Essentials of Mathematics

a one volume course for O' level

C. F. Chuter and R. W. Fox

Edward Arnold

First published 1981
by Edward Arnold (Publishers) Ltd.
41, Bedford Square, London WCIB 30P

British Library Cataloguing in Publication Data

Chuter, C F
A concise 'O' Level mathematics.
1. Mathematics—1961–
I. Title II. Fox, Ronald William
510 QA39.2

ISBN 0–7131–0418–X

Typeset by Macmillan India Ltd.
Bangalore 560 001
Printed in Hong Kong by
Wing King Tong Co. Ltd

Preface

This book is intended to provide a one volume course in O'level Mathematics. It is designed for pupils in their final year of such a course, or those re-sitting the examination; but should also be ideal for mature students attending a centre of Further Education or studying privately. Recent changes in the O'level syllabuses of the major examining boards (including the new London Syllabus 'B') have been covered, together with material which the authors anticipate will be required for future examinations at 16+.

Each topic in the course commences with notes and examples of 'lower level' work. These are intended to act as a reminder of material encountered by the student in earlier years, and to provide the basic notes which many pupils fail to retain. Complex terminology has been avoided where possible and wordy explanation reduced to a minimum. Much of the essential material to be learnt has been conveyed to the reader by means of worked examples. Each chapter contains sets of questions designed to cover the stages of progress, and concludes with a series of questions taken, in most cases, from past O'level examination papers.

Acknowledgements

The authors wish to thank the following examination boards for permission to use questions taken from their papers.

The University of London University Entrance and School Examinations Council
The Joint Matriculation Board
The University of Oxford Delegates of Local Examinations
The University of Cambridge Local Examinations Syndicate
The Associated Examining Board
The Welsh Joint Education Committee

C.F.C.
R.W.F.
January 1981

Contents

1

Number Systems and their Application

Types of Number

The set of counting numbers {1, 2, 3, ...} is known as the **natural** number system. The set {1, 2, 3, ...} can be defined also as the set of positive **integers**, an integer being a whole number. To enable the subtraction of one natural number from another, negative integers together with zero have been 'invented'. To enable the division of one natural number by another, **fractional** numbers are required. All these sets of numbers; all the integers together with the positive and negative fractions, are known as **rational** numbers.

A rational number therefore is one that can be expressed exactly in the form of a fraction.

$$\text{e.g.} \quad 200\left(=\frac{400}{2}\right), \frac{2}{3}, \frac{1}{100},$$

are all rational.

A number that cannot be expressed in this way is known as **irrational**.

e.g. $\sqrt{2} = 1.4142. \ldots$ (it has no exact answer)

is irrational.

The set of **real** numbers is the combination of the sets of rational and irrational numbers. This means that any point along the number line can be represented by a real number.

Products of a number by itself are usually written in **index** notation.

e.g. $2 \times 2 \times 2 \times 2 = 2^4$ (two to the power of four)

The 4 is called the index (plural indices).

A **prime** number is a number that can only be divided exactly by itself.

e.g. {2, 3, 5, 7}

is the set of the first four prime numbers.

Common Fractions

A common fraction has two parts, a **numerator** and a **denominator**. In the fraction $\frac{1}{2}$, the 1 is the numerator (upper part), the 2 is the denominator (lower part). A mixed number consists of a whole number and a fraction. $2\frac{1}{2}$ is a mixed number, $\frac{5}{2}$ is an example of an **improper** fraction.

Before adding or subtracting fractions their denominators must be made the same. This is done by multiplying or dividing the numerator and denominator by the same amount.

Example

To change the fractions $\frac{1}{2}$, $\frac{3}{4}$, $\frac{5}{8}$ all into eighths, ask the question 'what must I multiply the first denominator (2) by in order to get 8?' The answer is obviously 4, therefore both the numerator (1) and the denominator (2) must be multiplied by 4.

$$\therefore \quad \frac{1}{2} = \frac{1 \times 4}{2 \times 4} = \frac{4}{8}$$

Asking the same question with the second fraction indicates that multiplication by 2 is required.

$$\therefore \quad \frac{3}{4} = \frac{3 \times 2}{4 \times 2} = \frac{6}{8}$$

The three fractions now become $\underline{\frac{4}{8}, \frac{6}{8}, \frac{5}{8}}$ ANS

Addition and Subtraction of Fractions

Examples

1. Simplify $\frac{1}{2} + \frac{3}{4} - \frac{3}{8}$

[Change into eighths] $\quad \frac{(1 \times 4)}{(2 \times 4)} + \frac{(3 \times 2)}{(4 \times 2)} - \frac{3}{8}$

$$= \frac{4 \quad + \quad 6 \quad - 3}{8}$$

$$= \frac{10 - 3}{8} = \underline{\frac{7}{8}} \quad \text{ANS}$$

2. Simplify $\quad 1\frac{3}{4} + 2\frac{5}{6} - 3\frac{1}{12}$

[Deal with whole numbers first: $1 + 2 - 3 = 0$] $\quad = \frac{3}{4} + \frac{5}{6} - \frac{1}{12}$

[Change into twelths] $\quad = \frac{9 + 10 - 1}{12}$

$$= \frac{19 - 1}{12} = \frac{18}{12} = 1\frac{6}{12} = \underline{1\frac{1}{2}} \quad \text{ANS*}$$

* (Always change your answer into its lowest terms.)

Exercise 1 In questions **1** to **10** change each fraction into the denominator indicated within the brackets.

1. $\frac{3}{8}$ (16) **2.** $\frac{2}{3}$ (6) **3.** $\frac{4}{9}$ (27) **4.** $\frac{1}{5}$ (20)
5. $\frac{6}{9}$ (3) **6.** $\frac{4}{8}$ (2) **7.** $\frac{10}{12}$ (6) **8.** $\frac{8}{12}$ (4)
9. $\frac{3}{17}$ (51) **10.** $\frac{3}{51}$ (17)

Simplify:

11. $\frac{3}{4}+\frac{5}{8}$ **12.** $\frac{3}{4}+\frac{5}{8}-\frac{1}{2}$ **13.** $\frac{2}{3}-\frac{1}{6}$ **14.** $\frac{2}{3}+\frac{5}{12}-\frac{1}{6}$
15. $\frac{4}{7}-\frac{9}{14}+\frac{5}{14}$ **16.** $1\frac{3}{4}-\frac{1}{2}$ **17.** $1\frac{7}{8}+2\frac{3}{4}$ **18.** $3\frac{9}{10}-4\frac{17}{20}+1\frac{3}{5}$
19. $2\frac{7}{8}+3\frac{5}{16}-\frac{3}{4}$ **20.** $4\frac{11}{12}-\frac{3}{8}-2\frac{7}{24}$

Multiplication and Division of Fractions

When multiplying fractions together first try to cancel any common factors then multiply the numerators together followed by the denominators.

Example Simplify $\frac{2}{3}\times\frac{9}{10}$

[The 2 and the 10 have 2 as a common factor] $=\frac{\not{2}^1}{3}\times\frac{9}{\not{10}_5}$

[The 3 and the 9 have 3 as a common factor] $=\frac{1}{\not{3}_1}\times\frac{\not{9}^3}{5}$

$=\frac{1}{1}\times\frac{3}{5}=\underline{\frac{3}{5}}$ ANS

Mixed numbers should first be changed into improper fractions.

Example Simplify $1\frac{3}{4}\times 2\frac{2}{7}$

$=\frac{7}{4}\times\frac{16}{7}$

[The 7s are obviously common] $=\frac{\not{7}^1}{4}\times\frac{16}{\not{7}_1}$

[The 4 and the 16 have 4 as a common factor] $=\frac{1}{\not{4}_1}\times\frac{\not{16}^4}{1}$

$=\frac{1}{1}\times\frac{4}{1}=\underline{4}$ ANS

To divide by a fraction, turn the fraction by which we are dividing upside-down and then multiply by that fraction.

Example Simplify $\dfrac{2}{3} \div \dfrac{5}{12}$

$$= \frac{2}{3} \times \frac{12}{5}$$

[The 3 and the 12 have 3 as a common factor]

$$= \frac{2}{\cancel{3}_1} \times \frac{\cancel{12}^4}{5}$$

$$= \frac{2}{1} \times \frac{4}{5} = \frac{8}{5} = \underline{1\tfrac{3}{5}} \quad \text{ANS}$$

Exercise 2 Simplify:

1. $\frac{2}{5} \times \frac{5}{8}$	**2.** $\frac{3}{7} \times \frac{14}{15}$	**3.** $\frac{4}{9} \div \frac{2}{3}$	**4.** $\frac{7}{10} \div \frac{2}{5}$
5. $1\frac{1}{4} \times 2\frac{2}{5}$	**6.** $1\frac{1}{5} \div \frac{3}{10}$	**7.** $2\frac{1}{3} \div 1\frac{5}{9}$	**8.** $4\frac{1}{8} \div 5\frac{1}{2}$
9. $3\frac{1}{4} \times 2\frac{3}{8} \div 4\frac{3}{4}$	**10.** $7\frac{1}{5} \div 1\frac{1}{3} \times 2\frac{1}{12}$		

Decimal Fractions

If you find yourself doing simple addition on your fingers you may realise that this is the **denary** (counting in tens) system. Numbers less than unity in this system are called **decimal** fractions. The denary number 605 means 6 hundreds, no tens and 5 units. Similarly, the number 605.43 means 6 hundreds, no tens, 5 units, 4 tenths and 3 hundredths.

Using common fractions this would appear as $605 + \dfrac{4}{10} + \dfrac{3}{100}$.

To convert a common fraction into a decimal fraction divide the numerator by the denominator.

e.g. $\dfrac{2}{5} = 5\overline{)2.0}^{\,0.4} \quad = \underline{0.4} \quad \text{ANS}$

$\dfrac{2}{3} = 3\overline{)2.000}^{\,0.666}$

$= 0.666$ (the 6 is said to recur)
$= \underline{0.67}$ correct to 2 decimal places ANS

In an examination question you may be asked to give such an answer correct to a specified number of *decimal places.* In the above example the answer is given to *two decimal places.* To do this, look at the third place, if it is equal to 5 or more the second place is increased by 1.

A similar process is carried out when giving an answer correct to a specified number of *significant figures.* The first figure of a whole number is the first significant figure, in a decimal fraction with no integers to the left of the point the first significant figure is the first figure other than a 0 following the point.

e.g. the number 54536 correct to 3 s.f. would be 54500
the decimal fraction 0.0345 correct to 2 s.f. would be 0.035

Exercise 3 Correct the following to the number of decimal places indicated in the brackets.

1. 4.342 (2)	**2.** 43.336 (2)	**3.** 0.3567 (3)
4. 0.46 (1)	**5.** 11.0474 (3)	**6.** 10.007 (2)
7. 18.046 (1)	**8.** 0.000 537 (5)	**9.** 0.3349 (2)
10. 45.3416 (1)		

Exercise 4 Correct the following to the number of significant figures indicated in the brackets.

1. 45.44 (3)	**2.** 4536 (3)	**3.** 5896.37 (5)	**4.** 1189.037 (6)
5. 49.05 (3)	**6.** 467.81 (3)	**7.** 0.00467 (2)	**8.** 20.0047 (3)
9. 41.348 (3)	**10.** 1.0006 (2)		

Multiplication and Division of Decimals

Example 23.12×8.6

[Converting to vulgar fractions] $= \frac{2312}{100} \times \frac{86}{10}$

$= \frac{2312 \times 86}{100 \times 10}$

$= \frac{198832}{1000} = \underline{198.832}$ ANS

Alternative solution Multiply 2312×86 using the process of long multiplication. The result will be 198832. There are 3 places of decimals in the question, therefore there must be 3 places in the answer = $\underline{198.832}$.

Example $\frac{198.832}{8.6}$

[Make the *divisor* (i.e. 8.6) a whole number by multiplying both parts by 10] $= \frac{1988.32}{86}$

$$\begin{array}{rl} & 23.12 \\ \text{[By long division]} \quad = & 86\overline{)1988.32} \\ & \underline{172} \\ & 268 \\ & \underline{258} \\ & 103 \\ & \underline{86} \\ & 172 \\ & \underline{172} \\ = & \underline{23.12} \quad \text{ANS} \end{array}$$

Exercise 5 Simplify:

1. 24.6×9.3 **2.** 0.6×181.41 **3.** 2.305×230.5

4. $\dfrac{208.12}{24.2}$ **5.** $\dfrac{11402.24}{102.4}$

Percentages

A percentage is a type of fraction, measured in parts of 100.

e.g. $1\% = \dfrac{1}{100} = 0.01.$

It follows that $100\% = \dfrac{100}{100} = 1$ (whole one)

To express a common fraction or a decimal as a percentage multiply by $\dfrac{100}{1}$.

e.g. $\dfrac{1}{5} \times \dfrac{100}{1} = 20\%$, $0.2 \times \dfrac{100}{1} = 20\%$

To express a percentage as a fraction divide by 100.

e.g. $25\% = \dfrac{25}{100} = \dfrac{1}{4}$ (or 0.25)

Exercise 6 Change these fractions to percentages:

1. $\frac{1}{2}$ **2.** $\frac{3}{4}$ **3.** $\frac{5}{8}$ **4.** $\frac{1}{3}$ **5.** $\frac{5}{12}$

6. 0.4 **7.** 0.04 **8.** 0.45 **9.** 0.666 **10.** 1.75

Change these percentages to common fractions:

11. 70% **12.** 10% **13.** $12\frac{1}{2}\%$ **14.** 95% **15.** 99%

Standard Form

A number with one figure, other than 0, before the decimal point is in **standard form**. Numbers greater than 10 can be written in standard form and multiplied by a power of 10.

e.g. **1.** 200 $= 2 \times 10^2$ in standard form
2. 395.8 $= 3.958 \times 10^2$ in standard form
3. 10096 $= 1.0096 \times 10^4$ in standard form.

Numbers less than 1 can be written in standard form by dividing by a power of 10.

e.g. **1.** 0.2 $= \frac{2}{10^1} = \quad = 2 \times 10^{-1}$ in standard form

2. 0.002 $= \frac{2}{1000} \quad = 2 \times 10^{-3}$ in standard form

3. 0.0369 $= \frac{3.69}{100} \quad = 3.69 \times 10^{-2}$ in standard form.

Exercise 7 Write the following numbers in standard form:

1. 300 **2.** 496 **3.** 25 **4.** 4006
5. 23109 **6.** 0.3 **7.** 0.003 **8.** 0.00167
9. 4961.43 **10.** 300.006

Ratio and Proportion

The **ratio** of pupils to teachers in a secondary school may be 18 : 1. This means that if there are 180 pupils there are 10 teachers, and so on. The ratio of pupils to teachers (18 : 1) can also be expressed as a ratio of teachers to pupils to give the ratio 1 : 18. This ratio can also be expressed as the common fraction $\frac{1}{18}$.

Ratios should always be expressed in their simplest terms. For example, in a school of 600 pupils with 30 teachers the pupil-teacher ratio would be 600 : 30, cancelling through by 30 to give the ratio 20 : 1.

Before expressing quantities in ratio, check that their units are the same.

Example A map has a scale of 2cm to 1km. Express this as a ratio.

The units used are centimetres and kilometres. It is therefore necessary to convert the larger unit to the smaller.

1km = 1000m = 1000 × 100cm = 100000cm.
i.e. 1cm to 1km = 1 : 100000
Therefore 2cm to 1km = 2 : 100000
∴ 2cm to 1km = <u>1 : 50000</u> ANS

When expressing one quantity as a **proportion** of another quantity, an ordinary fraction or a percentage is used. As with a ratio, the units must be the same.

Example — What proportion of £5 is 50p?

$$£5 = 500\text{p}$$

[Expressing as a fraction] $\frac{50}{500} = \frac{1}{10}$ ANS

Exercise 8 Express each of these ratios in their lowest terms:

1. 12 hours to 24 hours **2.** 30 minutes to 2 hours
3. 5cm to 1m **4.** 5cm to 1km
5. 40g to 2kg

Express the second quantity as a proportion of the first:

6. £1, 10p **7.** 10m, 10cm **8.** 1km, 160m
9. 2kg, 50g **10.** 5m^2, 125cm^2

Changing the Base

As explained earlier, our usual system of counting is in the denary (or decimal) scale. This system is said to have a **base** of ten, the reason can be seen in this denary abacus.

......	10^3 (10 × 10 × 10) 1000	10^2 (10 × 10) 100	10^1 (10) 10	U
	3	4	0	1

The example 3401 means 3000 + 400 + 1, or 3401_{10}.
(Small figures are used to indicate the base of the number, although we do not usually show the 10.)

It is possible to construct a counting system using any base.

The Binary Scale In this system the base is two. This means only the digits 0 and 1 are used. (Compared with our denary base of ten using the digits 0, 1, 2, 3, 4, 5, 6, 7, 8, 9). The binary sequence is 0, 1, 10, 11, 100, 101, 110, 111. . . . .

Each column of the binary abacus has twice the value of the column to its right.

......	2^4 (2 × 2 × 2 × 2) 16	2^3 (2 × 2 × 2) 8	2^2 (2 × 2) 4	2^1 (2) 2	U
	1	0	0	1	1

The example 10011 means (1 × 16) + (1 × 2) + 1, or 10011_2 (base two). The denary equivalent to this number would, of course, be $16 + 2 + 1 = 19_{10}$.

The Octal Scale

In this system the base is eight. The sequence is therefore 0, 1, 2, 3, 4, 5, 6, 7, 10, 11, 12, ...

We use only the digits 0, 1, 2, 3, 4, 5, 6, 7.
The octal abacus would be

......	8^3 (8 × 8 × 8) 512	8^2 (8 × 8) 64	8^1 (8) 8	U
	4	0	7	0

The example 4070_8 would be $(4 \times 512)+(7 \times 8) = 2104_{10}$.

Other Scales

The above procedures apply to all bases.

Examples

1. 234_6 can be written in abacus form as:

6^2 36	6^1 6	U
2	3	4

$= (2 \times 36)+(3 \times 6)+4 = 94_{10}$

2. 412_5 can be written in abacus form as:

5^2 25	5^1 5	U
4	1	2

$= (4 \times 25)+(1 \times 5)+2 = 107_{10}$

The processes of addition, subtraction, multiplication and division may all be performed in any of the scales, care being taken to remember the values of each column on the abacus.

Exercise 9

Evaluate the following:

1. $23_{10}+19_{10}$ **2.** 10_2+11_2 **3.** 101_2+111_2
4. 15_6+23_6 **5.** 45_7+36_7 **6.** 13_4+22_4
7. $18_{10}-9_{10}$ **8.** 11_2-1_2 **9.** 111_2-10_2
10. 43_5-24_5 **11.** 27_8-11_8 **12.** 21_3-12_3
13. $25_{10}\times 11_{10}$ **14.** $11_2\times 11_2$ **15.** $101_2\times 100_2$
16. $14_5\times 22_5$ **17.** $36_8\times 41_8$ **18.** $99_{10}\div 11_{10}$
19. $1100_2\div 10_2$ **20.** $44_5\div 11_5$

Examination Standard Questions

Worked Examples

The worked examples which follow are intended as a guide to setting out solutions of examination standard questions. Hints and comments are contained within brackets, such notes would not be required in your solutions. However, your attempted solutions in an examination must show clearly your essential working, an incorrect answer in some papers can still earn a large proportion of the marks if your approach is clear and correct.

1. Calculate the value of $\left(3-\frac{1}{5}\right)\div\frac{9}{11}$

[Brackets must always be evaluated first] $= 2\frac{4}{5}\div\frac{9}{11}$

[Changing the mixed number into a fraction] $= \frac{14}{5}\div\frac{9}{11}$

[Invert the divisor and multiply] $= \frac{14}{5}\times\frac{11}{9}$

[There are no common factors] $= \frac{14\times 11}{5\times 9} = \frac{154}{45} = 3\frac{19}{45}$ ANS

2. Find the exact value of $\frac{45\times 0.55}{0.011\times 2.25}$

$$= \left(\frac{45}{1}\times\frac{55}{100}\right)\div\left(\frac{11}{1000}\times\frac{225}{100}\right)$$

[Invert the whole divisor and multiply] $= \frac{45}{1}\times\frac{55}{100}\times\frac{1000}{11}\times\frac{100}{225}$

[The 45 and 225 have 45 as a common factor, 55 and 11 have 11 as a common factor, the 100s are common.] $= \frac{\cancel{45}^{1}}{1}\times\frac{\cancel{55}^{5}}{\cancel{100}_{1}}\times\frac{1000}{\cancel{11}_{1}}\times\frac{\cancel{100}^{1}}{\cancel{225}_{5}}$

[The 5s are common] $= \underline{1000}$ ANS

3. After an increase of 12% the price of an article is £45.92. Calculate the price before the increase.

An increase of 12% gives 112% of the original price.

i.e. $(\text{Previous price})\times\frac{112}{100} = £45.92$

[Dividing both sides of the equation by $\frac{112}{100}$] Previous price $= \frac{£45.92}{1}\times\frac{100}{112}$

[Cross-cancel] $= \frac{£\cancel{45.92}^{41}}{1}\times\frac{100}{\cancel{112}_{1}}$

$= \underline{£41}$ ANS

4. In a general election only 70% of the electors actually voted, and the ratio of the votes cast for the two candidates was 3 : 2. If the winning candidate secured 980 votes more than the loser, find the total number of electors.

Winning candidate received $\frac{3}{5}$ of the votes, loser received $\frac{2}{5}$

Therefore the additional $\frac{1}{5}$ received by the winner was 980 votes

$$\text{total votes} = \frac{5}{5} = 980 \times 5 = 4900 \text{ votes}$$

this represents 70% of the electrotate

$$\therefore \text{ total electroate} = \frac{4900^{70}}{1} \times \frac{100}{\cancel{70}_1}$$

$$= \underline{7000} \quad \text{ANS}$$

5. Simplify $17_{10} + 14_8$, giving your answer in the octal scale.

$$14_8 = (1 \times 8) + 4 = 12_{10}$$
$$17_{10} + 12_{10} = 29_{10}$$

Using the octal abacus

8^2	8^1	
64	8	U
	3	5

$= (3 \times 8) + 5 = 29_{10}$

$\therefore$ required answer is $\underline{35_8}$

Exercise 10

1. Find the greatest $\frac{3}{4}, \frac{10}{13}, \frac{12}{17}, \frac{32}{43}$ (O)

2. Evaluate $(3\frac{3}{4} \div 2\frac{1}{3}) - \frac{1}{4}$ (C)

3. Find the exact value of $\frac{14\frac{3}{8}}{(2\frac{1}{2})^2}$ (L)

4. Find the exact value of $3.432 \div 16.5$ (C)
5. Find the exact value of $(0.6)^3 - 0.0179$ (O)
6. An article is sold for £3.08 at a profit of 12% on the cost price. Calculate the profit which is made. (C)
7. A chef uses 14kg of fruit and 33kg of other ingredients to make Christmas puddings. Express the weight of the fruit as a percentage of the weight of the puddings. (JMB)
8. A man's annual salary of £2400 is increased by $7\frac{1}{2}$% and then by an extra £60. Calculate the total percentage increase in his salary. (L)
9. The perimeter of a triangle is 64cm and the sides are in the ratio 8:15:17. Calculate the length of the longest side. (C)
10. In a recent village election the total number of votes cast was 2772. There were two candidates A and B. The ratio of the votes cast for A to the votes cast for B was 7:5. If 105 of the voters had given their votes to B instead of A and the remainder had voted as before,

calculate the new ratio of votes cast for A to the votes cast for B. Give your answer in its lowest terms.

(*Hint:* First calculate the original number of votes cast for A.) (C)

11. A legacy was divided between two nephews in proportion to their ages which were 29 and 13 years. If the older received £736 more than the younger find how much the younger received. (O)

12. Write down (in binary notation) the number which is four times the binary number 1011. (JMB)

13. Given that $0.0431 = 4.31 \times 10^n$, state the value of n. (JMB)

14. If all numbers are given in binary notation, calculate the following, leaving your answer in binary form. (WJEC)
(i) 110×101 (ii) $10001 - 111$

15. Express the number 111_8 in the binary scale. (C)

16. Calculate $37_8 + 64_8$, giving the answer in the same scale. (C)

17. (i) Change 125_{10} to base 6, (ii) Change 125_6 to base 10. (O)

18. The number n is a positive integer which is a multiple of 6. Which of the following statements are true and which are false?
(a) In base 6, the representation of n ends with a zero.
(b) In base 10, the digits add together to give a multiple of 3.
(c) In base 7, the representation of n never ends with a zero.
(d) In base 9, the representation of n never ends with a 2. (O)

2
Logarithms

For our puposes at this level of mathematics, we need only master the mechanics of logarithms. The theory is best left to a more advanced stage for those who care to further their studies. In fact, the developers of logarithms, Napier and Briggs, produced tables as an aid to calculation without attempting to outline their theoretical basis.

Throughout this chapter, and in many of the topics that follow, you will need a set of four-figure tables.

In the following examples you may find that your tables provide an answer that has a difference of one in the fourth digit. (e.g. 2.304 instead of 2.305) This is simply because logarithmic tables provide an approximate solution in most cases, different editions of tables may contain slight variations. Both answers would be accepted by an examiner.

Turn to the double page in your book of tables headed 'Logarithms'. Consider the numbers 3.4, 34 and 340. In the tables you will find that all appear to have the same log. Look down the left hand column in heavy type to find the digits **34**, ignoring the decimal point. The log is read off as 5315. (This part of the log is called the **mantissa**). To distinguish between small and large numbers we place a **characteristic** in front of the mantissa, with a decimal point between.

To find the characteristic, count the number of figures to the left of the decimal point and subtract one.

e.g.	the characteristic of 3.4 is 0	
	the characteristic of 34 is 1	(think of 34.0)
	the characteristic of 340 is 2	(think of 340.0)
Therefore	the log of 3.4 is 0.5315	
	the log of 34 is 1.5315	
	the log of 340 is 2.5315	

When dealing with numbers less than unity, the characteristic will be negative. Following the above pattern the characteristic of 0.34 will be −1. In logs the minus sign is written above the characteristic and termed a **bar**. Therefore the log of 0.34 is $\bar{1}.5315$ (read 'bar-one point' 5315). Similarly the log of 0.034 would be $\bar{2}.5315$.

The rule with negative characteristics is to count the number of places from the decimal point up to and including the first significant figure.

This place is the required **bar** number.

e.g. 0.0009 will have a characteristic of $\bar{4}$ (the four places after the decimal point are numbered 1 2 3 4)

Examples

Find the logarithms of **1.** 35 **2.** 5.7 **3.** 6700 **4.** 0.89 **5.** 0.0009

1. The characteristic of 35.0 is 1
the mantissa of **35** from the tables is 0.5441
therefore, the log of 35 = 1.5441 ANS

2. The characteristic of 5.7 is 0
the mantissa of **57** from the tables is 0.7559
therefore, the log of 5.7 = 0.7559 ANS

3. The characteristic of 6700.0 is 3
the mantissa of **67** from the tables is 0.8261
therefore, the log of 6700 is 3.8261 ANS

4. The characteristic of 0.89 is $\bar{1}$
the mantissa of **89** is 0.9494
therefore, the log of 0.89 is $\bar{1}.9494$ ANS

5. The characteristic of 0.0009 is $\bar{4}$
the mantissa of **9** is 0.9542 (look up 90)
therefore, the log of 0.0009 is $\bar{4}.9542$ ANS

Exercise 11

Find the logarithms of the following numbers:

1. 29	**2.** 2.9	**3.** 290	**4.** 570	**5.** 0.57
6. 8.8	**7.** 880	**8.** 1000	**9.** 0.001	**10.** 99 000

To find the log of a three-figure number look down the left hand column for the first two figures, then read carefully across to a position beneath the third figure.

e.g. the log of 34.5 is 1.5378

The characteristic is found as usual, the mantissa is found as above by starting from the **34** position (0.5315) then moving horizontally along the **34** line to arrive in the '5' column (0.5378). A rule placed under the line of figures helps to keep the eyes from reading above or below the required line.

Examples

Find the logarithms of **1.** 12.3 **2.** 1.34 **3.** 0.933.

1. The characteristic of 12.3 is 1
the mantissa of 123 is 0.0899
therefore, the log of 12.3 is 1.0899 ANS

2. The characteristic of 1.34 is 0
the mantissa of 134 is 0.1271
therefore, the log of 1.34 is 0.1271 ANS

3. The characteristic of 0.933 is $\bar{1}$
the mantissa of 933 is 0.9699
therefore, the log of 0.933 is $\underline{\bar{1}.9699}$ ANS

To find the log of a four-figure number continue horizontally across the page of tables to the 'difference' columns.

e.g. the log of 34.56 is 1.5386

Follow the process as before for 34.5 (1.5378), then continue horizontally across to the difference columns (the narrower columns on the right of the same page) and read off the value beneath the '6'. In this example it is 8, and this 8 is added to the fourth digit of the mantissa (0.5378 + 0.0008). The result is 1.5386.

Exercise 12 Find the logarithms of:

1. 23.5	**2.** 4.26	**3.** 83 900	**4.** 0.4125	**5.** 29.41
6. 397.8	**7.** 8396	**8.** 417 800	**9.** 0.003 333	**10.** 0.3004

Antilogarithms

To convert a logarithm back into an ordinary number we use the table of antilogarithms, found on the next double page of your book of tables.

Examples Find the numbers whose logs are **1.** 2.3456 **2.** $\bar{3}.4444$.

1. Only the mantissa (0.3456) is taken from the antilog tables. As before, look down the left-hand column for the 0.34 (gives 2188), go horizontally across beneath the '5' column (gives 2213), travel further across to the '6' column in the difference section (a value of 3), add this on (= 2216).

The decimal point *always* starts *between* the first two digits, then moves the number of places indicated by the characteristic.

i.e. Putting point between first two digits gives 2.216, now move the point the number of places (2) indicated by the characteristic:

2 . 2 1 6 becomes $\underline{221.6}$ ANS

2. The mantissa of 0.4444 from the antilog tables gives 2783. Putting point between first two digits gives 2.783, now move the point the number of places ($\bar{3}$ means 3 places to the *left*) indicated by the characteristic.

0 0 0 0 2 . 7 8 3 becomes $\underline{0.002783}$ ANS

Care must be taken when adding or subtracting logarithms.

Examples

Add together these logarithms: **1.** $\bar{1}.3031 + \bar{2}.9131$ **2.** $0.6416 + \bar{1}.496$

1.

$$\begin{array}{r} \bar{1}.3031 \\ +\,\bar{2}.9131 \\ \hline \bar{2}.2162 \\ \hline 1 \end{array}$$

['Carry-one' gives $(-1)+(-2)+(+1) = -2$]

2.

$$\begin{array}{r} 0.6416 \\ +\,\bar{1}.4967 \\ \hline 0.1383 \\ \hline 1\;1\;1 \end{array}$$

['Carry-one' gives $(0)+(-1)+(+1) = 0$]

Subtract the second log from the first: **3.** $3.416 - \bar{1}.322$ **4.** $\bar{2}.499 - \bar{1}.267$ **5.** $\bar{3}.4772 - \bar{1}.5681$.

3.

$$\begin{array}{r} 3.416 \\ -\,\bar{1}.322 \\ \hline 4.094 \\ \hline \end{array}$$

[$(3)-(-1) = 3+1 = 4$]

4.

$$\begin{array}{r} \bar{2}.499 \\ -\,\bar{1}.267 \\ \hline \bar{1}.232 \\ \hline \end{array}$$

[$(-2)-(-1) = (-2)+1 = -1$]

5.

[By the 'borrowing' method, subtract 1 from $\bar{3}$, i.e. $(-3)-1 = -4$]

[$(-4)-(-1) = (-4)+1 = -3$]

$$\begin{array}{r} \overset{\bar{4}}{\bar{3}}.4\overset{6}{7}72 \\ -\,\bar{1}.5681 \\ \hline \bar{3}.9091 \\ \hline \end{array}$$

Exercise 13 Simplify the following – do not antilog:

1. $1.3459 + 0.2397$ **2.** $5.3324 + 1.0004$ **3.** $4.35 + 1.4496$
4. $2.8115 + \bar{1}.3496$ **5.** $\bar{3}.0408 + \bar{2}.0737$ **6.** $3.2176 - 1.4533$
7. $2.816 - \bar{1}.399$ **8.** $\bar{2}.467 - \bar{3}.238$ **9.** $\bar{3}.2968 - \bar{4}.5784$
10. $\bar{3}.699 - \bar{1}.877$

Multiplication

To multiply numbers together *add* their logarithms, then find the antilog.

Examples

Evaluate **1.** 24.91×3.405 **2.** 0.35×0.04176.

1.

[Write the numbers beneath each other then look up their logs.]

[Add their logs]

[Antilog of 1.9285 gives the answer]

No	Log
24.91	1.3964
3.405	0.5321 +
	1.9285
	= 84.82 ANS

2.

[Remember to turn back to the log pages]

[*Note*: When adding, there is a 'carry-one' from the mantissa which is positive. The characteristic therefore becomes $(-1)+(-2)+(+1) = -2$]

No	Log
0.35	$\bar{1}.5441$
0.04176	$\bar{2}.6207$ +
	$\bar{2}.1648$
	1

[Using antilog tables] $= \underline{0.01462}$ ANS

Exercise 14 Evaluate using tables of logarithms:

1. 3.491×24.35 **2.** 456.6×2970 **3.** 41.66×0.3516
4. 0.047×0.003526 **5.** 8100×0.03175

Division

Examples Evaluate using tables of logarithms **1.** $\frac{234.7}{8.39}$ **2.** $0.041 \div 0.00356$.

1. In this situation *subtract* the log of the lower number from the log of the upper number then antilog, (or log of upper number *minus* log of lower number).

[Subtract]

No	Log
234.7	2.3705
8.39	0.9238 −
	1.4467

$= \underline{27.97}$ ANS

2.

[Log of first number *minus* log of second number]
[Subtract]
$[(-2)-(-3) = (-2)+3 = +1]$

No	Log
0.041	$\bar{2}.6128$
0.00356	$\bar{3}.5514$ −
	1.0614

$= \underline{11.52}$ ANS

Exercise 15 Evaluate using tables of logarithms:

1. $234.6 \div 12.38$ **2.** $12\,960 \div 1.35$ **3.** $0.369 \div 0.0048$
4. $41.68 \div 0.359$ **5.** $21.67 \div 485$

Powers

To raise a number to a given power, *multiply* its logarithm by the value of this power, then find the antilog.

Examples Evaluate using tables of logarithms **1.** $(23.45)^2$ **2.** $(0.034)^5$

1.

No	Log
23.45	1.3701
	× 2
	2.7402

[Antilog] = <u>549.8</u> ANS

2.

No	Log
0.034	$\bar{2}.5315$
	× 5
	$\bar{8}.6575$
	2

['Carry-two' is positive, i.e. $-10+2+.6575 = -8+.6575$]

[Antilog] = <u>0.000 000 045 44</u> ANS

Exercise 16 Evaluate using tables of logarithms:

1. $(24)^2$ **2.** $(24.38)^2$ **3.** $(4960)^3$
4. $(0.0396)^4$ **5.** $(0.145)^5$

Roots

To find the given root of a number, divide its logarithm by the value of this root, then find the antilog.

Examples Evaluate using tables of logarithms **1.** $\sqrt{24.5}$ **2.** $\sqrt[3]{5976}$ **3.** $\sqrt[3]{0.03986}$ **4.** $\sqrt[4]{0.0047}$.

1.

[*Note.* No root value indicated is an accepted way of writing a square root, i.e. $\sqrt[2]{\ }$]

No	Log
24.5	1.3892 ÷ 2
	0.6946

[By short div.]
[Antilog] = 4.950
= <u>4.95</u> ANS

2.

No	Log
5976.0	3.7766 ÷ 3
	1.2589

[1.2588 remainder $\frac{2}{3}$ = 1.2589 to 4 d.p.]

= <u>18.15</u> ANS

3.

No	Log
0.03986	$\bar{2}.6006 \div 3$
	$= \bar{3} + 1.6006 \div 3$
	$\bar{1}.5335$
	$= 0.3416$ ANS

[We cannot divide $\bar{2}$ by 3 exactly, we therefore increase the negative characteristic to a multiple of 3. To do this we add $\bar{1}$ to the $\bar{2}$. To maintain the true value we must also add 1 to the decimal part of the log]

[ignore remainder]

4.

No	Log
0.0047	$\bar{3}.6721 \div 4$
	$= \bar{4} + 1.6721 \div 4$
	$\bar{1}.4180$
	$= 0.2618$ ANS

[$\bar{3} + \bar{1} = \bar{4}$, therefore add 1 to the decimal part of the log to retain the true value
ignore remainder $\frac{1}{4}$]

Exercise 17

Evaluate using tables of logarithms:

1. $\sqrt{49}$ **2.** $\sqrt{568.3}$ **3.** $\sqrt[3]{0.0698}$ **4.** $\sqrt[4]{0.005996}$
5. $\sqrt[5]{0.00789}$ (Hint: $\bar{3} + \bar{2} = \bar{5}$)

Examination Standard Questions

Worked examples

1. Evaluate $\sqrt{\frac{593}{409}}$, correct to one place of decimals.

Using logs

No	Log
593.0	2.7731
409.0	2.6117 −
	$0.1614 \div 2$
	= 0.0807
	= 1.204
	= 1.2 ANS

[First, perform the division by subtracting the log of the lower number from the log of the upper number.]

[Secondly, divide by 2 to find the square root]

[Antilog]

[Correct to 1 d.p.]

2. Use tables to evaluate to 3 significant figures $\frac{15.99 \times 8.23}{107.4}$.

Using logs

No	Log
15.99	1.2039
8.23	0.9154 +
	2.1193
107.4	2.0310 −
	0.0883
	= 1.226
	= 1.23 ANS

[First, perform the multiplication by adding the logs]

[Secondly, perform the division by subtracting the log of the lower number from the log of the upper number.]

[Antilog]

[Correct to 3 s.f.]

Exercise 18

1. Use tables to evaluate $\sqrt[3]{\dfrac{3.264}{2761}}$. (L)
2. Use logarithms to calculate $289.5 \div 8.32$ (WJEC)
3. Use tables to evaluate, to three significant figures $\left(\dfrac{20.56}{47.83}\right)^3$. (O)
4. Evaluate $\sqrt{\dfrac{707}{404}}$, correct to two places of decimals. (JMB)
5. Use tables to evaluate $\sqrt{\dfrac{193.9}{0.07046}}$. (C)
6. Find the positive square root of 236, giving your answer correct to 3 significant figures. (JMB)
7. Which of the following numbers is the greatest? 4^{15}, 16^5, 2^{32} 32^6. (JMB)
8. Perform the following calculations and correct the answers to 1 significant figure. State which of your answers are equal to 60. (O)
   (a) $44.7 + 24.7$ (b) 0.0324×1950 (c) $2.4 \div 0.043$ (d) $\sqrt{(38\,000)}$.
9. The value, corrected to the nearest integer if necessary, of $\sqrt{1600}$ $+ \sqrt{9000}$ is equal to (a) 43, (b) 70, (c) 107, (d) 108, (e) 135. State which answer is correct. (O)
10. Use tables to evaluate (a) $\sqrt[5]{0.0068}$, (b) $\dfrac{1}{\sqrt{(14.3^2 + 8.7^2)}}$. (C)

3

Area & Volume

Area of plane figures

An area is measured in square units, e.g. cm^2, m^2, . . .

This can be illustrated by considering the square.

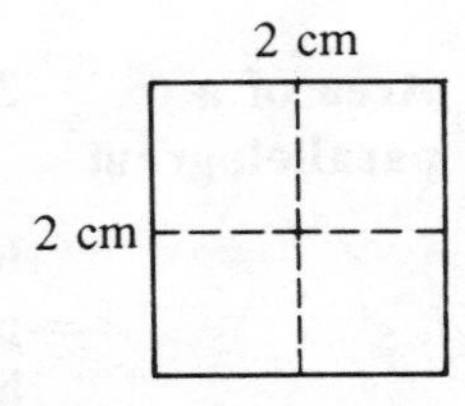

This square has length of 2cm and breadth of 2cm. It obviously contains 4 one centimetre squares, this fact being referred to as its area.
Therefore, the area is $2\,\text{cm} \times 2\,\text{cm} = \underline{4\text{cm}^2}$

The same applies to the rectangle.

breadth

length

area of a rectangle = length × breadth

Example Find how many square metres of carpet need to be purchased to cover a rectangular floor, measuring 6.5 metres long and 3.9 metres wide.

$$\begin{aligned} \text{area} &= \text{length} \times \text{breadth} \\ &= \quad 6.5\ \text{m} \\ &\times \quad \underline{3.9}\ \text{m} \\ &\quad 1950 \\ &\quad \underline{585} \\ &\quad \underline{25.35}\ \text{m}^2 \end{aligned}$$

thus requiring 26 square metres of carpet to cover the floor.

It can be shown that a triangle is one-half of a rectangle.

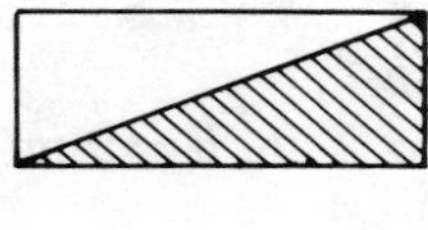

$\therefore$ **area of a triangle** = $\frac{1}{2}$(length × breadth)
or as usually expressed, **$\frac{1}{2}$ base × perpendicular height**

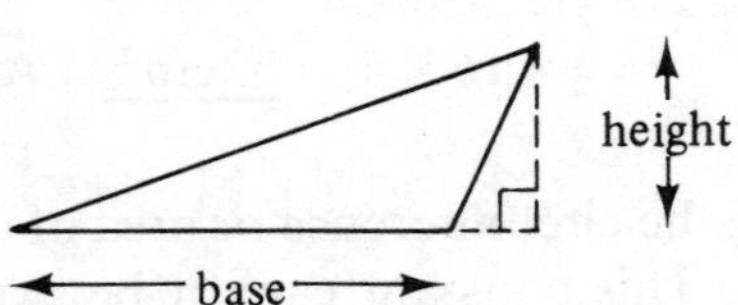

Example Find the area of this triangular plot of land.

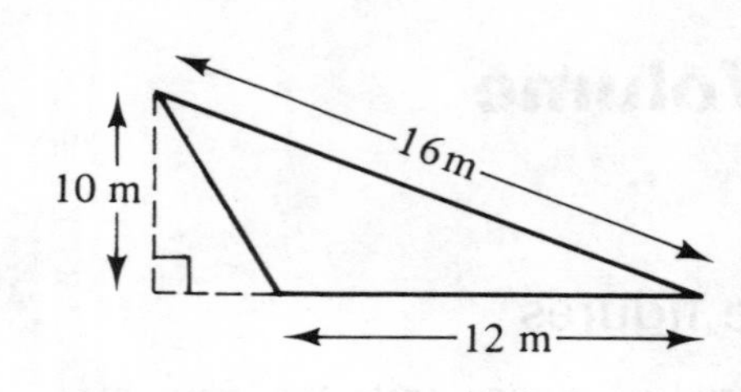

area $= \frac{1}{2}$ base $\times$ perpendicular height

In order to use the perpendicula height of 10m shown, the base mus be taken as the side of length 12m

$\therefore$ area $= \frac{1}{2} \times 12 \times 10$

$= 6 \times 10 = \underline{60\text{m}^2}$ ANS

A parallelogram consists of two equal triangles.

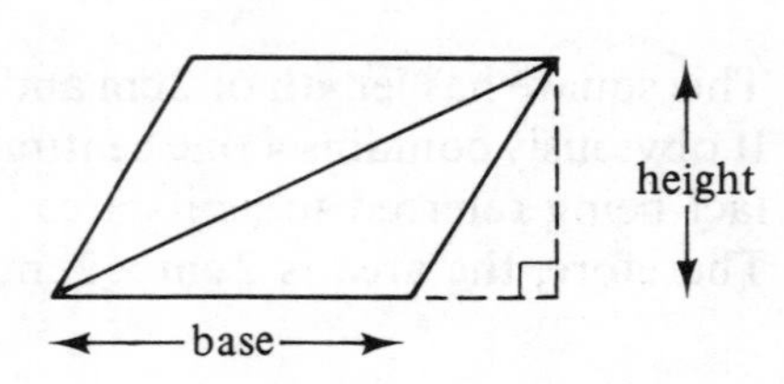

Area of a parallelogram $=$ 2 $\times$ area of one triangle

$=$ **base $\times$ perpendicular height**

Example Find the area of the parallelogram shown.

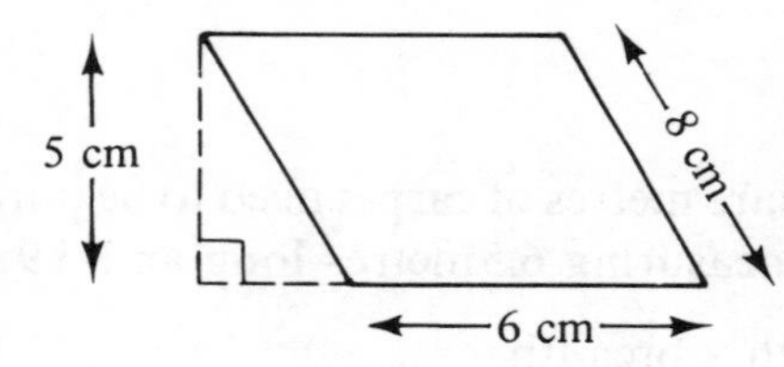

area $=$ base $\times$ perpendicular heigh

In order to use the perpendicula height of 5cm shown, the base mus be taken as the side of length 6cm

$\therefore$ area $= 6 \times 5 = \underline{30\text{cm}^2}$ ANS

A trapezium is a quadrilateral with one pair of parallel sides.

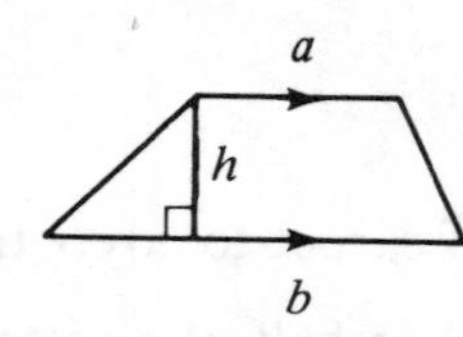

Area of a trapezium $= \frac{1}{2}(a+b)h$,
where a and b are the lengths of the parallel side and h is the perpendicular height (or distanc between the parallel sides).

Example Find the area of the trapezium shown.

7 cm

4 cm

9 cm

area $= \frac{1}{2}(a+b)h$

In this case, $a = 7$cm, $b = 9$cm and $h = 4$cm.

area $= \frac{1}{2}(7+9) \times 4$

$= \frac{1}{2} \times 16 \times 4 = \underline{32\text{cm}^2}$ ANS

Calculations involving the circumference or area of a circle require the use of the constant π. This constant cannot be expressed exactly as either a decimal or a fraction, the approximate value to be used will be given in the question or examination paper involved.

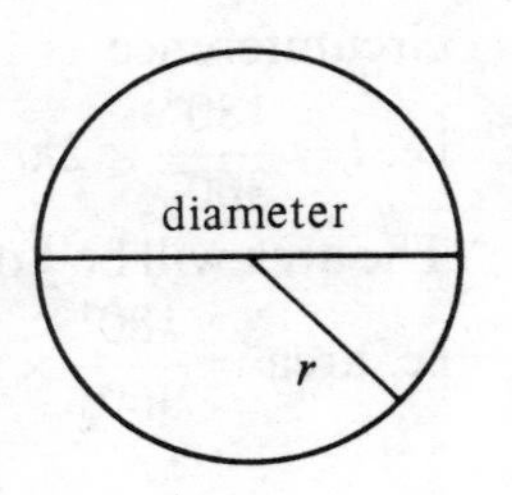

circumference of a circle = $2\pi r$ or πd (where r is the radius and d the diameter)

xample

Find the circumference of a circular pond of diameter 14 metres. Take π to be $\frac{22}{7}$.

circumference $= \pi d$

$= \frac{22}{7} \times \frac{14}{1}$

[Cross-cancel by 7] $= \frac{22}{\not{7}_1} \times \frac{\not{14}^2}{1} = \underline{44\,\text{m}}$ ANS

area of a circle = πr^2

Example

Find the area of the circular pond in the above example.

area $= \pi r^2$

[Radius = ½ of the diameter] $= \frac{22}{7} \times \frac{7}{1} \times \frac{7}{1}$

[Cross-cancel by 7] $= \frac{22}{\not{7}_1} \times \frac{\not{7}^1}{1} \times \frac{7}{1}$

$= \underline{154\,\text{m}^2}$ ANS

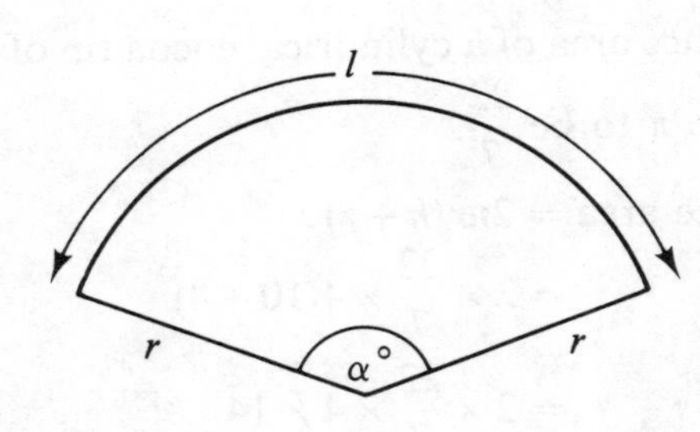

Length of a circular arc = $\frac{\alpha}{360} \times 2\pi r$

Area of a circular sector = $\frac{\alpha}{360} \times \pi r^2$

where α is the angle subtended by the arc at the centre of the circle.

Example Let $\alpha = 180^\circ$, then the sector is $\frac{1}{2}$ of the circle (i.e. a semi-circle).

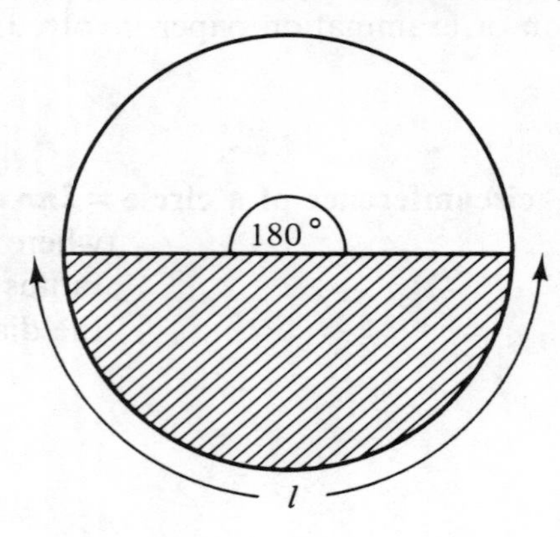

The length (l) will be $\frac{1}{2}$ of the circumference

i.e. $l = \frac{\cancel{180}^1}{\cancel{360}_2} \times 2\pi r = \underline{\pi r}$ ANS

The area will be $\frac{1}{2}$ of the circle's area

i.e. area $= \frac{\cancel{180}^1}{\cancel{360}_2} \times \pi r^2$

$= \underline{\frac{\pi r^2}{2}}$ ANS

Note

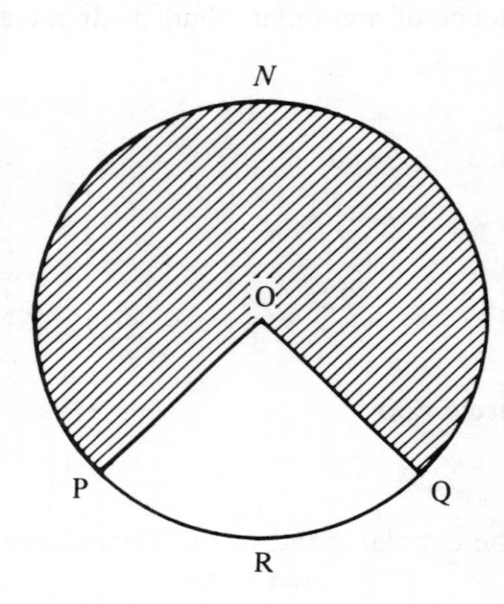

The **arc** PNQ is called the **major arc**.

The **arc** PRQ is called the **minor arc**.

(A major arc is greater than half the circle's circumference, a minor arc is less than half the circle's circumference.)

The area bounded by the major arc PNQ and the radii OP and OQ is called the **major sector** (shaded).

The area bounded by the minor arc PRQ and the radii OP and OQ is called the **minor sector** (unshaded).

Total surface area

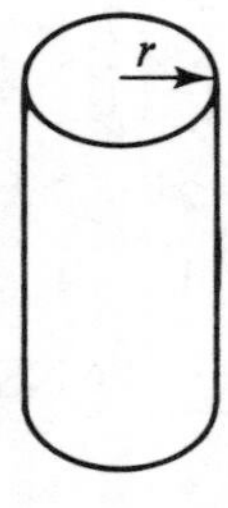

Total surface area of a cylinder

= area of curved surface + area of both ends

$= 2\pi rh + 2\pi r^2$ OR $2\pi r(h+r)$

Example Find the *total* surface area of a cylindrical cocoa tin of radius 4cm and height 10cm. Take π to be $\frac{22}{7}$.

$$\text{Total surface area} = 2\pi r(h+r)$$
$$= 2 \times \frac{22}{7} \times 4(10+4)$$
$$= 2 \times \frac{22}{7} \times 4 \times 14$$

[Cross-cancel by 7] $= \frac{2}{1} \times \frac{22}{\cancel{7}_1} \times \frac{4}{1} \times \frac{\cancel{14}^2}{1}$

$= \underline{352\text{cm}^2}$ ANS

Volume

Volume of a cylinder $=$ **area of base × height.**

$= \pi r^2 h$, where h is the height.

Volume of a cuboid or prism = area of base × height.

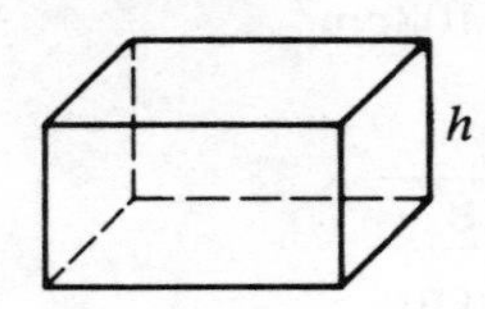

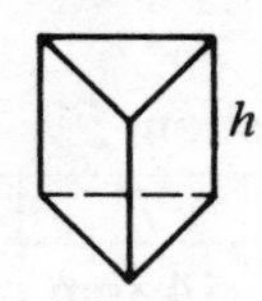

Example Find the volume of the cylindrical cocoa tin mentioned above, radius 4cm, height 10cm. Take π to be 3.14.

Volume $= \pi r^2 h$

$= 3.14 \times 4 \times 4 \times 10$

[$3.14 \times 10 = 31.4$] $= 31.4 \times 16$

$= \underline{502.4\text{cm}^3}$ ANS

Volume of a cone or pyramid $= \frac{1}{3}$ **area of base × perpendicular height**

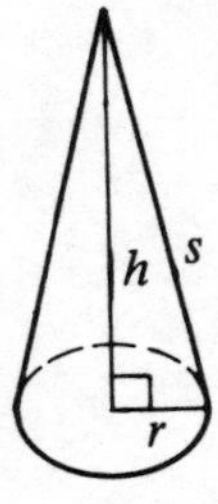

∴ **Volume of cone** $= \frac{1}{3}\pi r^2 h$ or $\frac{\pi r^2 h}{3}$

Curved surface area $= \pi rs$, (s is the slant height)

Example Find the volume of a pyramid base area 6cm², height 10cm.

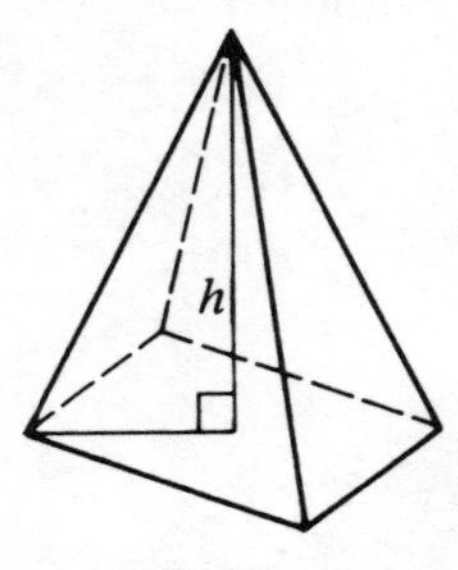

Volume $= \frac{1}{3}$ area of base × perpendicular height

$= \frac{1}{3} \times 6 \times 10 = \underline{20\text{cm}^3}$ ANS

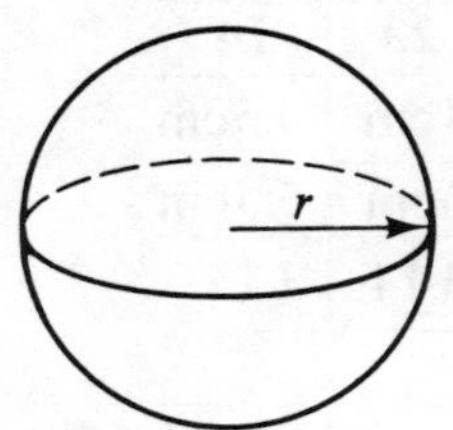

Surface area of a sphere $= 4\pi r^2$

Volume of a sphere $= \frac{4}{3}\pi r^3$

Exercise 19 Give the missing entries in the following tables:

Rectangle or Parallelogram

Question

	1	2	3	4
Length	5cm	10cm	8m	9cm
Breadth	8cm	4.9cm	50cm	
Area				108cm^2

Triangle Question

	5	6	7	8
Base	8cm	9m	14.8cm	6cm
Height	5cm	7m	10cm	
Area				27cm^2

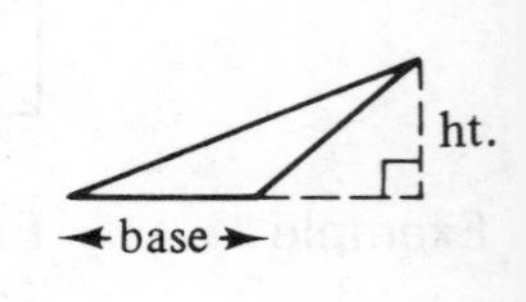

Trapezium Question

	9	10	11	12
Length a	4cm	3cm	0.5m	14cm
Length b	8cm	4cm	1.5m	16cm
Height h	5cm	4cm	80cm	
Area				75cm^2

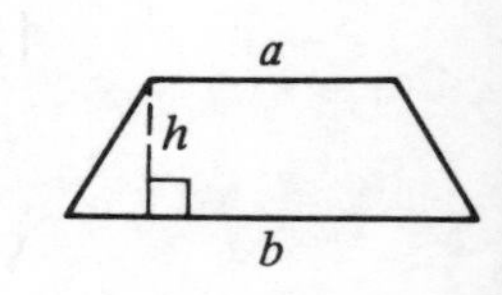

Circle Question

	13	14	15	16
Radius	7cm	$10\frac{1}{2}$cm	10cm	
π	$\frac{22}{7}$	$\frac{22}{7}$	3.14	3.14
Circumf				15.7cm
Area				

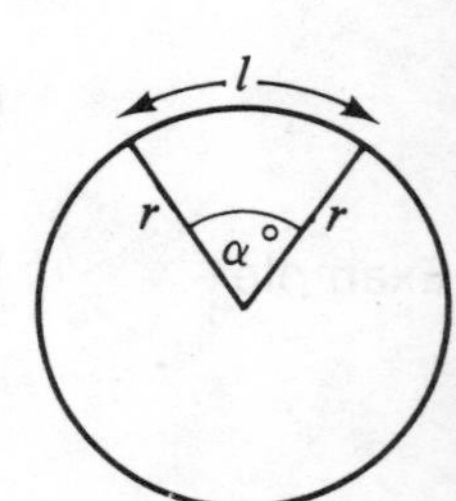
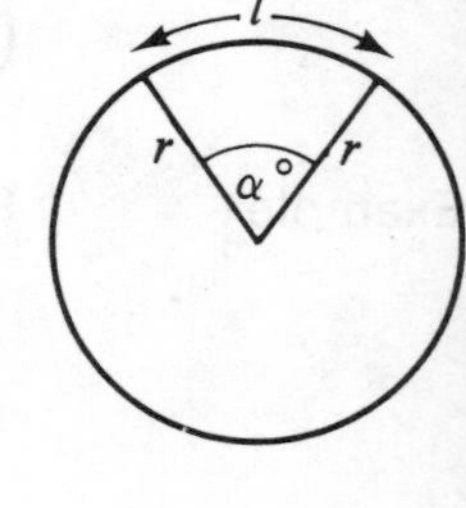

Sector (Leave answer in terms of π and r.)

Question

	17	18	19	20
Angle α	90°	120°		
Length l			$3r$	
Area				$\frac{\pi r^2}{5}$

Cylinder Question

	21	22	23	24
Radius	7cm	21cm	4cm	4.5cm
Height	17cm	14cm	6cm	5.5cm
π	$\frac{22}{7}$	$\frac{22}{7}$	3.14	3.14
Total sur. area				
Volume				

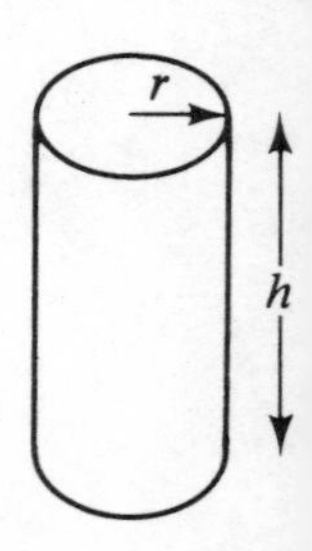

Other solids

	Question 25	26	27	28
Base area	10m^2	15cm^2	$3\frac{1}{2}\text{cm}^2$	34.5cm^2
Height	5cm	7cm	$5\frac{1}{2}$cm	0.3m
Cuboid or Prism Volume				
Cone or Pyramid Volume				

Sphere

	Question 29	30
Radius	7cm	9cm
π	$\frac{22}{7}$	3.14
Surface Area		
Volume		

Examination Standard Questions

Worked examples

1. Find the area, in cm^2, of a circle whose circumference is 12πcm.

[From the formulae] circumference of a circle $= 2\pi r$

[Substituting] $12\pi = 2\pi r$

[Divide both sides by 2π] $\frac{12\pi}{2\pi} = r$

$\underline{6\text{cm}} = \text{radius}$

[From the formulae] Area of a circle $= \pi r^2$

$= \pi(6^2)$

$= \underline{36\pi\text{cm}^2}$ ANS

2.

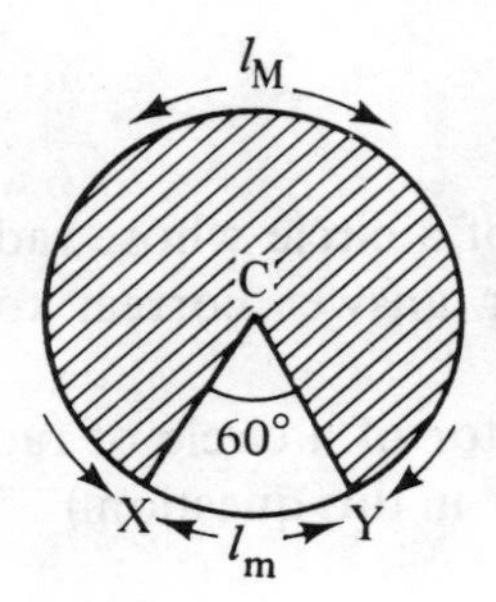

In the diagram, C is the centre of the circle. Find (a) the ratio of the lengths of the minor and major arcs XY, (b) the area shaded as a fraction of the area of the circle.

(a) Let l_m represent the minor arc and l_M the major arc XY.

[From the formulae]

$$\text{Length of circular arc } (l_m) = \frac{60}{360} \times 2\pi r$$

$$l_m = \frac{1}{6} \times 2\pi r$$

$$= \frac{\pi r}{3}$$

$$\text{and } l_M = \frac{300}{360} \times 2\pi r$$

$$= \frac{5}{6} \times 2\pi r$$

$$= \frac{5\pi r}{3}$$

$$l_m : l_M = \frac{\pi r}{3} : \frac{5\pi r}{3} = \underline{1:5} \quad \text{ANS}$$

(b) [From the formulae] area of major sector is shaded

$$\text{Therefore, the required fraction} = \frac{\text{area of sector}}{\text{area of circle}}$$

$$= \frac{\frac{300}{360}\pi r^2}{\pi r^2}$$

$$= \frac{300}{360} = \underline{\frac{5}{6}} \quad \text{ANS}$$

Exercise 20 1.

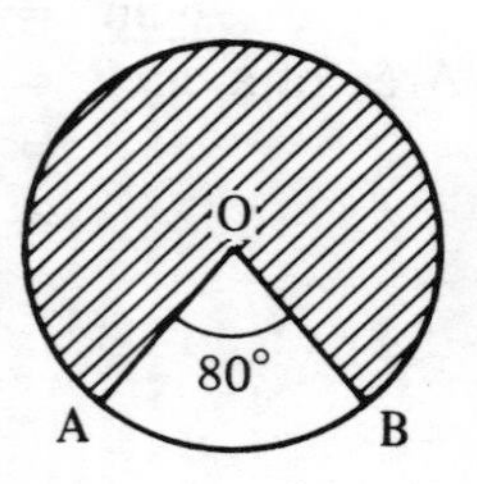

In this diagram, O is the centre of the circle. Find (a) the ratio of the lengths of the minor and major arcs AB, (b) the area shaded as a fraction of the area of the circle. (JMB)

2. Calculate the circumference of a circle whose radius is 3cm. (Take π as 3.14 and give your answer correct to two significant figures.) (JMB)

3. The diagram represents a sector of a circle of radius 14cm. Angle AOB is a right angle. (Take π as $\frac{22}{7}$ in this question.)

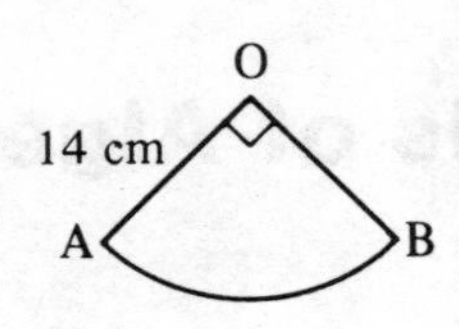

(a) Calculate the length of the arc AB

(b) The edges OA and OB are brought together so that the sector forms the curved surface of a cone. Calculate the radius of the base of this cone and the height of the cone.

(c) Another cone is formed from a sector of the same circle with angle 270°. Calculate the radius of its base. (JMB)

4. An arc LM of a circle, centre O, radius 6cm, is of length 8.8cm. Calculate the size in degrees of angle LOM. (Take $\pi = \frac{22}{7}$) (L)

5. A cylindrical garden roller has a diameter of 40cm and is 70cm wide. Calculate the area, in m^2, rolled during 100 revolutions of the roller. (Take $\pi = \frac{22}{7}$.) (L)

6. A cubical box has each of its internal edges of length 90cm. Find the volume of the inside of the box in cubic centimetres, giving your answer in standard index form. (O)

7. A right circular cone has a height of 4cm and a base whose radius is 3cm. Determine whether the following statements are true or false.

(a) The total surface area is $5\pi cm^2$

(b) The circular cylinder which has the same height and volume as the cone has diameter 1cm. (O)

8. A hollow tin cylinder is closed at both ends. Its length is 13 cm and its radius is 7cm. Taking $\pi = 3\frac{1}{7}$, calculate the total surface area of the tin cylinder. (WJEC)

9. A cylinder is 10.5cm high and has a circular base of radius 6cm. Taking π as $3\frac{1}{7}$, find the volume of the cylinder. Find also the length of an edge of a cube which has the same volume. Give your answer correctly rounded off to two significant figures. (O)

10. Calculate the radius of the circle whose area is $154cm^2$. (Take π as $\frac{22}{7}$.) (JMB)

4
The Fundamentals of Algebra

Definitions

When letters are used to represent numbers, this branch of mathematics is called Algebra. The symbols $+, -, \times, \div, =$, have the same meanings in Algebra as they do in Arithmetic.

Note

$$x + x = 2x$$
$$x + y = y + x \quad \text{(Just as } 2 + 3 = 3 + 2\text{)}$$
$$3x - x = 2x$$

Example

Simplify $2x + y - 3x + 2y + 4x$

First rearrange $2x - 3x + 4x + y + 2y$

This gives $\underline{3x + 3y}$ ANS

The collection of symbols in the above example is called an algebraic **expression**. The components $2x$, $+y$, $-3x$, $+2y$, $+4x$ are all called **terms** in that expression.

Rules for directed numbers and algebraic terms are as follows:

$$++ = +, \quad +- = -+ = -, \quad -- = +.$$

e.g.

$$+(+2) = +2 = 2 \qquad +(+x) = +x = x$$
$$+(-2) = -2 \qquad +(-x) = -x$$
$$-(+2) = -2 \qquad -(+x) = -x$$
$$-(-2) = +2 = 2 \qquad -(-x) = +x = x$$

$$(+2) \times (+3) = +6 = 6, \quad (+6) \div (+2) = +3 = 3, \quad (+x) \times (+y) = +xy = xy,$$
$$(+2) \times (-3) = -6, \quad (+6) \div (-2) = -3, \quad (+x) \times (-y) = -xy,$$
$$(-2) \times (+3) = -6, \quad (-6) \div (+2) = -3, \quad (-x) \times (+y) = -xy,$$
$$(-2) \times (-3) = +6 = 6, \quad (-6) \div (-2) = +3 = 3, \quad (-x) \times (-y) = +xy = xy.$$

Note

Any term without a $+$ or a $-$ sign in front of it is accepted as being positive. The rules for division are the same as those for multiplication.

Examples

1. Find the sum of $3x$ and $-2x$.

Sum means add i.e. $(+3x) + (-2x)$

$$= +3x - 2x$$
$$= \underline{x} \quad \text{ANS}$$

2. Subtract $-4y$ from $6y$.

This means $(+6y)-(-4y)$

$= +6y+4y$

$= \underline{10y}$ ANS

Exercise 21 Find the sum of:

1. $+12$ and $+13$
2. -12 and -13
3. -12 and $+13$
4. x and $2x$
5. x and $-x$
6. $2x$ and $-7x$
7. $-x$ and $-3x$
8. $-a$, $2a$, $4a$
9. $7a$, $-3a$, $-9a$
10. $-b$, $-14b$, $11b$

Subtract:

11. $+3$ from $+5$
12. -3 from $+5$
13. $+3$ from -5
14. x from $2x$
15. x from $-x$
16. $2x$ from $-7x$
17. $-2x$ from $-7x$
18. $-11a$ from $8a$
19. $12a$ from $6a$
20. $-12a$ from $-6a$

When two numbers are multiplied together in Arithmetic we call the result the **product**.

e.g. the product of 3 and 5 is $3 \times 5 = \underline{15}$

In Algebra the product of x and y would be written simply as xy. (The multiplication sign is left out.) We also say that x and y are the **factors** of the product xy.

e.g. the factors of 15 are 3 and 5.
the factors of $2pq$ are 2, p and q.

Note

$a \times a = a^2$ (a squared)

$a \times a \times a = a^3$ (a cubed)

The small numbers 2 and 3 in the above note are called **index** numbers, (plural indices).

e.g. $2^2 = 2 \times 2 = \underline{4}$
$2^3 = 2 \times 2 \times 2 = \underline{8}$

Higher powers (another word for indices) are expressed as follows:

a^4 is read 'a to the fourth',
a^5 is read 'a to the fifth', etc.

Substitution

Examples

1. Find the volume of a cube, each edge being of length 4 cm. Give also an expression for the volume of *any* cube, whose edges are of length x cm.

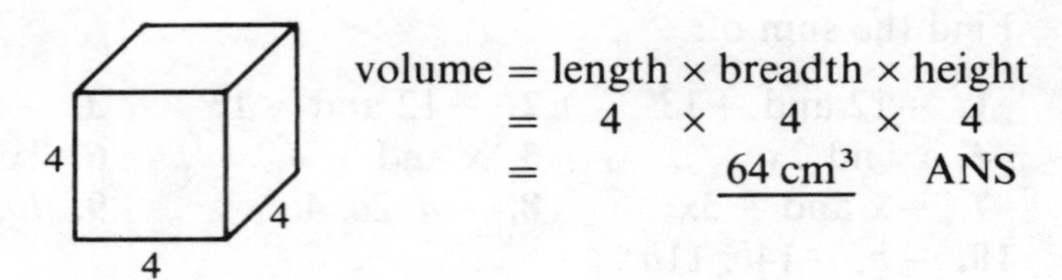

$$\begin{aligned} \text{volume} &= \text{length} \times \text{breadth} \times \text{height} \\ &= 4 \times 4 \times 4 \\ &= \underline{64 \text{ cm}^3} \quad \text{ANS} \end{aligned}$$

(The above answer is read as 64 cubic centimetres.)

$$\begin{aligned} \text{volume of any cube} &= x \times x \times x \\ &= \underline{x^3 \text{ cm}^3} \quad \text{ANS} \end{aligned}$$

2. Find the cubic capacity of a refrigerator of internal dimensions 2 m by 1 m by 1 m. Give also an expression for the cubic capacity of any refrigerator of internal dimensions; height x metres, depth y metres, width y metres.

$$\begin{aligned} \text{volume} &= 2 \times 1 \times 1 \\ &= \underline{2 \text{ m}^3} \quad \text{ANS} \end{aligned}$$

$$\begin{aligned} \text{The volume of any such refrigerator} &= x \times y \times y \\ &= \underline{xy^2 \text{ m}^3} \quad \text{ANS} \end{aligned}$$

(Do not be confused by the m^3, this simply refers to the units being used, cubic metres in this case.).

3. If $a = 3$, $b = 4$, $c = 1$, find: (a) a^3, (b) ab^2, (c) $\frac{a}{b}$, (d) ac, (e) $\frac{2}{3}a^2$.

(a) Substitute the value 3 for a,

i.e. $a^3 = 3^3 = 3 \times 3 \times 3 = \underline{27}$ ANS

(b) Substitute for a and b,

i.e. $ab^2 = 3 \times (4)^2 = 3 \times 16 = \underline{48}$ ANS

Note $ab^2 = a \times b \times b$ but $(ab)^2 = (a \times b) \times (a \times b)$.

(c) $\frac{a}{b} = \underline{\frac{3}{4}}$ ANS

(d) $ac = 3 \times 1 = \underline{3}$ ANS

(e) $\frac{2}{3}a^2 = \frac{2}{3} \times 3^2 = \frac{2}{3} \times 9 = \underline{6}$ ANS

Exercise 22 If $x = 2$, $y = 3$, $z = 1$, find the value of:

1. x^2 **2.** x^2y **3.** xyz **4.** $\frac{x}{y}$

5. $\frac{xy}{z}$ **6.** $\frac{xy^2}{z}$ **7.** $\frac{(xy)^2}{z}$ **8.** $x^2y - xy^2$

9. $x^4y - xy^3$ **10.** $xy^3 + x^2y$ **11.** x^y **12.** z^x

13. $\frac{3}{4}x^2$ **14.** $\frac{1}{3}y^3$ **15.** $\frac{y^3z}{9}$

Examples

1. Find the product of $-12x$ and $-2x$.

[This means] $(-12x) \times (-2x)$

$= \underline{24x^2}$ ANS

2. Divide $18y$ by $-3y$

[This means] $(+18y) \div (-3y)$

[The *y*s cancel] $= \frac{+18\not{y}}{-\ 3\not{y}}$

$= \underline{-6}$ ANS

Note

$x^2 + x$ cannot be simplified

$x^2 \times x$ however is x^3 (because $x^2 \times x = x \times x \times x$)

Do not confuse these two conditions.

Examples

1. Find the product of y and $-y$.

i.e. $(+y) \times (-y) = \underline{-y^2}$ ANS

2. Find the sum of y and $-y$.

[In the same way that $(+2) + (-2) = 0$] $(+y) + (-y) = \underline{0}$ ANS

Exercise 23 Find the products of:

1. 7 and 5, **2.** $7x$ and 5 **3.** $7x$ and $5x$

4. $5x$ and $7x$ **5.** $7x$ and $-5x$ **6.** $-7x$ and $5x$

7. $-7x$ and $-5x$ **8.** $+7x$ and $+5x$ **9.** $2x$ and $-9x$

10. $-11x$ and $-10x$

Divide:

11. 35 by 7 **12.** $35x$ by 7 **13.** $35x^2$ by $5x$

14. $35x^2$ by $7x$ **15.** $-35x^2$ by $7x$ **16.** $-35x^2$ by $-7x$

17. $35x^2$ by $-5x$ **18.** $35x^2$ by $35x$ **19.** $-35x^2$ by x

20. $-35x^2$ by $-x$.

Bracketed Quantities

Examples

1. $(3+7)+(2+4)$
$= (10) + (6)$
$= \underline{16}$ ANS

Note

Always deal with the bracketed terms first.

2. $(3+7)+(2-4)$
$= (10) + (-2)$
$= 10 - 2$
$= \underline{8}$ ANS

3. $(3+7)-(2+4)$
$= (10) - (6)$
$= \underline{4}$ ANS

4. $3+7-2+4$
$= +14-2$
$= \underline{12}$ ANS

It can be seen that (4) does not equal (3) and therefore it is important to note that if there is a minus sign outside a bracket the numbers inside change their sign, as shown in the rules at the beginning of this chapter. The same rule applies in Algebra.

Examples

1.

$(2x+1)+(3x+3)$
[No minus sign outside means no change] $= 2x+1 + 3x+3$
$= \underline{5x+4}$ ANS

2.

$(2x+1)-(3x+3)$
[The minus sign outside the second bracket changes the signs within] $= 2x+1-3x-3$
$= \underline{-x-2}$ ANS

Exercise 24

Simplify:

1. $(1+3)-(1+2)$
2. $(2+3)-(2+1)+(4-2)$
3. $(a+3a)-(a+2a)$
4. $(2a+3a)-(2a+a)+(4a-2a)$
5. $(3a+2)+(a-2)$
6. $(3a+2)-(a-2)$
7. $(2a+2b)+(a+b)$
8. $(2a+2b)-(a+b)$
9. $(4x-2y)+(2x-y)$
10. $(4x-2y)-(2x-y)$
11. $(x+y+z)-(x+y-z)$
12. $(x^2+2)+(2x^2+3)$
13. $(x^2+2x)+(2x^2-3x)$
14. $(4x^3+y^2)-(2x^3-y)$
15. $(3a^2-a)-(a^3+1)$

A value immediately in front of a bracket multiplies the terms within.

e.g. $3(2x+1) = 6x+3$
also $-2(x^2-3) = -2x^2+6$

Examples

Simplify **1.** $3(2x-2y)$

[The terms in the bracket are multiplied by the 3] $= \underline{6x-6y}$ ANS

2. $x(x^3+x^2+x+1)$

[Multiply by the x] $= \underline{x^4+x^3+x^2+x}$ ANS

3. $12x^2-3(x-4)+7(x^2+x)$

[The first bracket is multiplied by -3, the second by $+7$] $= 12x^2-3x+12+7x^2+7x$

[Like terms are grouped together] $= 12x^2+7x^2-3x+7x+12$

$= \underline{19x^2+4x+12}$ ANS

Note

The type of expression shown in the answer to **3.** above is known as a **quadratic.**

Exercise 25

Simplify:

1. $2(2x+1)$ **2.** $3(x^2+x-2)$
3. $4(a+b)-2(a-b)$ **4.** $a(a+b)$
5. $x(x+y)+3x(x-z)-y(x-y)$ **6.** $3(2a+3b+c)-2(a-4b+2c)$
7. $3a-2(a-b)+3(a+b)$ **8.** $pq(r+s)+rs(p+q)$
9. $3x^2-2x(x+1)+3$ **10.** $9x+3x(x^2+2x)-7(x+3)+x^3$

The next situation involves the multiplication of two or more bracketed quantities.

e.g. Simplify $(a+b)(c+d)$
The answer is $ac+ad+bc+bd$
Each term in the second bracket is multiplied by each term in the first.

Examples

Simplify
1. $(x+y)(x+z)$.

This is done in four steps

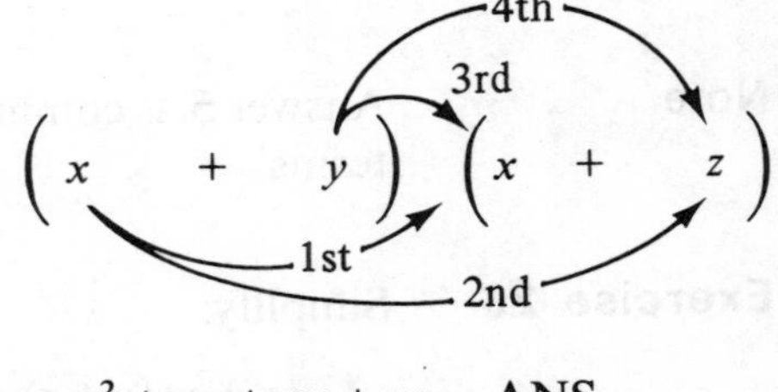

$= \underline{x^2+xz+xy+yz}$ ANS

Note

You may think that the third term in the above answer should have been written as yx, but it is conventional to write such terms in alphabetical order.

2. $(x+2)(x^2+x-1)$

Three terms multiplied by two terms means six steps

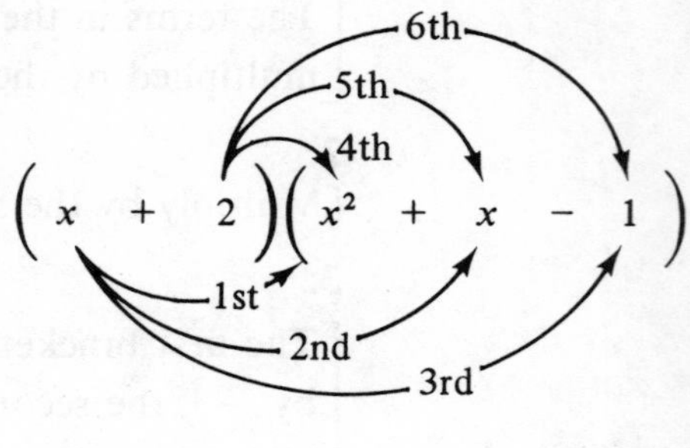

$$= x^3+x^2-x+2x^2+2x-2$$
$$= \underline{x^3+3x^2+x-2} \quad \text{ANS}$$

3. $(a+b)^2$

[This means] $(a+b)(a+b)$

$$= a^2+ab+ab+b^2$$
$$= \underline{a^2+2ab+b^2} \quad \text{ANS}$$

Note $(a+b)^2$ is not the same as a^2+b^2. You can prove this for yourself by substituting simple numbers for a and b in both situations.

e.g. $(2+3)^2 = 5^2 = \underline{25}$
but $2^2+3^2 = 4+9 = \underline{13}$

4. $(a-b)^2$

$$= (a-b)(a-b)$$
$$= a^2-ab-ab+b^2$$
$$= \underline{a^2-2ab+b^2} \quad \text{ANS}$$

5. $(a+b)(a-b)$

$$= a^2-ab+ab-b^2$$
$$= \underline{a^2-b^2} \quad \text{ANS}$$

Note Answer **5.** is commonly referred to as the 'difference of the two squared terms'.

Exercise 26 Simplify:

1. $(x+y)(y+z)$ **2.** $(x+2)(x+3)$ **3.** $(x^2+x)(x+1)$
4. $(x^2+2x+3)(x-2)$ **5.** $(x+y)^2$ **6.** $(x-y)^2$
7. $(x+y)(x-y)$ **8.** $3(p+q)^2$ **9.** $4(p+q)(p-q)$
10. $8(p^2+q^2)^2$

Factorisation

Factorisation can be described as the reverse of the previous procedures. Earlier we simplified $3(2x+1)$ as an example. The answer was $6x+3$. To reverse this process, if we were asked to factorise $6x+3$, the answer would be $3(2x+1)$. The 3 is the factor common to both terms in the question.

Examples

Factorise:

1. $x^4+x^3+x^2+x$

[x is common to all terms] $= x(x^3+x^2+x+1)$ ANS

This can be checked by multiplying out the answer. You should get back to the question.

2. $3x^2+6x^4$

[Both 3 and x^2 are common to both terms] $= 3x^2(1+2x^2)$ ANS

3. $4a^2-9b^2$

This should be recognised as the 'difference of two squares', the squared terms being $2a$ and $3b$.

[Compare with example **5** from the previous section] $= (2a+3b)(2a-3b)$ ANS

4. $ac+ad+bc+bd$

Two steps are required as there is no factor common to all four terms.
First step: take a out as the common factor in terms 1 and 2
take b out as the common factor in terms 3 and 4

$$= a(c+d)+b(c+d)$$

Second step: take the common factor of $(c+d)$ out

$$= (c+d)(a+b) \quad \text{ANS}$$

5. $pr+ps-qr-qs$

Two steps are again required.
First step: take out p as the common factor in terms 1 and 2
take out q as the common factor in terms 3 and 4

[Note that $(-q)(+s) = -qs$] $= p(r+s)-q(r+s)$
Second step: take out the common factor of $(r+s)$

$$= (p-q)(r+s) \quad \text{ANS}$$

6. x^2+4x+4
This quadratic type usually factorises into two pairs of brackets; ( )( ). The factors of the third term, $+4$, could be $+1$ and $+4$, -1 and -4, $+2$ and $+2$, -2 and -2.
The factors of the first term x^2, can only be $+x$ and $+x$ or $-x$ and $-x$. By inspection it can be seen that the correct combination is $(x+2)(x+2)$. Multiply out the brackets to prove this for yourself. Two more examples now follow, the procedure is also dealt with fully in Chapter 5.

7. $3x^2-7x-20$
The factors of $3x^2$ are $3x$ and x. (Either both plus or both minus.) The factors of 20 can be 20 and 1, 5 and 4 or 10 and 2. We must find which combination is correct, noting that the two signs within the brackets must be different in order to get -20.

Trials: $(3x+20)(x-1) = 3x^2+17x-20$ (incorrect)
$(3x+10)(x-2) = 3x^2+4x-20$ (incorrect)
$\underline{(3x+5)(x-4)} = 3x^2-7x-20$ (correct)

8. $35a^2+3a-2$
The factors of $35a^2$ can be $35a$ and a, or $7a$ and $5a$. The factors of 2 can only be 2 and 1. To obtain -2 the signs within the brackets must be different.

Trials: $(35a+2)(a-1) = 35a^2-33a-2$ (incorrect)
$\underline{(7a+2)(5a-1)} = 35a^2+3a-2$ (correct)

Exercise 27 Factorise:

1. x^3+x^2+2x	**2.** $4x^2+12x^3$	**3.** $3a+3b$
4. ab^2+ac^2	**5.** $abc-abd$	**6.** $24xy-8x^2y$
7. $pq-pqr$	**8.** $80a-40b$	**9.** $18xy-3x$
10. x^2+xy	**11.** $5x+5y+ax+ay$	
12. $3pq+5q+5r+3pr$	**13.** $ac-bc+2ad-2bd$	
14. $x^2-xy+xz-yz$	**15.** $20+5x+4y+xy$	
16. $x^2+9x+20$	**17.** $x^2+17x+66$	**18.** x^2+2x+1
19. $x^2+18x+80$	**20.** x^2+4x+3	**21.** $x^2+15x+54$
22. $x^2+11x+24$	**23.** $x^2+20x+96$	**24.** $x^2-14x+48$
25. $x^2-9x+20$	**26.** $x^2-6x-27$	**27.** $x^2-3x-70$
28. $2x^2+19x+35$	**29.** $4x^2+13x+10$	**30.** $21x^2+x-2$
31. $6x^2-5x-6$	**32.** x^2-49	**33.** a^2-c^2
34. b^2-100	**35.** $4x^2-y^2$	**36.** $9x^2-16y^2$
37. $9-x^2$	**38.** $36x^2-25$	**39.** a^2-1
40. 17^2-13^2.		

Formulae and Change of Subject

In the formula $v = u + ft$, the letter v is the **subject**. This same formula could be written in three other ways:

$$u = v - ft$$
$$f = \frac{v-u}{t}$$
$$t = \frac{v-u}{f}$$

The subject has been changed in each case.

Examples

1. Make B the subject of the formula $A = LB$.

This means the formula has to be rearranged to leave B by itself. The rule to remember is that we may add, subtract, multiply or divide one side of an equation by any amount, provided we do the same to the other side. (I refer here to both 'sides' of the equals sign.)

In this example LB means L multiplied by B.
Therefore, in order to leave B by itself we must divide by L.

The equation now becomes $\frac{A}{L} = \frac{LB}{L}$

[The Ls on the right cancel] $\underline{\frac{A}{L} = B}$ ANS

[*Note:* This is the same as saying $B = \frac{A}{L}$]

2. Make A the subject of the formula $V = \frac{Ah}{3}$

[Multiply both sides by 3 to clear the fraction] $3V = Ah$

[Divide both sides by h] $\underline{\frac{3V}{h} = A}$ ANS

3. Make a the subject of the formula $A = \frac{h(a+b)}{2}$

[Multiply both sides by 2] $\therefore 2A = h(a+b)$

[Divide through by h] $\therefore \frac{2A}{h} = \frac{{}^1\cancel{h}(a+b)}{\cancel{h}_1}$

[There is no need for the brackets now that the h has gone on that side] $\therefore \frac{2A}{h} = a+b$

[Subtract b from both sides] $\therefore \underline{\frac{2A}{h} - b = a}$ ANS

4. Make r the subject of the formula $\quad V = \pi r^2 h$

[Divide through by π] $\quad \therefore \frac{V}{\pi} = \frac{{}^1\not{\pi} r^2 h}{\not{\pi}_1}$

$$\therefore \frac{V}{\pi} = r^2 h$$

[Divide through by h] $\quad \therefore \frac{V}{\pi h} = \frac{r^2 \not{h}^1}{\not{h}_1}$

$$\therefore \frac{V}{\pi h} = r^2$$

[Take the square root of each side] $\quad \therefore \sqrt{\frac{V}{\pi h}} = r$ ANS

Exercise 28 Rearrange the following formulae to make the indicated component the new subject:

1. $A = LB \ (L)$ 2. $V = \frac{1}{3}Ah \ (h)$ 3. $A = \frac{h(a+b)}{2} \ (h)$

4. $V = LBW \ (B)$ 5. $A = 2\pi r(r+h) \ (h)$ 6. $C = \frac{5}{9}(F-32) \ (F)$

7. $I = \frac{PRT}{100} \ (T)$ 8. $A = P + \frac{PRT}{100} \ (R)$ 9. $x = u + at \ (a)$

10. $v^2 = u^2 + 2ax \ (u)$ 11. $v^2 = u^2 + 2ax \ (x)$ 12. $\frac{1}{u} + \frac{1}{v} = \frac{1}{f} \ (f)$

Indices

Positive Indices

When multiplying terms of the same base, such as x^3 and x^2, we add the index numbers. In this case we would get x^5.

Examples

1. $y^2 \times y^3 = y^{(2+3)} = \underline{y^5}$
2. $2y^2 \times 3y^3 = 2 \times 3 \times y^{(2+3)} = \underline{6y^5}$
3. $3p^2q^3 \times 4pq^4 = \underline{12p^3q^7}$
4. $2x^2 \times 3x^3 \times 4x^4 = \underline{24x^9}$

When **dividing** one term by another, if the bases are the same, such as $x^5 \div x^2$, we subtract the second index number from the first. In this case we get x^3.

Examples

1. $y^5 \div y^2 = y^{(5-2)} = \underline{y^3}$
2. $10y^5 \div 2y^2 = \frac{10}{2}\left[y^{(5-2)}\right] = \underline{5y^3}$

3. $12p^2q^3 \div 3pq^4 = \dfrac{12p^2q^3}{3pq^4} = \underline{\dfrac{4p}{q}}$

Note — If terms are NOT of the same base these methods cannot be used:

$$a^4 \times b^2 = \underline{a^4b^2} \quad \text{NOT } ab^6$$

$$a^4 \div b^2 = \underline{\frac{a^4}{b^2}}$$

Exercise 29 — Simplify:

1. $a^2 \times a^3$ 2. $a^5 \times a^4$ 3. $a^7 \times a$ 4. $2a^3 \times 3a^2$
5. $5a^5 \times 4a^4$ 6. $a^2b \times ab^2$ 7. $2a^2b \times 3ab$ 8. $a^4 \div a^2$
9. $30a^5 \div 10a$ 10. $4a^3b^2 \div 2ab^4$

Negative Indices — A negative index number indicates a **reciprocal.**

Note — A **reciprocal** is an '*upside-down*' form of the number or term.

e.g. the reciprocal of 2 is $\frac{1}{2}$

the reciprocal of 10 is $\frac{1}{10}$

the reciprocal of x is $\frac{1}{x}$

the reciprocal of $\frac{3}{4}$ is $\frac{4}{3} = 1\frac{1}{3}$

Examples

1. $2^{-1} = \dfrac{1}{2}$

2. $a^{-1} = \dfrac{1}{a}$

3. $3^{-2} = \dfrac{1}{3^2} = \dfrac{1}{9}$

4. $a^{-2} = \dfrac{1}{a^2}$

5. $a^{-2} \times a^{-3} = a^{(-2)+(-3)} = a^{-5} = \dfrac{1}{a^5}$

6. $a^{-4} \div a^2 = a^{(-4)-(2)} = a^{-6} = \dfrac{1}{a^6}$

Note — Any number or letter raised to the power zero equals one.

e.g. $a^0, x^0, 2^0, 3^0, \ldots = 1$

Double indices

Examples

1. $(x^2)^3 = x^{2\times 3} = x^6$ (This time we multiply the indices.)
2. $(a^{-2})^2 = a^{-4} = \dfrac{1}{a^4}$
3. $(3a^2)^3 = 3^3 \times a^{2\times 3} = 27a^6$
4. $(3a^{-2})^2 = 3^2 \times a^{(-2)\times(2)} = 9a^{-4} = \dfrac{9}{a^4}$

Exercise 30 Simplify, giving your answer with positive indices:

1. x^{-2}	**2.** b^{-2}	**3.** d^0
4. $a^4 \times a^{-2}$	**5.** $a^4 \div a^{-2}$	**6.** $b^{-2} \times b^{-5}$
7. $b^{-2} \div b^{-5}$	**8.** $a^2b^{-2} \times ab$	**9.** $\dfrac{1}{a} \times a^3$
10. $\dfrac{b}{a^2} \times a^3b$	**11.** $(a^2)^3$	**12.** $(a^4)^2$
13. $(a^{-2})^2$	**14.** $(a^2)^{-2}$	**15.** $(a^2b)^3$
16. $(a^2b^{-1})^3$	**17.** $(2a^{-3})^3$	**18.** $(3a^2b^{-2})^2$
19. $(3a^2)^2 \times (2a^{-1})^2$	**20.** $(4a^2b^3)^2 \times b^{-6}$	

Fractional indices

The roots of quantities can be expressed as a fractional index.

Examples

1. $4^{\frac{1}{2}} = \sqrt{4} = 2$ (*Note*: $2^2 = 2 \times 2 = 4$)
2. $a^{\frac{1}{2}} = \sqrt{a}$
3. $8^{\frac{1}{3}} = \sqrt[3]{8} = 2$ (*Note*: $2^3 = 8$)
4. $a^{\frac{1}{3}} = \sqrt[3]{a}$
5. $a^{\frac{1}{25}} = \sqrt[25]{a}$
6. $8^{\frac{2}{3}} = (\sqrt[3]{8})^2 = (2)^2 = 4$
7. $a^{\frac{2}{3}} = (\sqrt[3]{a})^2$
8. $25^{\frac{3}{2}} = (\sqrt{25})^3 = (5)^3 = 125$

Note The form $\sqrt[3]{a^2}$ is known as **surd** form.

Exercise 31 Write in **surd** form:

1. $a^{\frac{3}{2}}$ **2.** $b^{\frac{1}{4}}$ **3.** $b^{\frac{3}{4}}$ **4.** $(x^3)^{\frac{1}{2}}$ **5.** $(x^{\frac{1}{3}})^{-2}$

Evaluate (i.e. obtain a numerical answer as in examples **1**, **3**, **6** above.):

6. $25^{\frac{1}{2}}$ **7.** $27^{\frac{1}{3}}$ **8.** $27^{\frac{2}{3}}$ **9.** $16^{-\frac{3}{4}}$ **10.** $49^{\frac{3}{2}}$
11. $81^{\frac{3}{4}}$ **12.** $32^{-\frac{3}{5}}$ **13.** $64^{\frac{3}{2}}$ **14.** $8^{-\frac{4}{3}}$ **15.** $125^{\frac{2}{3}}$

Examination Standard Questions

Worked examples

1. Simplify $(2a+b)(a+2b)-2(a-b)^2$

$$= \overbrace{2a^2+4ab+ab+2b^2}^{\text{1st pair of brackets}}-2(a^2-2ab+b^2)$$
$$= 2a^2+4ab+ab+2b^2-2a^2+4ab-2b^2$$

[Grouping like terms] $= 2a^2-2a^2+2b^2-2b^2+4ab+ab+4ab$

$= \underline{9ab}$ ANS

2. Express as a product of factors (a) $pq+3p$, (b) p^2-49

(a)

['p' is the factor common to both terms] $\therefore$ $pq+3p$

$= \underline{p(q+3)}$ ANS

(b)

[Revise the earlier example referred to as the 'difference of two squared terms'. In this case it is the square of 'p' minus the square of '7'] $\therefore$ p^2-49

$= \underline{(p+7)\,(p-7)}$ ANS

3. Evaluate $4a^2+ab-2b^2$, when $a=\frac{1}{4}$, $b=-\frac{1}{2}$, giving your answer as simple as possible.

[Substitute the given values, remembering to use brackets] $\left[4\left(\frac{1}{4}\right)^2\right]+\left[\left(\frac{1}{4}\right)\left(-\frac{1}{2}\right)\right]-\left[2\left(-\frac{1}{2}\right)^2\right]$

[Simplify the brackets first] $=\left[4\left(\frac{1}{16}\right)\right]+\left[\left(-\frac{1}{8}\right)\right]-\left[2\left(+\frac{1}{4}\right)\right]$

$=\frac{4}{16}\quad -\frac{1}{8}\quad -\frac{2}{4}$

[The common denominator is 16] $=\frac{4-2-8}{16}=-\frac{6}{16}=\underline{-\frac{3}{8}}$ ANS

4. If $y=\frac{p}{2x^3+q}$, express x in terms of y, p and q.

[Multiply both sides by $(2x^3+q)$ to clear the fraction] $\therefore\ y(2x^3+q)=p$

[Divide through by y] i.e. $\frac{{}^1\!\not{y}(2x^3+q)}{\not{y}_1}=\frac{p}{y}$

$\therefore\ 2x^3+q=\frac{p}{y}$

[Subtract q from both sides] $\therefore\ 2x^3=\frac{p}{y}-q$

[Divide through by 2] $\therefore x^3 = \dfrac{p}{2y} - \dfrac{q}{2}$

[Common denominator is $2y$] $\therefore x^3 = \dfrac{p-qy}{2y}$

[Taking the cube root gives] $x = \sqrt[3]{\dfrac{p-qy}{2y}}$ ANS

5. Given that $\dfrac{a^7}{a^3 \times a^2} = a^y$ find the value of y.

[Simplifying the denominator gives] $\dfrac{a^7}{a^5} = a^y$

[Dividing, therefore subtract index 5 from index 7] $\therefore a^2 = a^y$

$\therefore y = \underline{2}$ ANS

Exercise 32

1. Factorise completely (a) $ab - 4a$, (b) $x^2 - 16$.
2. Simplify $(3x+y)(x+2y) - 3(x-y)^2 + y^2$.
3. Evaluate $3x^2 + xy$, when $x = \frac{1}{2}$, $y = -\frac{1}{3}$, simplify your answer.
4. Factorise completely (a) $18x^2 - 2$, (b) $a(b-3) + 6c - 2bc$ (WJEC)
5. Factorise completely (a) $2a^2 + 4ac - 6bc - 3ab$, (b) $12r^2 - 17rs + 6s^2$ (c) $(p+q)^2 - 4r^2$ (C)
6. Factorise completely (a) $45x^2 - 20$, (b) $6eg - 3eh + 4fg - 2fh$. (C)
7. Find the value of $p^3(t^2-1) + p^2(t-1)^2$, when $p = 30$ and $t = 7$. (JMB)
8. If $a = \dfrac{2b}{c^2 - d}$, express c in terms of a, b and d.
9. Given that $a = b^2\sqrt{(c+d)}$, express c in terms of a, b and d. (C)
10. Given that $H = a + \dfrac{b}{P}$, express P in terms of a, b and H. (JMB)
11. Without using tables, calculate (a) $9^{\frac{1}{2}}$, (b) $8^{\frac{2}{3}}$ (AEB)
12. Without using tables, calculate (a) $81^{\frac{3}{4}}$ (b) $125^{-\frac{2}{3}}$ (C)
13. Without using tables, evaluate (a) $(2^3)^{-2}$ (b) $(\frac{9}{25})^{\frac{3}{2}}$ (L)
14. Find the exact value of $x^2 - y^2$ when $x = 12.875$ and $y = 2.875$. (JMB)
15. Calculate the exact value of $\frac{30}{7}(4.25^2 - 2.75^2)$ (JMB)
16. Calculate x if $5^x = 25^3$ (L)
17. Find the numerical value of $a(b^2+c^2) + b(c^2+a^2) + c(a^2+b^2)$, when $a = 3$, $b = 2$ and $c = 0$. (JMB)
18. Find the real number 'n' such that $1^3 + 2^3 + 3^3 = (1+2+3)^n$. (JMB)
19. Factorise $(2x-1)^2 + 2x - 1$. (L)
20. Calculate $\left(\dfrac{8}{27}\right)^{-\frac{2}{3}}$ (L)

5

Further Algebraic Equations and Expressions

Simple Equations

An equation shows that two expressions are equal. We can multiply both sides of an equation by any equal amount; the two expressions will alter correspondingly but they will still be equal. We may also add, subtract or divide in the same way.

e.g. If $A = B$, then

$$2A = 2B, \quad 3A = 3B, \quad \ldots$$
$$A+1 = B+1, \quad A+2 = B+2, \quad \ldots$$
$$A-1 = B-1, \quad A-2 = B-2, \quad \ldots$$
$$\frac{A}{2} = \frac{B}{2}, \quad \frac{A}{3} = \frac{B}{3}, \quad \ldots$$

Examples

Solve the equations:

1.

$x+2 = 4$

[Subtract 2 from both sides] $\therefore x = 4-2$

$\therefore x = \underline{2}$ ANS

2.

$4x+1 = 2x+9$

[Subtract 1 from each side] $\therefore 4x+1-1 = 2x+9-1$

$\therefore 4x = 2x+8$

[Subtract $2x$ from each side] $\therefore 4x-2x = 2x-2x+8$

$\therefore 2x = 8$

[Divide each side by 2] $\therefore x = \underline{4}$ ANS

3.

$3(x-2)+2(1-x) = 27-4(2x+1)$

[Multiply out the brackets] $\therefore 3x-6+2-2x = 27-8x-4$

[Simplify each side] $\therefore x-4 = 23-8x$

[Add 4 to each side] $\therefore x = 27-8x$

[Add $8x$ to each side] $\therefore 9x = 27$

[Divide each side by 9] $\therefore x = \underline{3}$ ANS

Exercise 33

Solve these equations:

1. $x+3 = 6$

2. $x-3 = 6$

3. $x+3 = -6$

4. $x+\frac{1}{2} = \frac{3}{4}$

5. $2x = x+3$

6. $4x = 2x-6$

7. $2x+2 = x-3$

8. $5x-3 = 2x-6$

9. $8x+9 = 17+4x$

10. $7x-3x-2 = 14$

11. $14x-11 = 9-6x$

12. $6a+4 = 54-4a$

13. $8a - 6 = 2a - 36$

14. $6 - 15a = 6a + 48$

15. $11 - 7a = -2a - 9$

16. $2(x - 3) = 4$

17. $2(x - 6) = 3(x - 2)$

18. $4x + 4 = 2(x + 12)$

19. $3(7 - a) = 4(3a + 4) - 10$

20. $4(a - 1) - 3(2 - 2a) = 4 + 3(a + 7)$

Inequations

If we consider the set of integers on a number line it can be seen that the numbers get larger as we move to the right.

e.g. number line: $-3 \quad -2 \quad -1 \quad 0 \quad 1 \quad 2 \quad 3$

As 2 is to the right of 1 we can say that 2 'is greater than' 1, or that 1 'is less than' 2. The signs used to show these **inequalities** are $>$ for 'is greater than' and $<$ for 'is less than'.

i.e. $2 > 1$ or $1 < 2$

Note

The smaller end of the 'arrow head' is always nearer the smaller number. The inequalities $0 < 1$ and $1 < 2$ can be combined in the form $0 < 1 < 2$, or $2 > 1 > 0$. If we consider the number line below and are told that x is an integer with the condition that $0 < x < 2$, then x must equal 1.

i.e. number line: 0, 1, 2 with x at 1

Should the condition read $0 \leqslant x \leqslant 2$ (meaning 0 'is less than or equal to' x which is 'less than or equal to' 2) the value of x could be 0, 1 or 2.

i.e. number line: 0, 1, 2 with x or x or x above them

Statements of the form $2x > 6$ are termed **inequations.** The rules for simplifying such statements are the same as those considered earlier for simple equations.

Examples

1. Simplify $\quad 2x > 6$

[Divide each side by 2] $\quad \therefore \ \underline{x > 3}$ ANS

2. Find the simplest equivalent inequation to $x + 1 < 6$

Given $\quad x + 1 < 6$

[Subtract 1 from each side] $\quad \therefore \ \underline{x < 5}$ ANS

3. Find the simplest statement equivalent to $3x - 2 \geqslant 7$

Given $\quad 3x - 2 \geqslant 7$

[Add 2 to each side] $\quad \therefore 3x \geqslant 9$

[Divide each side by 3] $\quad \therefore \ \underline{x \geqslant 3}$ ANS

Exercise 34 Find the simplest statements equivalent to:

1. $2x > 4$ **2.** $7x > 5$ **3.** $5x < 10$
4. $x+3 > 5$ **5.** $2+x > 8$ **6.** $x-2 < 7$
7. $4-x < 2$ **8.** $3x+4 > 13$ **9.** $7+4x > 19$
10. $2x-7 \leqslant 5$ **11.** $8+3x \leqslant 20$ **12.** $4-3x \geqslant -17$
13. $11 < 4x-5$ **14.** $17 > 2-5x$ **15.** $5-3x > 14$

Sequences

A set of numbers arranged in order according to a rule is called a **sequence** e.g.

1. 1, 2, 3, 4, 5, ... is the sequence of natural numbers.
2. 1, 4, 9, 16, 25, ... is the sequence of square numbers.
3. 0, 1, 1, 2, 3, 5, 8, 13 is an example of the Fibonacci sequence named after the man who discovered it. The rule is that each term is the sum of the preceeding two terms.

Five of the terms in sequence (2) above have been listed, it is of course possible to evaluate any term. Let n represent the term we wish to find, then n^2 gives the value of that term.

e.g. Let $n = 3$ (the 3rd. term), then $n^2 = 3^2 = \underline{9}$
Let $n = 4$ (the 4th. term), then $n^2 = 4^2 = \underline{16}$

It is said that the nth. term of the sequence is n^2.

Examples

1. Find a formula for the nth. term of the sequence 4, 7, 10, 13, ...

Let $n = 2$ (i.e. the 2nd. term), which is 7,
then $n+5$ or $2n+3$ or $3n+1$ or $4n-1$ . . . . all equal 7.

Let $n = 3$ (i.e. the 3rd. term), which is 10,
then $n+7$ or $2n+4$ or $3n+1$ or $4n-2$ . . . . all equal 10.
The common element is $3n+1$. This formula works for $n = 2$ and $n = 3$, now check with $n = 4$.

Let $n = 4$, then $3n+1 = 13$
$3n+1$ is the formula for the nth. term of the sequence.

2. Find the fifth term of a sequence whose nth term is $2^n + 3$.

Here, $n = 5$. Substituting for n gives:

$$2^n + 3 = 2^5 + 3$$
$$= (2 \times 2 \times 2 \times 2 \times 2) + 3 = \underline{35} \quad \text{ANS}$$

Exercise 35 In questions **1** to **10** find a formula for the nth. term of these sequences:

1. 3, 4, 5, 6, . . . **2.** 3, 5, 7, 9, . . . **3.** 3, 7, 11, 15, . . .
4. 2, 5, 10, 17, . . . **5.** 1, 8, 27, 64, . . . **6.** 3, 9, 19, 33, . . .
7. 3, 24, 81, . . . **8.** 5, 16, 39, . . . **9.** $\frac{1}{4}$, 1, $2\frac{1}{4}$, . . .
10. -1, $+2$, -3, $+4$, . . .

In questions **11** to **20** find the third terms of the sequences whose nth. terms are:

11. $n+1$ **12.** $2n+3$ **13.** n^2-1 **14.** n^3+2n-1

15. $4n+2^n$ **16.** $\dfrac{1}{n}+2^{-n}$ **17.** $2n+(-1)^n$ **18.** $\dfrac{3}{n^2}-n^{-1}$

19. $n^{-2}-(2n)^2$ **20.** $\dfrac{n+3}{n^2}+3^{-n}$

Highest Common Factor and Lowest Common Multiple

In Arithmetic the H.C.F. of a pair of numbers is found in the following way.

Consider the numbers 36 and 42:
Factorising fully gives: $36 = 2 \times 2 \times 3 \times 3$
$42 = 2 \times 3 \times 7$

The H.C.F. $= 2 \times 3 = \underline{6}$, this being the largest factor common to both 36 and 42. (i.e. 6 is the largest number that divides 36 and 42 without remainder.)

In Arithmetic, again consider the pair of numbers 36 and 42.

Factorising fully gives: $36 = 2^2 \times 3^2$
$42 = 2 \times 3 \times 7$

The L.C.M. $= 2^2 \times 3^2 \times 7 = \underline{252}$, this being the smallest number divisible by both 36 and 42 without remainder.

Examples

1. Find the H.C.F. and L.C.M. of $15a^3b^3c$, $25ab^3c^2$.

$$15a^3b^3c = 3 \times 5 \times a \times a \times a \times b \times b \times b \times c$$
$$25ab^3c^2 = 5 \times 5 \times a \times b \times b \times b \times c \times c$$

H.C.F. = the largest expression that divides each exactly
$= 5 \times a \times b \times b \times b \times c = \underline{5ab^3c}$ ANS

L.C.M. = the smallest expression divisible exactly by both
$= 3 \times 5 \times 5 \times a \times a \times a \times b \times b \times b \times c \times c$
$= \underline{75a^3b^3c^2}$ ANS

2. Find the H.C.F. and L.C.M. of $a(a-1)$, $(a-1)^2$

$$a(a-1) = a \times (a-1)$$
$$(a-1)^2 = (a-1) \times (a-1)$$
$$\therefore \text{H.C.F.} = \underline{(a-1)} \quad \text{ANS}$$
$$\text{and L.C.M.} = \underline{a(a-1)^2} \quad \text{ANS}$$

3. Find the H.C.F. and L.C.M. of a^2+4a, a^2+6a+8

$$a^2+4a = a(a+4)$$
$$a^2+6a+8 = (a+2)(a+4)$$
$$\therefore \text{H.C.F.} = \underline{(a+4)} \quad \text{ANS}$$
$$\text{and L.C.M.} = \underline{a(a+2)(a+4)} \quad \text{ANS}$$

Exercise 36 Find the H.C.F. and L.C.M. of:

1. 12, 38 **2.** $3x^2$, $5x$
3. $6xy^2$, $8x^2y$ **4.** $x^3y^2z^4$, $z^4x^2y^5$
5. $3(x+2)$, $9(x+2)^2$ **6.** $x(x-2)$, $(x-2)^3$
7. x^2+7x, $x^2+9x+14$ **8.** x^2+3x+2, x^2+5x+6
9. $x^2-8x+16$, x^2-16 **10.** x^2-5x, x^2-6x+5, x^2-25

Simplification of Fractions

In Algebra, as in Arithmetic, fractions can often be simplified by dividing the numerator and the denominator by the highest common factor. This is known as **cancelling**.

e.g. To simplify $\frac{24}{56}$

$$24 = 2 \times 2 \times 2 \times 3 = 2^3 \times 3$$
$$56 = 2 \times 2 \times 2 \times 7 = 2^3 \times 7$$
$$\therefore \text{H.C.F.} = 2^3 = 8,$$
$$\therefore \frac{24}{56} = \frac{3 \times 8}{7 \times 8}$$

[The 8s cancel] $= \underline{\frac{3}{7}}$ ANS

Examples Simplify the expressions **1.** $\frac{7a^2bc}{28a^3bc^2}$ **2.** $\frac{x^2-3x-4}{x^2-16}$

3. $\frac{21x^2y^3}{13ab^2} \times \frac{39a^2b^3}{28x^3y^2}$

1.
$$7a^2bc = 7 \times a^2 \times b \times c$$
$$28a^3bc^2 = 2^2 \times 7 \times a^3 \times b \times c^2$$
$$\therefore \text{H.C.F.} = 7a^2bc$$

[Dividing numerator and denominator by $7a^2bc$ gives]
$$\frac{7a^2bc}{28a^3bc^2} = \frac{1 \times 7a^2bc}{4ac \times 7a^2bc} = \underline{\frac{1}{4ac}} \quad \text{ANS}$$

2. $$x^2-3x-4=(x-4)(x+1)$$
$$x^2-16=(x-4)(x+4)$$
$$\therefore \text{H.C.F.}=(x-4)$$
$$\frac{x^2-3x-4}{x^2-16}=\frac{(x+1)\cancel{(x-4)}^1}{(x+4)\cancel{(x-4)}_1}=\frac{x+1}{x+4} \quad \text{ANS}$$

3. H.C.F. of $21x^2y^3$ and $28x^3y^2=7x^2y^2$
H.C.F. of $39a^2b^3$ and $13ab^2=13ab^2$

$$\frac{21x^2y^3}{13ab^2}\times\frac{39a^2b^3}{28x^3y^2}=\frac{3y(7x^2y^2)}{(13ab^2)}\times\frac{3ab(13ab^2)}{4x(7x^2y^2)}$$

[Cancelling the common factors]
$$=\frac{3y\times 3ab}{4x}=\frac{9aby}{4x} \quad \text{ANS}$$

Exercise 37 Simplify the following expressions:

1. $\dfrac{3a}{6a}$ 2. $\dfrac{4a^2b^3}{2ab^2}$ 3. $\dfrac{11a^2b^4cd^2}{33a^3bd^2}$

4. $\dfrac{x^2+7x+12}{x^2-9}$ 5. $\dfrac{x^2-x-42}{x^2+8x+12}$ 6. $\dfrac{3a}{b}\times\dfrac{b}{a}$

7. $\dfrac{16a^2}{3b^3}\times\dfrac{6b}{4a^3}$ 8. $\dfrac{13a^3b}{39c}\times\dfrac{5c^2}{25b}$ 9. $\dfrac{a^2+a}{a^2}\times\dfrac{a+2}{a^2+3a+2}$

10. $\dfrac{7x^2y^4}{9xz^3}\times\dfrac{27z^2p^4}{11p^3q}$

Addition and Subtraction of Fractions

Before fractions can be added or subtracted, in Algebra or in Arithmetic, their **lowest common denominator** must be found. This common denominator is the L.C.M. of the original denominators.

Examples Simplify **1.** $\dfrac{1}{3}+\dfrac{3}{5}$, **2.** $\dfrac{x}{3}+\dfrac{x}{5}$, **3.** $\dfrac{x+3}{4}+\dfrac{x-1}{3}$,
4. $\dfrac{2}{a-7}+\dfrac{2a+3}{a^2-3a-28}$.

1. $$\frac{1}{3}+\frac{3}{5}=\frac{(1\times 5)+(3\times 3)}{15}=\frac{5+9}{15}=\frac{14}{15} \quad \text{ANS}$$

2. $$\frac{x}{3}+\frac{x}{5}=\frac{(x\times 5)+(x\times 3)}{15}=\frac{5x+3x}{15}=\frac{8x}{15} \quad \text{ANS}$$

3. $$\frac{x+3}{4}+\frac{x-1}{3}=\frac{3(x+3)+4(x-1)}{12}=\frac{3x+9+4x-4}{12}$$
$$=\frac{7x+5}{12} \quad \text{ANS}$$

4. $\dfrac{2}{a-7}+\dfrac{2a+3}{a^2-3a-28}$

[L.C.M. of denominators is $(a-7)(a+4)$] $=\dfrac{2(a+4)+(2a+3)}{(a-7)(a+4)}$

$=\dfrac{2a+8+2a+3}{(a-7)(a+4)}$

$=\dfrac{4a+11}{(a-7)(a+4)}$ ANS

Exercise 38 Simplify:

1. $\dfrac{1}{4}+\dfrac{2}{5}$ **2.** $\dfrac{a}{4}+\dfrac{a}{5}$

3. $\dfrac{a}{5}+\dfrac{3a}{7}$ **4.** $\dfrac{a+1}{2}+\dfrac{a+2}{3}$

5. $\dfrac{2(a-3)}{4}-\dfrac{a+1}{5}$ **6.** $\dfrac{3}{a+2}+\dfrac{4a-1}{a^2-a-6}$

7. $\dfrac{3a}{a-3}-\dfrac{2}{a^2-6a+9}$ **8.** $\dfrac{2(a+1)}{a^2-9}+\dfrac{3a+6}{a^2+6a+9}$

9. $\dfrac{12a}{a^2+a-2}+\dfrac{3}{a^2+3a-4}$ **10.** $\dfrac{11(a+1)}{a^2+4a-77}-\dfrac{2a-3}{a^2+a-56}$

Simple Equations Involving Fractions

Example Solve $\dfrac{3a+2}{3}=\dfrac{3a-1}{6}+\dfrac{a-4}{9}$

[Common denominator = 18] $\dfrac{6(3a+2)}{18}=\dfrac{3(3a-1)}{18}+\dfrac{2(a-4)}{18}$

[Multiply through by 18] $6(3a+2)=3(3a-1)+2(a-4)$

$18a+12=9a-3+2a-8$

[Group 'a's on the left numbers on the right] $18a-9a-2a=-3-8-12$

$\therefore\ 7a=-23$

$\therefore\ a=-3\frac{2}{7}$ ANS

Exercise 39 Solve:

1. $\dfrac{x}{3}=6$ **2.** $\dfrac{x+1}{2}=\dfrac{x-2}{5}$ **3.** $\dfrac{x}{3}+\dfrac{3x}{7}=4$

4. $\dfrac{4x}{9}+7=\dfrac{x}{4}$ **5.** $\dfrac{5x-1}{2}=\dfrac{3x-3}{3}$

Proportion and Variation

This topic was introduced in Chapter 1 where associated problems were of an arithmetical nature. Questions requiring an algebraic solution will now be considered.

Direct proportion

If a man walks at a steady speed, the distance that he travels depends upon the time he has been walking. It is said that his distance travelled is **directly proportional** (or varies directly) to the time taken. If d represents the distance travelled, and t the time taken it can be said that $d \propto t$, (i.e. d is directly proportional to t)

If he walks for two hours at a constant speed of 6km/h he travels a distance of $d = 6 \times 2 = \underline{12\text{km}}$. If he walks for four hours at the same speed he travels $d = 6 \times 4 = \underline{24\text{km}}$. It can be seen that the constant speed of 6km/h has turned the statement $d \propto t$ into the equation $d = 6t$.

In problems of this type we usually employ the letter k to represent the constant that equates the two variables.

Example

Two variables, a and b are directly proportional to each other. If $a = 6$ when $b = 3$, find the value of a when $b = 7\frac{1}{2}$.

Given $\qquad a \propto b$

[Let k be the constant] $\qquad \therefore a = kb,$

Substituting $a = 6$ and $b = 3$ in this equation gives:

$$6 = 3k$$

$$\therefore \frac{6}{3} = k = \underline{2}$$

In the second case when $b = 7\frac{1}{2}$, $a = kb$ becomes

$$a = 7\tfrac{1}{2}k$$

[We now know that $k = 2$] $\quad \therefore a = 7\frac{1}{2} \times 2 = \underline{15}$ ANS

Exercise 40

In questions **1** to **5** x is directly proportional to y

1. If $x = 2$ when $y = 4$, find x when $y = 6$.
2. If $x = 3$ when $y = 1\frac{1}{2}$, find x when $y = 7$.
3. If $x = 7$ when $y = 14$, find y when $x = 9$.
4. If $x = \frac{1}{2}$ when $y = \frac{1}{8}$, find y when $x = \frac{3}{4}$.
5. If $x = 17$ when $y = 4$, find x when $y = 11$.

Inverse proportion

Considering the earlier example of the man walking, if he had to travel a distance of 8km in two hours he obviously would not need to walk as fast as a friend who was trying to do the same distance in one hour. In this case it is said that the speed of walking a given distance is **inversely proportional** (varies inversely) as the time allowed.

If s represents the speed required and t the time available, then it is said that $s \propto \frac{1}{t}$, (i.e. s is inversely proportional to t).

The equation would be $s = \frac{k}{t}$, where k is constant.

Example Two variables a and b are inversely proportional to each other. If $a = 4$ when $b = 2$, find a when $b = 7$.

Given $$a \propto \frac{1}{b}$$

[Let k be the constant] $$\therefore a = \frac{k}{b}$$

Substituting $a = 4$ and $b = 2$ in this equation gives:

$$4 = \frac{k}{2}$$

$$\therefore 2 \times 4 = k = \underline{8}$$

In the second case when $b = 7$, $a = \frac{k}{b}$ becomes

$$a = \frac{k}{7}$$

[We now know that k = 8] $\therefore a = \frac{8}{7} = \underline{1\frac{1}{7}}$ ANS

Exercise 41 If x is inversely proportional to y:

1. When $x = 2$, $y = 4$. Find x when $y = 6$.
2. When $x = 3$, $y = 2$. Find x when $y = 7$.
3. When $x = 12$, $y = 3$. Find y when $x = 4$.
4. When $x = \frac{1}{2}$, $y = \frac{1}{4}$. Find y when $x = \frac{3}{4}$.
5. When $x = 21$, $y = 4$. Find x when $y = 7$.

Joint Variation

The area of a rectangle varies directly as the length and directly as the width of the rectangle. This is called **joint variation**, and the area is to vary jointly as the length and the width.

i.e. $A \propto LW$

$\therefore A = kLW$

Examples **1.** If a varies jointly to b and to c, and $a = 12$ when $b = 3$, $c = 2$, find the value of a when $b = \frac{1}{4}$, and $c = \frac{1}{2}$.

Given $$a \propto bc$$
[Let k be the constant] $$\therefore a = kbc$$

Substituting $a = 12$, $b = 3$ and $c = 2$ gives:

$$12 = k \times 3 \times 2$$

$$\therefore \frac{12}{6} = k = \underline{2}$$

In the second case when $b = \frac{1}{4}$ and $c = \frac{1}{2}$, $a = kbc$ becomes

$$a = 2 \times \frac{1}{4} \times \frac{1}{2} = \underline{\frac{1}{4}} \quad \text{ANS}$$

2. If a varies directly as b and inversely as c, and if $a = 16$ when $b = 4$, $c = 6$, find the value of a when $b = 7$, $c = 2$.

Given $\qquad a \propto \frac{b}{c}$

$$\therefore a = \frac{kb}{c}$$

Substituting $a = 16$, $b = 4$ and $c = 6$ gives:

$$16 = \frac{4k}{6}$$

$$\therefore 6 \times 16 = 4k$$

$$\therefore k = \underline{24}$$

In the second case when $b = 7$ and $c = 2$, $a = \frac{kb}{c}$ becomes

$$a = \frac{24 \times 7}{2} = 12 \times 7 = \underline{84} \quad \text{ANS}$$

Exercise 42

1. If $x \propto yz$, and $x = 8$ when $y = 2, z = 1$, find x when $y = 3, z = 2$.
2. If $x \propto yz$, and $x = 10$ when $y = 4, z = 5$, find x when $y = \frac{1}{2}, z = 4$.
3. If $x \propto \frac{y}{z}$, and $x = 6$ when $y = 2, z = 3$, find x when $y = 5, z = 3$.
4. If $x \propto \frac{y}{z}$, and $x = 9$ when $y = 15, z = 20$, find x when $y = 5, z = 3$.
5. If $x \propto \frac{y}{z}$, and $x = 4$ when $y = 2, z = 3$, find x when $y = 7, z = 3$.

Simultaneous Equations

As the name implies, these are two or more equations that are solved simultaneously (i.e. together).

Examples

1. Solve the equations $2x + y = 5$, $x + y = 3$.

It is necessary to eliminate one of the variables (x or y) so that we may find the value of the other.

[Number the equations] i.e. $\quad 2x + y = 5 \qquad (1)$

$\qquad x + y = 3 \qquad (2)$

[Subtract] $\qquad x \quad = \underline{2} \quad$ ANS

[i.e. $2x - x = x$, $y - y = 0$,
$5 - 3 = 2$.]

[Now use the value we have found for x in order to find y]

Substitute $x = 2$ in equation (2)

$$x + y = 3$$

becomes $\quad 2 + y = 3$

[Subtract 2 from each side] $\quad y = \underline{1}$ ANS

[A check may be made of our answers by substituting for x and for y in the other equation.]

Substitute $x = 2$, $y = 1$ in equation (1)

$$2x + y = 5$$

becomes $\quad (2 \times 2) + 1 = 5$

i.e. $\quad 5 = 5$

which proves our answer to be correct

2. Solve the equations $5x - 3y = 29$, $2x + 4y = 22$.

[Number the equations] $\quad 5x - 3y = 29 \quad (1)$

$$2x + 4y = 22 \quad (2)$$

Neither variable can be eliminated immediately by addition or subtraction. We must multiply one or both equations by a suitable number so that either the xs or the ys have equal values.

[Multiply (1) by 4 and (2) by 3]

i.e. $\quad 20x - 12y = 116 \quad (1) \times 4$

$$6x + 12y = 66 \quad (2) \times 3$$

[Add] $\quad 26x = 182$

[Divide both sides by 26] $\quad x = \underline{7}$ ANS

Substitute $x = 7$ in equation (2)

$$2x + 4y = 22$$

becomes $\quad 14 + 4y = 22$

$$4y = 8$$

$$y = \underline{2} \text{ ANS}$$

[Check by substituting for x and y in equation (1)]

i.e. $\quad 5x - 3y = 29$

becomes $\quad 35 - 6 = 29$

i.e. $\quad 29 = 29$ hence correct

3. Solve the equations $3x = 4 - 4y$, $y = 5x - 22$.

Both equations must be rearranged so that like terms are in the same positions.

[Rearrange and number equations] i.e. $\quad 3x + 4y = 4 \quad (1)$

$$-5x + y = -22 \quad (2)$$

[Multiply (2) by 4] i.e. $\quad 3x + 4y = 4 \quad (1)$

$$-20x + 4y = -88 \quad (2) \times 4$$

[Subtract (2) from (1)] $\quad 23x = 92$

$$\left[\begin{array}{c} 3x-(-20x) = 3x+20x \\ 4-(-88) = 4+88 \end{array}\right] \qquad x = \frac{92}{23}$$

$$= \underline{4} \quad \text{ANS}$$

Substitute $x = 4$ in equation (1)

i.e. $\quad 3x+4y = 4$

becomes $\quad 12+4y = 4$

$\quad 4y = -8$

$\quad y = \underline{-2}$ ANS

Check in equation (2)

i.e. $\quad -5x+y = -22$

becomes $\quad -20-2 = -22$

i.e. $\quad -22 = -22$ hence correct

Exercise 43 Solve the following pairs of simultaneous equations:

1. $3x+y = 10,\ x-y = 2$ **2.** $4x+2y = 32,\ 3x-2y = 10$
3. $2x+4y = 26,\ x+y = 9$ **4.** $5x+3y = -6,\ x-y = -6$
5. $2x+3y = 5,\ 4x-5y = 21$ **6.** $x = 1-y,\ 2x = 11-y$
7. $3x-4 = 2y,\ 2x+3y = 7$ **8.** $3y = -7-11x,\ 2x = 21-5y$
9. $12 = 3x-y,\ 2x = 13-y$ **10.** $5x-4y-4 = 0,\ 8x-3y+14 = 0$

Quadratic Equations

A method for factorising a quadratic expression was given in Chapter 4. This method can now be used to solve a quadratic equation. The solutions to such an equation are called the **roots** of that equation. A quadratic equation contains $x^{②}$ and therefore has ② roots, a simple equation contains $x^{①}$ and therefore has only ① root.

e.g. If $x^2 = 9$, then $x = \underline{\pm 3}$ (the 2 roots)

If $4x^2 = 9$, then $x^2 = \frac{9}{4}$, $\therefore\ x = \sqrt{\frac{9}{4}} = \pm\frac{3}{2} = \underline{\pm 1\frac{1}{2}}$

Examples

1. Solve the equation $x^2-3x+2 = 0$.

The factors of x^2 are x and x, either both plus or both minus, the factors of $+2$ are 2 and 1, either both plus or both minus.

Trials: $(x+2)(x+1) = x^2+3x+2$ (incorrect)
$(x-2)(x-1) = x^2-3x+2$ (correct)

Given $\quad x^2-3x+2 = 0$
$(x-2)(x-1) = 0$

We know that any number multiplied by zero equals zero, therefore either $(x-2) = 0$ or $(x-1) = 0$.

If $x-2=0$ or if $x-1=0$
then $x=2$ or $x=1$
The roots are $\underline{x=2}$ or $\underline{x=1}$ ANS

2 Solve the equation $6x^2+x-2=0$

Factors of $6x^2$ are $6x$ and x or $3x$ and $2x$, $\pm$

Trials:

$$(6x+2)(x-1) = 6x^2-4x-2 \quad \text{(incorrect)}$$
$$(3x-1)(2x+2) = 6x^2+4x-2 \quad \text{(incorrect)}$$
$$(3x+2)(2x-1) = 6x^2+x-2 \quad \text{(correct)}$$

Given $6x^2+x-2=0$

$(3x+2)(2x-1)=0$

either $3x+2=0$ or $2x-1=0$

$3x=-2$ or $2x=1$

$x=-\frac{2}{3}$ or $x=\frac{1}{2}$

Roots are $\underline{x=-\frac{2}{3}}$ or $\underline{x=\frac{1}{2}}$ ANS

Exercise 44 Solve the following equations:

1. $x^2+3x+2=0$ **2.** $x^2-7x+12=0$ **3.** $x^2-x-6=0$
4. $x^2-2x-63=0$ **5.** $2x^2+7x+3=0$ **6.** $6x^2+13x+6=0$
7. $6x^2-5x-6=0$ **8.** $6x^2-38x+56=0$ **9.** $10x^2-26x=56$
10. $16x^2=24x-9$

All the equations handled so far had rational roots that are easily found. An equation with fractional or irrational roots may be solved by use of the formula

$$x=\frac{-b\pm\sqrt{b^2-4ac}}{2a}$$

which applies to the general quadratic equation $ax^2+bx+c=0$.

Example Solve the equation $2x^2-6x+3=0$, giving the roots correct to two decimal places.

Here, $a=2$, $b=-6$, $c=3$.
Substituting in the above formula gives:

$$x=\frac{-(-6)\pm\sqrt{(-6)^2-(4\times2\times3)}}{2\times2}$$

$$=\frac{+6\pm\sqrt{+36-24}}{4}$$

$$=\frac{6\pm\sqrt{12}}{4}.$$

[From square root tables] $= \dfrac{6 \pm 3.464}{4}$

[Using the + sign] $x = \dfrac{6+3.464}{4} = \dfrac{9.464}{4} = 2.366$

[Using the − sign] $x = \dfrac{6-3.464}{4} = \dfrac{2.536}{4} = 0.634$

$x = \underline{2.37}$ or $\underline{0.63}$ (each correct to 2 d.p) ANS

Exercise 45 Solve the following equations, giving the roots correct to two decimal places:

1. $x^2 - 4x + 1 = 0$ **2.** $2x^2 - 3x - 1 = 0$ **3.** $3x^2 - 4x - 5 = 0$
4. $5x^2 + 2x - 1 = 0$ **5.** $3x^2 + 7x - 6 = 0$ **6.** $4x^2 - 5x - 1 = 0$
7. $7x^2 + 12x + 4 = 0$ **8.** $5x^2 + 8x - 5 = 0$ **9.** $6x^2 - 3x - 2 = 0$
10. $(x-2)(x+3) = 7$

Note When questions on quadratic equations include a reference to an 'answer correct to decimal places', this is a good indication that the quadratic expression will not factorise. The formula should be used.

Examination Standard Questions

Worked examples

1. Find the least integer x which satisfies the inequality:

$$2x - 26 > 5 - x$$

[Add x to each side] $\therefore 3x - 26 > 5$
[Add 26 to each side] $\therefore 3x > 31$
[Divide both sides by 3] $\therefore x > 10\frac{1}{3}$

The nearest integer (i.e. whole number) greater than $10\frac{1}{3}$ is 11.

$\therefore x = \underline{11}$ ANS

2. The nth term of a sequence is $4n + \frac{1}{4}(-1)^n$. Find the tenth term.

Substituting $n = 10$ gives:

$$4 \times 10 + \frac{1}{4}(-1)^{10}$$

[-1 to any even power $= +1$] $= 40 + \dfrac{1}{4} = \underline{40\frac{1}{4}}$ ANS

3. Express as a single fraction in its simplest form

$$\frac{1}{p^2 + pq} + \frac{1}{q^2 + pq}$$

[Factorise denominators] $\quad = \dfrac{1}{p(p+q)} + \dfrac{1}{q(q+p)}$

[Common factor is $(p+q)$] $\quad = \dfrac{(1 \times q) + (1 \times p)}{pq(p+q)}$

[Cancel $(p+q)$] $\quad = \dfrac{(q+p)}{pq(p+q)} = \dfrac{(\cancel{p+q})^1}{pq(\cancel{p+q})_1} = \underline{\dfrac{1}{pq}}$ ANS

4. Solve the equation $\dfrac{4p-2}{3} - \dfrac{2p-3}{5} = \dfrac{9}{5}$

[Common denominator = 15] $\quad \dfrac{5(4p-2) - 3(2p-3)}{15} = \dfrac{9 \times 3}{15}$

[Multiply through by 15] i.e. $\quad 5(4p-2) - 3(2p-3) = 27$

$$20p - 10 - 6p + 9 = 27$$

$$14p - 1 = 27$$

[Add 1 to both sides] i.e. $\quad 14p = 27 + 1$

$$14p = 28$$

$$p = \underline{2} \quad \text{ANS}$$

5. P is proportional to a^2 and $P = 54$ when $a = 3$. Calculate P when $a = 4$.

Given that $P \propto a^2$

$\therefore \; P = \mathrm{k}a^2$

Substituting $P = 54$ and $a = 3$ gives:

$54 = \mathrm{k} \times 3^2$

$\therefore \; \dfrac{54}{9} = \mathrm{k} = \underline{6}$

Substituting $\mathrm{k} = 6$ and $a = 4$ in $P = \mathrm{k}a^2$ gives:

$P = 6 \times 4^2$

$= 6 \times 16 = \underline{96}$ ANS

6. Numbers x, t and v are such that x varies directly as t and inversely as the square of v. Given that when $x = 12$, $t = 4$ and $v = 2$, find the value of x when $t = 8$ and $v = 4$.

Given that $x \propto \dfrac{t}{v^2}$

$\therefore \; x = \dfrac{\mathrm{k}t}{v^2}$

Substituting $x = 12$, $t = 4$ and $v = 2$ gives:

$12 = \dfrac{4\mathrm{k}}{4}$

$\therefore \; k = \underline{12}$

In the second case when $t = 8$ and $v = 4$,

$x = \dfrac{12 \times 8}{4^2} = \dfrac{96}{16} = \underline{6}$ ANS

7. Partial variation

When q is expressed in terms of p it consists of the sum of two terms, the first directly proportional to p and the second inversely proportional to p. If $q = 6$ when $p = 1$, and $q = 9$ when $p = 2$, find the precise expression for q in terms of p.

Let $q = k_1 p + \frac{k_2}{p}$ where k_1 and k_2 are two constants.

Substituting $q = 6$ and $p = 1$ gives:

$$6 = k_1 + k_2$$

$$\therefore k_2 = 6 - k_1 \qquad (1)$$

Substituting $q = 9$ and $p = 2$ in the initial equation gives:

$$9 = 2k_1 + \frac{k_2}{2}$$

Substituting for k_2 from equation (1) above gives:

$$9 = 2k_1 + \frac{(6 - k_1)}{2}$$

[Multiply both sides by 2] $\quad \therefore 18 = 4k_1 + 6 - k_1$

$$12 = 3k_1$$

$$\therefore \underline{k_1 = 4}$$

and from (1) above $\quad k_2 = 6 - 4 = 2$

Expression becomes $\quad \underline{q = 4p + \frac{2}{p}}$ ANS

8. Solve the equation $x(x+3) = 5$, giving your answer correct to two decimal places.

Multiplying out $\quad x^2 + 3x = 5$

Rearranging $\quad x^2 + 3x - 5 = 0$

By formula $\quad x = \dfrac{-b \pm \sqrt{b^2 - 4ac}}{2a}$

in the quadratic $ax^2 + bx + c = 0$.

Here, $a = 1$, $b = +3$, $c = -5$

$$x = \frac{-3 \pm \sqrt{9 - 4(-5)}}{2}$$

$$x = \frac{-3 \pm \sqrt{29}}{2}$$

From square root tables $\quad x = \dfrac{-3 \pm 5.385}{2}$

$$x = \frac{-3 + 5.385}{2} \text{ or } \frac{-3 - 5.385}{2}$$

$$x = \frac{2.385}{2} \text{ or } \frac{-8.385}{2}$$

$$x = 1.193 \text{ or } -4.193$$

$= \underline{1.19}$ or $\underline{-4.19}$ (each correct to 2 d.p.) ANS

Exercise 46

1. If x, y and z are consecutive even integers such that $x < y < z$, calculate $(x - z)$.
2. Find the least integer x which satisfies the inequality $3x - 40 > 10 - x$. (JMB)
3. Express $x + 3 < 9 < 4x + 1$ in the form $a < x < b$. (C)
4. Rewrite the statement $7 - 3x > 16$ in the form $x < a$.
5. The nth term of a sequence is $\frac{(-1)^n}{3} + n^2$. Find the fifth term.
6. Find the eighth term of a sequence whose nth term is $n^2 + 2n - n^{-1}$.
7. Find the nth term of the sequence $-3, 4\frac{1}{2}, -8\frac{1}{3}, 16\frac{1}{4}, \ldots$
8. Simplify $\frac{x-1}{x^2+3x+2} + \frac{2}{x+1}$, giving your answer as a single fraction in its lowest terms. (C)
9. Express $\frac{a-b}{3} - \frac{a+b}{5}$ as a single fraction. (O)
10. Write $\frac{x-1}{2} - \frac{x+3}{5}$ as a single fraction and simplify your answer. (JMB)
11. Simplify $\frac{1}{x} + \frac{1}{2x} + \frac{1}{3x}$. (L)
12. Solve the equation $\frac{3x-2}{5} - \frac{2x-3}{4} = \frac{1}{2}$. (L)
13. Solve the equation $\frac{2x-1}{3} + \frac{1-3x}{2} = \frac{3}{4}$. (JMB)
14. Solve the equation $\frac{x-3}{2} + \frac{x-4}{5} = 4$. (O)
15. Solve the equation $\frac{x}{3} - \frac{x-1}{4} = \frac{5}{12}$, where x is a real number. (JMB)
16. The real variables F and r satisfy the relationship $F = \frac{k}{r^2}$, where k is a constant.
    When $r = 2$, and $F = 25$. Calculate (a) the value of k, (b) the value of F when $r = 4$. (JMB)
17. Given that a is proportional to b and that $a = 3$ when $b = 4\frac{1}{5}$, calculate the value of a when $b = 8$.
18. A is proportional to r^2 and $A = 81$ when $r = 3$. Calculate A when $r = 5$.
19. Given that q varies inversely as p, and that $q = 9$ when $p = 2$, find the value of q when $p = 3$.
20. F varies inversely as the square of d. If $F = 9$ when $d = 4$, calculate the value of F when $d = 10$. (WJEC)

21. When y is expressed in terms of x it consists of the sum of two terms, the first directly proportional to x and the second inversely proportional to x. If $y = 11$ when $x = 1$, and $y = 10$ when $x = \frac{1}{2}$, find the precise expression for y in terms of x. Use your expression to find (i) the value of y when $x = -3$, (ii) the values of x for which $y = 14$. (O)

22. At time t seconds the height h metres of a stone thrown vertically upwards, is given by the formula $h = at + bt^2$ where a and b are constants. It is known that the stone is 136 m high when $t = 2$ and 156 m high when $t = 3$. Calculate the values of a and b. Hence calculate the values of t for which the stone is 24 m high. (JMB)

23. The sag in a plank supported at its two ends is found to vary directly as the cube of its length and inversely as the square of its thickness. A plank 10 m long and 2 cm thick is found to sag 2.5 cm. Calculate the sag in a similar plank 15 m long and 3 cm thick. (C)

24. Numbers x, t and u are such that x varies directly as t and inversely as the square of u. Given that when $x = 27$, $t = 36$ and $u = 2$, find the value of x when $t = 64$ and $u = 4$. (C)

25. When a current passes through a wire, the rate at which heat is produced, W watts, is directly proportional to the square of the voltage V volts, and inversely proportional to the lengths of the wire l centimetres. Express W in terms of V, l and a constant of variation k. Hence calculate (i) the value of k given that $W = 800$ when $V = 200$ and $l = 40$, (ii) the value of l given that $W = 800$ when $V = 240$. (AEB)

26. Given that y varies as x^n, write down the value of n in each of the following cases:
(a) y is the area of a circle of radius x,
(b) y is the volume of a cylinder of given base area and height x,
(c) y and x are the sides of a rectangle of given area. (C)

27. Find the positive number x which satisfies the equation $2x^2 - x - 3 = 0$. (JMB)

28. Solve the equation $(x-5)^2 = 9$. (C)

29. Solve the simultaneous equations $x - y = 7$, $2x + 3y = 4$. (JMB)

30. Solve the simultaneous equations $2x + 7y = 0$, $3x + 5y = 11$. (JMB)

31. Solve the simultaneous equations $6x - 10y = 3$, $5x - 6y = 6$. (JMB)

32. Solve the equation $3x^2 - 5x - 4 = 0$, giving the roots correct to two decimal places. (L)

33. Solve the equation $2x^2 - 13x + 5 = 0$, giving the roots correct to one place of decimals. (L)

34. Solve the equation $x(x+1) = 7$, giving your answer correct to two decimal places. (JMB)

35. Express in the form $a \leqslant x \leqslant b$ and $c \leqslant y \leqslant d$ the inequalities,

$$1 - x \leqslant x - 6 \leqslant 9 - x$$
$$2(15 - y) \leqslant 4(y + 3) \leqslant 3(y + 18)$$

Hence write down the least and greatest possible values of $x + 3y$ and of $3y - x$. (O)

6

Functional Notation

The expression $2x^3+6x^2+x-4$ contains x as the only variable (or unknown) and is therefore described as a **function of x**. A shorthand notation that may be used to represent this function is **f(x)**. i.e. **f(x)** is read **a function of x**.

When $x = 1$ the above expression becomes

$$\begin{aligned} &2(1)^3+6(1)^2+(1)-4 \\ &= 2 \quad +6 \quad +1 \; -4 \\ &= \underline{5} \end{aligned}$$

It can therefore be said that $f(1) = 5$, i.e. when $x = 1$ the value of the function is 5.

Example

If $f(x) = 3x^2 - 7x - 6$, find $f(-1)$, $f(0)$, $f(3)$.

$$\begin{aligned} f(-1) &= 3(-1)^2-7(-1)-6 = 3+7-6 = \underline{4} \\ f(0) &= 0 \quad -0 \quad -6 = \underline{-6} \\ f(3) &= 3(3)^2 \quad -7(3) \quad -6 = 27-21-6 = \underline{0} \end{aligned}$$

Factor Theorem

In the above example $f(3) = 0$, which means that $x = 3$ is a **root** of the equation $f(x) = 0$. i.e. $x-3$ is a **factor** of the expression $3x^2-7x-6$. The other factor, by inspection, is $3x+2$.

i.e. $3x^2-7x-6 = (x-3)(3x+2)$.

Example

Factorise fully the cubic expression $f(x) = x^3-x^2-4x+4$.

If $f(x) = 0$, when x is an integer, then this integer must be a factor of 4.

Trials
$$\begin{aligned} f(1) &= (1)^3 - (1)^2 - 4(1) + 4 = 1-1-4+4 = \underline{0} \\ f(-1) &= (-1)^3-(-1)^2-4(-1)+4 = -1-1+4+4 = \underline{6} \\ f(2) &= (2)^3 - (2)^2 - 4(2) + 4 = 8-4-8+4 = \underline{0} \\ f(-2) &= (-2)^3-(-2)^2-4(-2)+4 = -8-4+8+4 = \underline{0} \end{aligned}$$

$f(1) = f(2) = f(-2) = 0$
$(x-1)$, $(x-2)$ and $(x+2)$ are all factors of the expression.

i.e. $x^3-x^2-4x+4 = \underline{(x-1)\,(x-2)\,(x+2)}$ ANS

Exercise 47

1. If $f(x) = 4x^2 + x - 1$; find $f(1)$, $f(-1)$.
2. If $f(x) = 3x^3 + 2x^2 + 4x - 6$; find $f(1)$, $f(-1)$, $f(2)$, $f(-2)$, $f(3)$, $f(-3)$.
3. Show that $(x+2)$ is a factor of $2x^2 - x - 10$.
4. Show that $(x-3)$ is *not* a factor of $2x^2 - x - 10$.
5. Given that $(3x+1)$ is a factor of $3x^2 - 8x - 3$, find the other factor.
6. Factorise fully $x^3 - 2x^2 - x + 2$.
7. Factorise fully $x^3 - 5x^2 - 49x + 245$.
8. Given that $(3x-2)$ is a factor of $3x^3 - 8x^2 - 5x + 6$ find the other factors.

Functions as Mappings

Domain and range

The expression $f: x \to x^2$ means that a function f maps x onto x^2.

i.e. $f(x) = x^2$

The function $x \to x^2$ maps the set of integers $\{1, 2, 3, 4\}$ onto the set $\{1, 4, 9, 16\}$. The first set is called the **domain** of the function and the second set the **range** of the function.

Example

If $f: x \to x^2 + 1$, find the range of $f(x)$ with domain $\{-2, -1, 0, 1, 2\}$.

Mapping each member in turn gives

$$\begin{aligned} f(-2) &= (-2)^2 + 1 = 5 \\ f(-1) &= (-1)^2 + 1 = 2 \\ f(0) &= (0)^2 + 1 = 1 \\ f(1) &= (1)^2 + 1 = 2 \\ f(2) &= (2)^2 + 1 = 5 \end{aligned}$$

The range of $f(x)$ is therefore $\underline{\{1, 2, 5\}}$ ANS

Note

The above is an example of a 'many to one' correspondence, more than one member of the domain is mapped onto the same member of the range. Functions are either 'many to one' or 'one to one' mappings; $f: x \to 3x$ is an example of a 'one to one' mapping, each member of the domain maps onto a different member of the range.

Compound Functions

Consider the two functions $f: x \to x^2$ and $g: x \to x + 1$ with domain $\{1, 2, 3\}$. The **compound**, or **composite**, function $gf(x)$ means $g[f(x)]$.

i.e. 'do f first then g'.

$$\begin{aligned} g[f(1)] &= g[(1)^2] = g[1] = 1 + 1 = 2 \\ g[f(2)] &= g[(2)^2] = g[4] = 4 + 1 = 5 \\ g[f(3)] &= g[(3)^2] = g[9] = 9 + 1 = 10 \end{aligned}$$

The range of $gf(x)$ is therefore $\{2, 5, 10\}$.

Examples 1. Let $f: x \rightarrow x+3$ and $g: x \rightarrow x-2$. Find (a) $fg(2)$, (b) $gf(2)$, (c) $fg(-4)$, d) $gf(-4)$.

(a) $f[g(2)] = f[2-2] = f[0] = 0+3 = \underline{3}$ ANS
(b) $g[f(2)] = g[2+3] = g[5] = 5-2 = \underline{3}$ ANS
(c) $f[g(-4)] = f[-4-2] = f[-6] = -6+3 = -\underline{3}$ ANS
(d) $g[f(-4)] = g[-4+3] = g[-1] = -1-2 = -\underline{3}$ ANS

Note $fg(x) = gf(x)$ in these cases.

2. Let $f: x \rightarrow 2x$ and $g: x \rightarrow x^2$. Find (a) $fg(2)$, (b) $gf(2)$, (c) $fg(-4)$, (d) $gf(-4)$.

(a) $f[g(2)] = f[(2)^2] = f[4] = 2 \times 4 = \underline{8}$ ANS
(b) $g[f(2)] = g[2(2)] = g[4] = 4^2 = \underline{16}$ ANS
(c) $f[g(-4)] = f[(-4)^2] = f[16] = 2 \times 16 = \underline{32}$ ANS
(d) $g[f(-4)] = g[2(-4)] = g[-8] = (-8)^2 = \underline{64}$ ANS

Note $fg(x) \neq gf(x)$ in these cases.

Exercise 48

1. If $f: x \rightarrow x+4$ find the range of the function given domain $\{1, 2, 3\}$.
2. If $g: x \rightarrow 3x$ find the range of the function given domain $\{-1, -2, -3\}$.
3. If $h: x \rightarrow 2x^2$ find the range of the function given domain $\{-2, -1, 0, 1, 2\}$.
4. State which of the functions in questions 1–3 above are 'one to one' mappings, and which, if any, are 'many to one'.
5. For $f: x \rightarrow x-2$, find (a) $f(1)$, (b) $f(3)$, (c) $f(-2)$, (d) $f(0)$.
6. For $g: x \rightarrow x^3+1$, find (a) $g(1)$, (b) $g(-1)$, (c) $g(2)$, (d) $g(0)$.
7. For $f: x \rightarrow 2x+2$ and $g: x \rightarrow x-1$, find (a) $fg(1)$, (b) $gf(1)$, (c) $fg(-1)$, (d) $gf(-1)$.
8. For $f: x \rightarrow 3x-2$ and $g: x \rightarrow x-6$, find (a) $fg(0)$, (b) $gf(0)$, (c) $fg(3)$, (d) $gf(-3)$.
9. For $h: x \rightarrow x^2$ and $j: x \rightarrow x+3$, find (a) $hj(1)$, (b) $jh(1)$, (c) $hj(-2)$, (d) $jh(-2)$.
10. For $f: x \rightarrow 2x$ and $g: x \rightarrow x+4$, find expressions which represent (a) $fg(x)$, (b) $gf(x)$. Are these expressions the same?
11. For $h: x \rightarrow x^2$ and $j: x \rightarrow x-1$, find expressions which represent (a) $hj(x)$, (b) $jh(x)$. Are these expressions the same?
12. For $m: x \rightarrow x^3$ and $p: x \rightarrow x^2$, find expressions which represent (a) $mp(x)$, (b) $pm(x)$. Are these expressions the same?

Inverse Functions

The inverse function of 'multiply by 2' is 'divide by 2' and the inverse function of 'add 2' is 'subtract 2'.

Using functional notation the **inverse function** of function f is written f^{-1}.

e.g. If $f: x \to 2x$, then $f^{-1}: x \to \frac{1}{2}x$
and if $g: x \to x+2$, then $g^{-1}: x \to x-2$

Example

If $f: x \to 2x$, for the set of integers, find (a) $f^{-1}(4)$, (b) $f^{-1}(-6)$.

Firstly $f^{-1}: x \to \frac{1}{2}x$.

(a) $f^{-1}(4) = \frac{1}{2}(4) = \underline{2}$ ANS
(b) $f^{-1}(-6) = \frac{1}{2}(-6) = \underline{-3}$ ANS

Note

The inverse of a 'many to one' function is *not* itself a function.
e.g. Consider the function $f: x \to x^2$ for the domain $\{-2, -1, 0, 1, 2\}$.

$f(-2) = (-2)^2 = +4$
and also $f(+2) = (+2)^2 = +4$

i.e. Two members of the function's domain map to the same member of the range ('many to one').

The inverse of 'many to one' is 'one to many', which is *not* a function as defined earlier.

Consider, however, the function $f: x \to x^2$ for the domain $\{0, 1, 2\}$.

The range would be $\{0, 1, 4\}$, which has 'one to one' correspondence with the domain.

$f: x \to x^2$ for domain $\{-2, -1, 0, 1, 2\}$ the inverse is *not* a function
for domain $\{0, 1, 2\}$ the inverse *is* a function

Example

If $f: x \to x-2$ and $g: x \to x^2$ for the set of positive integers, find (a) $fg^{-1}(9)$, (b) $f^{-1}g(5)$, (c) $g^{-1}f^{-1}(2)$.

Firstly, $f^{-1}: x \to x+2$ and $g^{-1}: x \to \pm\sqrt{x}$.

(a) $f[g^{-1}(9)] = f[\sqrt{9}]^* = f[3] = 3-2 = \underline{1}$ ANS
(b) $f^{-1}[g(5)] = f^{-1}[5^2] = f^{-1}[25] = 25+2 = \underline{27}$ ANS
(c) $g^{-1}[f^{-1}(2)] = g^{-1}[2+2] = g^{-1}[4] = \sqrt{4^*} = \underline{2}$ ANS

Note

(*)1. There is only one possible value as we are operating in the set of positive integers, negative values are excluded.

2. $g^{-1}f^{-1}(x)$ could also be written $(fg)^{-1}(x)$.

Exercise 49

1. If $f: x \to 3x$, find (a) $f^{-1}(6)$, (b) $f^{-1}(-11)$.
2. If $g: x \to x-3$, find (a) $g^{-1}(2)$, (b) $g^{-1}(-7)$.
3. If $h: x \to x^3$, find (a) $h^{-1}(27)$, (b) $h^{-1}(-27)$.
4. If $j: x \to \frac{1}{4}x$, find (a) $j^{-1}(1)$, (b) $j^{-1}(-4)$.
5. Using the functions f, g, h and j from questions **1** to **4** above, find:
(a) $fg^{-1}(6)$, (b) $f^{-1}g(6)$, (c) $f^{-1}g^{-1}(6)$, (d) $g^{-1}f^{-1}(6)$,
(e) $fh^{-1}(8)$, (f) $f^{-1}h(3)$, (g) $f^{-1}h^{-1}(27)$, (h) $h^{-1}f^{-1}(375)$,
(i) $j^{-1}h(2)$, (j) $h^{-1}j(32)$, (k) $h^{-1}j^{-1}(2)$, (l) $j^{-1}f^{-1}(9)$.
6. If $f: x \to x^2$, would its inverse also be a function for the domain, zero excluded, (a) the set of real numbers, (b) the set of positive rational numbers, (c) the set of positive integers.

Examination Standard Questions

Worked examples

1. If $x-3$ is a factor of $2x^2+x-p$, find the value of p and find the other factor.

Let $f(x) = 2x^2+x-p$

$f(+3) = 2(3)^2+(3)-p = 18+3-p = 21-p$

But $f(+3) = 0$ since $(x-3)$ is a factor of the expression.

$\therefore\ 21-p = 0$

$\therefore\ p = \underline{21}$ ANS

The expression is now $2x^2+x-21$

The other factor, by inspection is $(2x+7)$

i.e. $2x^2+x-21 = (x-3)\underline{(2x+7)}$ ANS

2. Show that $x-3$ is a factor of $6x^3-19x^2+2x+3$ and factorise fully this expression. Hence find the solutions of $6x^3-19x^2+2x+3=0$, (a) in the set of integers, (b) in the set of real numbers.

Trial: $f(+3) = 6(3)^3-19(3)^2+2(3)+3$

$= 162 \ -171 \ +6 \ +3 = 0$

$\therefore$ $\underline{x-3 \text{ is a factor of the expression}}$

$(x-3)(ax^2+bx+c) = 6x^3-19x^2+2x+3$ where a, b and c have now to be found.

$$\begin{bmatrix} 6x^3 = (x)(6x^2), +3 = (-3)(-1), \\ -19x^2 = (-3)(6x^2)+(x)(-x), \\ +2x = (x)(-1)+(-3)(-x) \end{bmatrix}$$

$\therefore\ 6x^3-19x^2+2x+3 = (x-3)(6x^2-x-1)$

Factorising the quadratic gives $(x-3)(3x+1)(2x-1)$

The roots are $+3, -\frac{1}{3}, \frac{1}{2}$.

(a) Integer root is $\underline{\{3\}}$ ANS

(b) Real roots are $\underline{\{3, -\frac{1}{3}, \frac{1}{2}\}}$ ANS

3. For which of the following pairs of functions f, g defined on the non-zero real numbers, is fg = gf?

1. $f: x \to 3x, \quad g: x \to x^2$

2. $f: x \to \dfrac{1}{x}, \quad g: x \to \dfrac{x}{5}$

3. $f: x \to x - 3, \quad g: x \to x + 5$

Trials: **1.** $fg(x) = f[g(x)] = f[x^2] = \underline{3x^2}$
$gf(x) = g[f(x)] = g[3x] = (3x)^2 = \underline{9x^2}$
$\therefore$ fg $\neq$ gf in this case

2. $$fg(x) = f[g(x)] = f\left[\frac{x}{5}\right] = \frac{1}{\frac{x}{5}} = \underline{\frac{5}{x}}$$

$$gf(x) = g[f(x)] = g\left[\frac{1}{x}\right] = \frac{\frac{1}{x}}{5} = \underline{\frac{1}{5x}}$$

$\therefore$ fg $\neq$ gf in this case

3. $fg(x) = f[g(x)] = f[x+5] = x + 5 - 3 = \underline{x+2}$
$gf(x) = g[f(x)] = g[x-3] = x - 3 + 5 = \underline{x+2}$
$\therefore$ fg = gf in this case

<u>Only pair **3** satisfies the condition fg = gf on the functions as defined ANS</u>

4. The function $f: x \to x^2 - 2x - 1$ is defined on the domain {0, 1, 2, 3}. Complete the mapping diagram for the function.

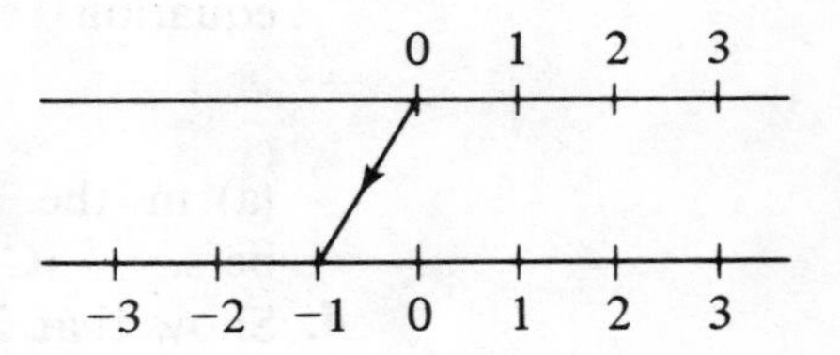

Mapping each member in turn gives

$$f(0) = (0)^2 - 2(0) - 1 = \underline{-1}$$
$$f(1) = (1)^2 - 2(1) - 1 = \underline{-2}$$
$$f(2) = (2)^2 - 2(2) - 1 = \underline{-1}$$
$$f(3) = (3)^2 - 2(3) - 1 = \underline{2}$$

The complete mapping diagram is therefore

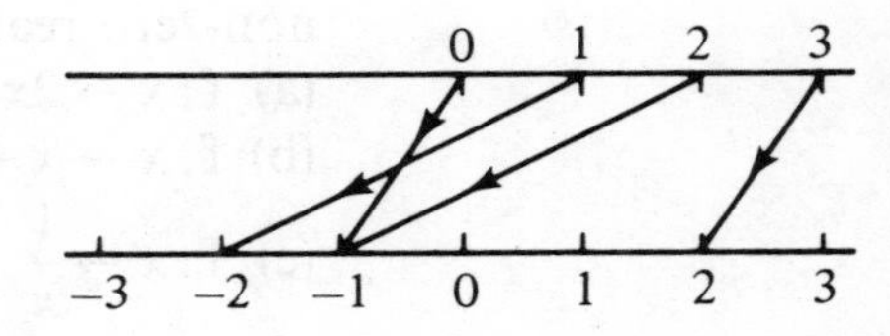

5. If $f : x \to 3x + 1$ and $g : x \to 2x + k$, find the value of k such that fg and gf are the same mapping.

If k has this value, find the vaue of x such that $fg(x) = 8\frac{1}{2}$.

$$fg(x) = f(2x+k) = 3(2x+k)+1 = 6x+3k+1$$
$$gf(x) = g(3x+1) = 2(3x+1)+k = 6x+k+2$$

If $fg(x) = gf(x)$, then $6x + 3k + 1 = 6x + k + 2$

$$\therefore\ 2k = 1$$
$$\therefore\ k = \tfrac{1}{2} \quad \text{ANS}$$

For $f : x \to 3x + 1$ and $g : x \to 2x + \frac{1}{2}$,
$fg(x) = f(2x + \frac{1}{2}) = 3(2x + \frac{1}{2}) + 1 = 6x + 2\frac{1}{2}$

Given $\quad fg(x) = 8\frac{1}{2}$,
Then $\quad 6x + 2\frac{1}{2} = 8\frac{1}{2}$

$$\therefore\ 6x = 8\tfrac{1}{2} - 2\tfrac{1}{2}$$
$$\therefore\ 6x = 6$$
$$\therefore\ x = \underline{1} \quad \text{ANS}$$

Exercise 50

1. If $x - 2$ is a factor of $3x^2 - x - a$, find the value of a and find the other factor. (L)
2. Calculate the values of a and b if $(x - 2)$ and $(x + 4)$ are factors of $x^3 + ax^2 + bx + 48$. (WJEC)
3. Show that $2x + 1$ is a factor of $2x^3 + 21x^2 + 52x + 21$ and factorise this expression completely. Hence write down the solution of the equation
$$2x^3 + 21x^2 + 52x + 21 = 0$$
(a) in the set of integers, (b) in the set of rational numbers. (C)
4. Show that $2a + 3$ is a factor of $2a^3 + 7a^2 - 9$ and use the factor theorem to find the missing factors. Use your results to solve the equation
$$2a^3 + 7a^2 - 9 = 0$$
(a) in the set of positive integers, (b) in the set of real numbers. (L)
5. Given that $f(x) = 2x^2 - 11x + k$ and that $f(3) = 0$, find the value of k and hence the value of $f(-1)$. (C)
6. For which of the following pairs of functions f, g defined on the non-zero real numbers, is $fg = gf$?
(a) $f : x \to 2x, \quad g : x \to x^2$
(b) $f : x \to x + 2, \quad g : x \to x - 1$
(c) $f : x \to \dfrac{1}{x}, \quad g : x \to \dfrac{x}{3}$. (L)

7. A function $x \to px^2 + qx$ where p and q are real numbers is defined on the set $\{0, 1, 2, 3\}$. A mapping diagram for the function is shown. Find the values of p and q. (JMB)

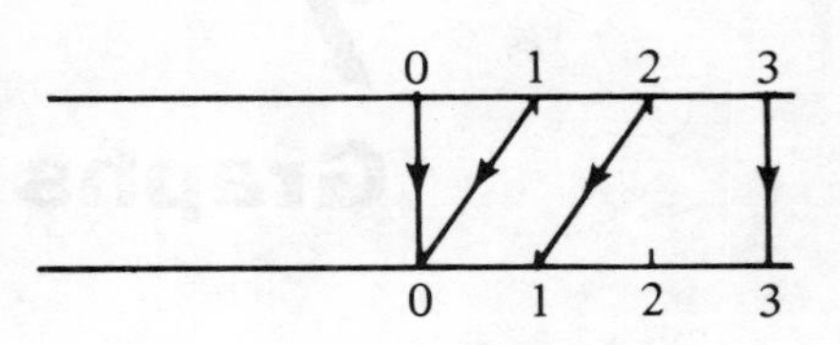

8. If $f : x \to 2x + 3$ and $g : x \to 4x + k$, find the value of k such that fg and gf are the same mapping.

If k has this value, find the value of x such that $fg(x) = 25$. (L)

9. $f(x) = 5 - 2x$.
(a) For what values of x is $f(x) < 3$?
(b) $f[f(x)] = a + bx$ where a and b are integers. Find the values of a and b.
(c) For what values of x is $f(x) + f(-x) = 10$? (O)

10. $f(x) = \dfrac{2x+3}{4}$. If $g[f(x)] = x$ (that is, g is the function which is the inverse of f), then find the function $g(x)$. (O)

11. The mappings f and g are defined as follows: $f : x \to x - 3$, $g : x \to x^2$.

Express in the form $x \to \cdots\cdots$
(a) the inverse mapping f^{-1}, (b) the inverse mapping g^{-1}, (c) gf, (d) fgf, (e) $(gf)^{-1}$

Note: (b) and (e) are 1-to-2 mappings. (L)

12. The function f maps x onto $f(x)$, where $f(x) = 5 - 3x$. Calculate (a) what f maps 4 onto, (b) the value of x if f maps x onto 17, (c) what f maps $f(x)$ onto (i.e. find $f[f(x)]$). (WJEC)

13. Given that $f(x) = 3^x$ where x is real find (a) $f(4)$, (b) x such that $f(x) = \dfrac{1}{27}$. (JMB)

14. Given that $f : x \to 3^x + 2^x$, write down and then evaluate (a) $f(3)$, (b) $f(-1)$. (JMB)

15. Draw a mapping diagram for the function $x \to 3^x$ defined on the domain $\{0, 1, 2\}$. (JMB)

7
Graphs and Calculus

Cartesian coordinates

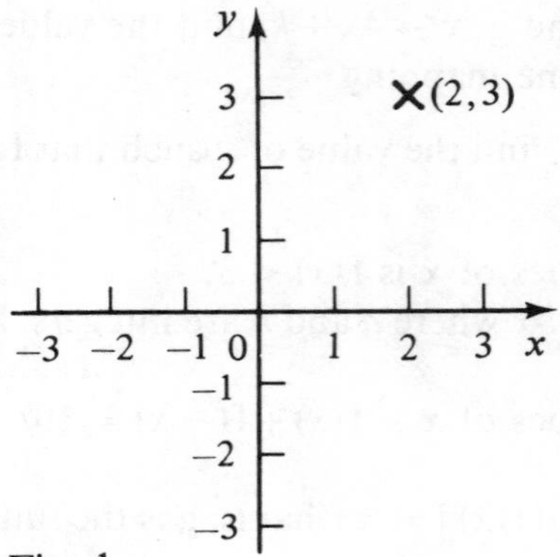

Fig. 1

Fig. (1) illustrates a pair of **coordinate axes**, the **origin** being the name given to the point where they meet, (O). Positive values are shown in the direction of the arrows to the right of and above, negative values to the left of and below. The horizontal axis is called the x-axis and the vertical axis the y-axis. The system we shall use to identify points on the paper is named after the French mathematician Descartes, and is known as the **Cartesian** system of **coordinates**. These are written in the form (x, y); the 'x' coordinate is always given first. The point marked in fig. (1) represents 2 steps along the x-axis followed by 3 steps vertically upwards parallel to the y-axis. The origin would be shown by the coordinates (0, 0).

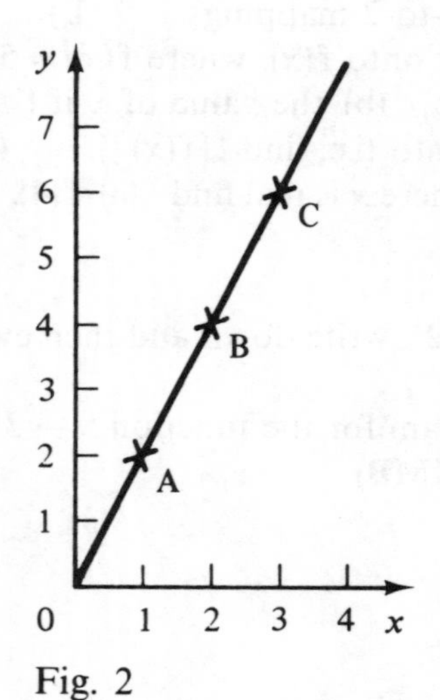

Fig. 2

In fig. (2) the points A(1, 2), B(2, 4), and C(3, 6) are marked. A straight line can be drawn through them. It can be seen that each y-coordinate is twice the corresponding x-coordinate. The line drawn is therefore referred to as the graph of $y = 2x$.

(The **gradient** of the line is 2.) All points on the line will satisfy the equation $y = 2x$. Therefore, if the x-coordinate is 1.5, then the y-coordinate will be $2(1.5) = 3$, etc.

Straight line graphs

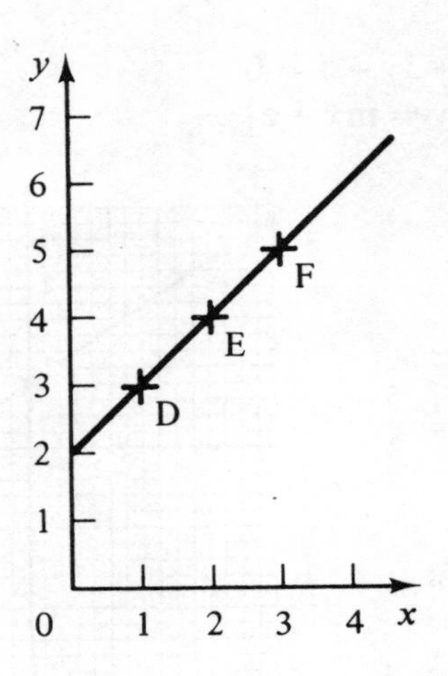

Fig. 3

In fig. (3) the points D(1, 3), E(2, 4), F(3, 5) are marked. A straight line drawn through them intersects the y-axis at the point $y = 2$. It can be seen that each y-coordinate is 2 more than the corresponding x-coordinate, i.e. $x+2$. The line drawn is therefore referred to as the graph of $y = x+2$. (The **gradient** is 1, as $y = 1x+2$.)

Equations such as $y = x+2$ are called linear equations, their graphs being straight lines. The general equation of a straight line graph is written in the form $y = mx+c$, where m is the gradient and c the intercept on the y-axis.

Examples

1. Draw the graph of $y = 3x+2$.

[Draw the axes near the centre of the paper.]

Substitute $x = 0$ in the equation.

$y = 3x+2$ becomes $y = 0+2 = \underline{2}$

Plot the coordinate $x = 0$, $y = 2$; i.e. (0, 2)

Substitute $x = -2$ in the equation

$y = 3(-2)+2 = -6+2 = \underline{-4}$

Plot the coordinate $x = -2$, $y = -4$; i.e. $(-2, -4)$

Substitute $x = 1$ in the equation

$y = 3(1)+2 = 3+2 = \underline{5}$

Plot the coordinate $x = 1$, $y = 5$; i.e. (1, 5)

Draw a line through the points plotted.

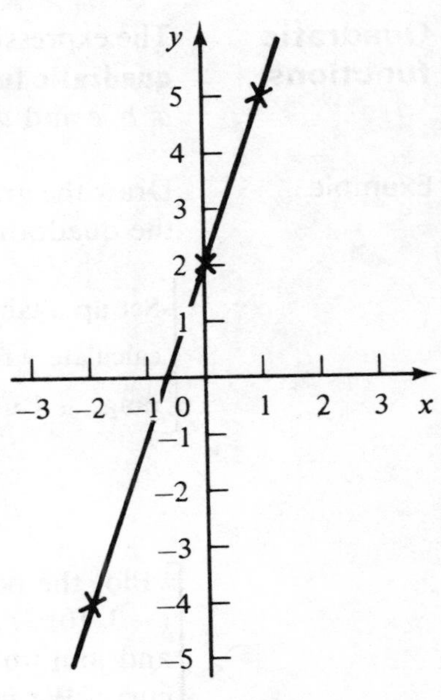

[The third point is calculated as a check on the arithmetical accuracy of the first two calculations.]

2. Draw the graph of $2x+3y-5=0$
[Rearrange into the form $y = mx + c$]

$$\therefore\ 3y = -2x+5$$
$$\therefore\ y = -\frac{2}{3}x+\frac{5}{3}$$

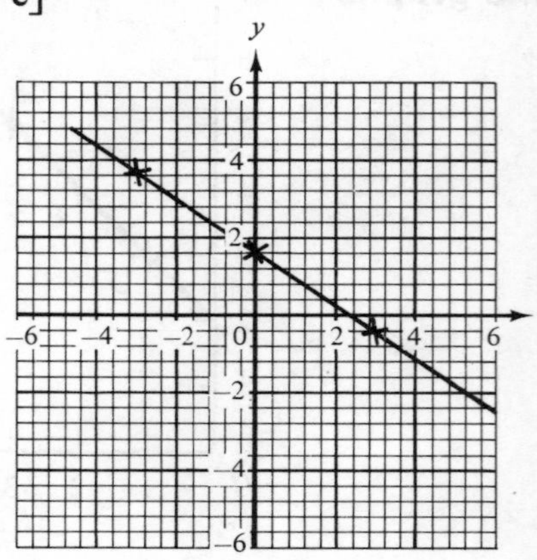

Substituting $x = 0$ gives $y = \frac{5}{3}$,

Substituting $x = -3$ gives
$y = 2 + \frac{5}{3} = 3\frac{2}{3}$,

Substituting $x = 3$ gives
$y = -2+\frac{5}{3} = -\frac{1}{3}$

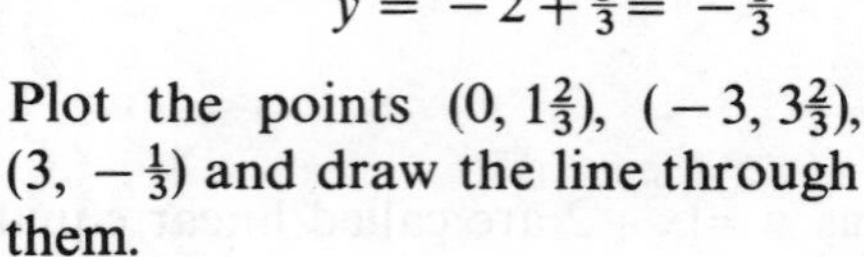

Plot the points $(0, 1\frac{2}{3})$, $(-3, 3\frac{2}{3})$, $(3, -\frac{1}{3})$ and draw the line through them.

Exercise 51 Draw the graphs of the following equations:

1. $y = x$, **2.** $y = x+3$, **3.** $y = \frac{1}{2}x$
4. $y = 2x+1$ **5.** $2y = 4x-1$, **6.** $3y+x = 6$,
7. $y+2x-3 = 0$, **8.** $14 = 3x-2y$,
9. What are the values of the gradients in questions **1** to **8** above?
10. In which direction does a negative gradient slope?

Quadratic functions The expression ax^2+bx+c, where a, b and c are constants is called a **quadratic function**. Similarly the expression ax^3+bx^2+cx+d, where a, b, c and d are constants is called a **cubic function**.

Example Draw the graph of $y = x^2-x-2$, when $-4 \leqslant x \leqslant 4$, and hence solve the quadratic equations **1.** $x^2-x-2 = 0$, **2.** $x(x-1) = 4$.

[Set up a table to calculate y for the range $-4 \leqslant x \leqslant 4$]

x	−4	−3	−2	−1	0	1	2	3	4
x^2	16	9	4	1	0	1	4	9	16
$-x$	4	3	2	1	0	−1	−2	−3	−4
-2	−2	−2	−2	−2	−2	−2	−2	−2	−2
y	18	10	4	0	−2	−2	0	4	10

[Plot the points $(-4, 18)$, $(-3, 10)$ . . . and join up with a smooth curve. We need to plot an extra point at $x = \frac{1}{2}$ to improve accuracy.]

When $x = \frac{1}{2}$,
$$y = x^2-x-2 = (\tfrac{1}{2})^2-\tfrac{1}{2}-2 = \tfrac{1}{4}-\tfrac{1}{2}-2 = -2\tfrac{1}{4}$$

Hence plot $(\frac{1}{2}, -2\frac{1}{4})$ before drawing the curve.

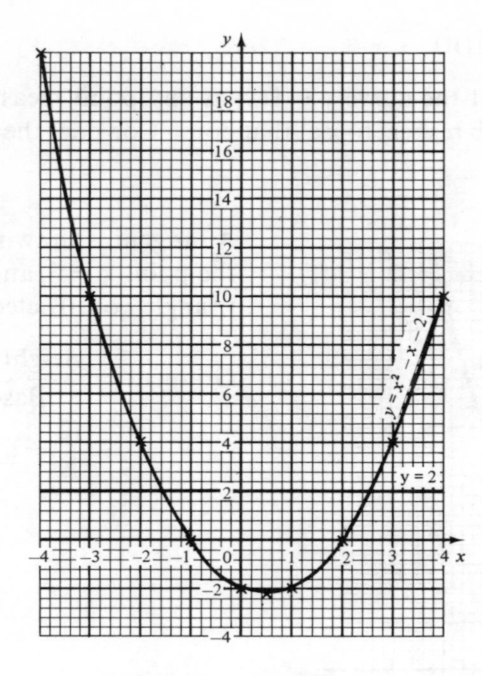

Note From the graph, the **minimum** value of y is $-2\frac{1}{4}$, that is $y = -2\frac{1}{4}$ when $x = \frac{1}{2}$.

1. $y = x^2 - x - 2$ becomes $x^2 - x - 2 = 0$ when $y = 0$.
The line $y = 0$ is the x-axis, and this cuts the curve at the points $x = -1$ and $+2$.
$\therefore$ The roots of $x^2 - x - 2 = 0$ are $\underline{-1}$ and $\underline{+2}$.

2. To solve $x(x-1) = 4$.
[Multiplying out the brackets] $\quad x^2 - x = 4$
[Subtract 2 from each side] $\quad x^2 - x - 2 = 4 - 2$

$$\therefore \; y = x^2 - x - 2 \text{ becomes } x^2 - x - 2 = 2 \text{ when } y = 2.$$

The horizontal line drawn through $y = 2$ cuts the curve at $x = \underline{-1.6}$ and $\underline{+2.6}$.

These are the approximate roots of the equation $x(x-1) = 4$.

Exercise 52

1. Draw the graph of $y = 2x^2 + 3x - 2$, for the range $-3 \leqslant x \leqslant 3$ and hence solve the quadratic equations (a) $2x^2 + 3x - 2 = 0$, (b) $2x^2 + 3x - 5 = 0$.
2. Draw the graph of $y = 3x^2 + 7x - 6 = 0$, for the range $-4 \leqslant x \leqslant 4$ and hence solve the quadratic equations (a) $3x^2 + 7x - 6 = 0$, (b) $3x^2 + 7x = -2$.
3. Draw the graph of $y = 6x^2 + x - 12$, for the range $-4 \leqslant x \leqslant 4$ and hence solve the quadratic equations (a) $6x^2 + x - 12 = 0$, (b) $x(6x+1) = 1$.

Differentiation

The gradient of the curve $y = f(x)$ at any point measures the rate of change of y with respect to x at that point. Consider the graph of $y = x^2$ shown here.

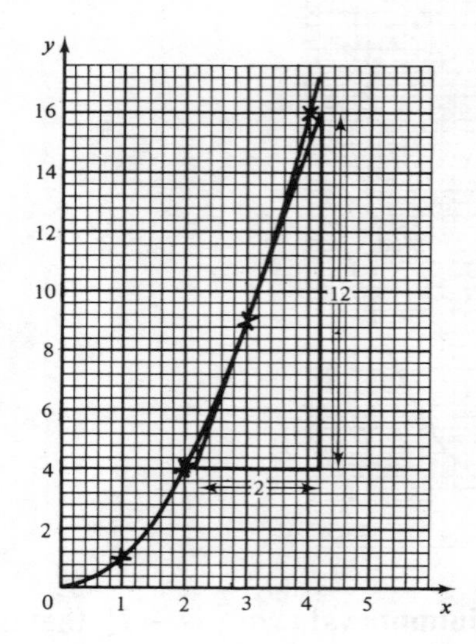

A tangent is drawn to the curve at the point $x = 3$, and a right-angled triangle constructed.

$$\text{Gradient} = \frac{\text{Height of triangle}}{\text{Base length}}$$

$$= \frac{12}{2} = \underline{6}$$

Calculating the gradient of a curve $y = f(x)$ is called **differentiating** y with respect to x. The gradient is usually represented by the symbols $f'(x)$ **or** $\frac{dy}{dx}$ (read dy by dx), called the **derived function** or **derivative**.

The rule for determining derivatives is:

$$\text{If } y = ax^n \qquad \text{then } \frac{dy}{dx} = anx^{n-1}$$

where ***a*** and ***n*** are ordinary numbers.
e.g. To differentiate $y = x^2$:
In the rule for $y = ax^n$; $a = 1$ and $n = 2$ in this case.

$$\frac{dy}{dx} = anx^{n-1} = 1 \times 2x^{2-1} = 2x^1 = \underline{2x} \quad \text{ANS}$$

Now let us consider the point on the curve $y = x^2$ where $x = 3$.

When $x = 3$, $\frac{dy}{dx} = 2(3) = \underline{6}$, which is the same value as that obtained from the graph by drawing the tangent at that point and finding its gradient.

Examples

Differentiate with respect to x **1.** $y = x^4$, **2.** $y = 2x^3$, **3.** $y = 2x^3 + 3x^2$, **4.** $y = 3x^3 + 2x^2 + x + 3$, **5.** $y = \frac{2}{x^3} - \frac{1}{x^2}$.

1. Given $y = x^4$; then in the rule for $y = ax^n$, $a = 1$ and $n = 4$.

$$\frac{dy}{dx} = anx^{n-1} = 1 \times 4x^{4-1} = \underline{4x^3} \quad \text{ANS}$$

2. Given $y = 2x^3$, then in the rule for $y = ax^n$, $a = 2$ and $n = 3$.

$$\therefore \quad \frac{dy}{dx} = anx^{n-1} = 2 \times 3x^{3-1} = \underline{6x^2} \quad \text{ANS}$$

3. Differentiating $y = 2x^3$ first; then $\dfrac{dy}{dx} = \underline{6x^2}$ [from **2** above]

Differentiating $y = 3x^2$; then in the rule for $y = ax^n$, $a = 3$ and $n = 2$.

$$\therefore \frac{dy}{dx} = anx^{n-1} = 3 \times 2x^{2-1} = \underline{6x}$$

$\therefore$ Given $y = 2x^3 + 3x^2$

$$\frac{dy}{dx} = \underline{6x^2 + 6x} \quad \text{ANS}$$

4. Given $y = 3x^3 + 2x^2 + x + 3$

$$\text{then } \frac{dy}{dx} = (3 \times 3x^{3-1}) + (2 \times 2x^{2-1}) + (1 \times 1x^{1-1}) + 0$$

Note — The derivative of a constant term (e.g. 3) is always zero.

$$\therefore \quad \frac{dy}{dx} = \underline{9x^2 + 4x + 1} \quad \text{ANS}$$

Note — $x^{1-1} = x^0 = 1$

5. Given $y = \dfrac{2}{x^3} - \dfrac{1}{x^2}$,

rewriting with negative indices gives

$$y = 2x^{-3} - x^{-2}$$

$$\therefore \frac{dy}{dx} = (2 \times -3x^{-3-1}) - (1 \times -2x^{-2-1})$$

$$= (-6x^{-4}) - (-2x^{-3})$$

$$= \underline{-\frac{6}{x^4} + \frac{2}{x^3}} \quad \text{ANS}$$

Exercise 53

1. On graph paper draw the graph of $y = x^2$ for the range $0 \leqslant x \leqslant +3$.
   (a) Draw a tangent to the curve where $x = 2$, and hence calculate the gradient at that point.
   (b) Differentiate x^2 and substitute $x = 2$ in your answer. Compare your answers to parts (a) and (b).
2. On graph paper draw the graph of $y = x^3$ for the range $0 \leqslant x \leqslant +3$.
   (a) Draw a tangent to the curve where $x = 2$, and hence calculate the gradient at that point.

(b) Differentiate x^3 and substitute $x = 2$ in your answer. Compare your answers to parts (a) and (b).

3. Differentiate with respect to x (a) $y = x^5$, (b) $y = 2x^2$, (c) $y = 4x^3$

4. Find the derivatives of the functions (a) $x^4 + x^3 + x^2$, (b) $2x^4 + 3x^2 + x$, (c) $\frac{3}{x^4} - \frac{2}{x^2}$.

5. Find the gradients of the following curves at the point $x = 2$.
(a) $y = 2x^2 + 3x + 1$, (b) $y = 4x^3 + 3x^2 + x$, (c) $y = \frac{2}{x^3} + \frac{4}{x^2} - \frac{1}{x} + x$

Turning points

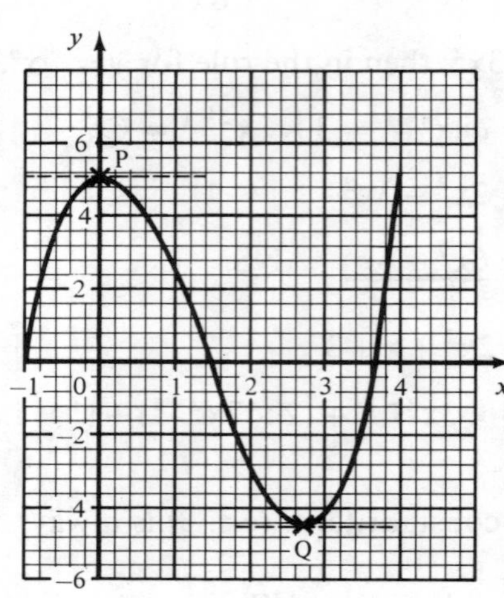

The graph of a cubic function $y = f(x)$ is shown here. At points P and Q a tangent drawn to the curve would be parallel to the x-axis, and therefore its gradient would be zero. These points are called **turning points**; P is also termed the **maximum point**, and Q is also termed the **minimum point**. At such turning points the value of $\frac{dy}{dx}$ (i.e. the gradient) is zero.

Examples

1. Find the coordinates of the turning point on the curve $y = 2x^2 + 2x$.

Given $\quad y = 2x^2 + 2x$

$$\therefore \frac{dy}{dx} = 4x + 2$$

$\left[\text{As } \frac{dy}{dx} = 0 \text{ at a turning point.}\right]$ $\therefore 0 = 4x + 2$

$$\therefore -2 = 4x$$

$$\therefore x = -\tfrac{1}{2}$$

Substitute for x in the equation of the curve:

i.e. $y = 2(-\tfrac{1}{2})^2 + 2(-\tfrac{1}{2})$
$= \tfrac{1}{2} - 1 = -\tfrac{1}{2}$

Coordinates are $(-\tfrac{1}{2}, -\tfrac{1}{2})$ ANS

2. Find the coordinates of the turning point of the curve $y = x^3 - 3x^2$.

Given $\quad y = x^3 - 3x^2$

$$\therefore \frac{dy}{dx} = 3x^2 - 6x$$

$\left[\text{At turning points } \frac{dy}{dx} = 0\right]$ $\therefore 3x^2 - 6x = 0$

[Factorising] $\quad \therefore \; 3x(x-2) = 0$
$\quad \therefore \; 3x = 0$ **or** $x - 2 = 0$; i.e. $x = \underline{0}$ **or** $\underline{+2}$

Substituting these values in the equation of the curve gives

$$\text{when } x = 0,\; y = (0)^3 - 3(0)^2 = \underline{0}$$
$$\text{when } x = 2,\; y = (2)^3 - 3(2)^2 = 8 - 12 = \underline{-4}$$

$\therefore$ The coordinates of the turning points are

$\underline{(0, 0)}$ and $\underline{(2, -4)}$ ANS

We are often required to identify whether the turning points are **maxima** (maximum values) or **minima** (minimum values). This can be done by differentiating $y = f(x)$ a second time, i.e. finding $f''(x)$ or $\dfrac{d^2y}{dx^2}$.

If $\dfrac{d^2y}{dx^2}$ is *positive* for the determined value of x, then this value gives a *minimum* value for y.

If $\dfrac{d^2y}{dx^2}$ is *negative* for the determined value of x, then this value gives a *maximum* value for y.

Examples

1. Find the maxima and minima of $y = \dfrac{2x^3}{3} + \dfrac{x^2}{2} - 6x$.

[To find turning points] $\quad \therefore \; \dfrac{dy}{dx} = 2x^2 + x - 6$

$$\text{and } \frac{d^2y}{dx^2} = 4x + 1$$

[Factorising] $\quad \dfrac{dy}{dx} = (2x-3)(x+2)$

When $\dfrac{dy}{dx} = 0 \qquad 0 = (2x-3)(x+2)$

i.e. $2x - 3 = 0$ **or** $x + 2 = 0$
i.e. $x = \frac{3}{2}$ **or** $\underline{-2}$

Substituting $x = \dfrac{3}{2}$ in $\dfrac{d^2y}{dx^2} = 4x + 1$ gives $4\left(\dfrac{3}{2}\right) + 1 = 7$ (postive)

$\therefore$ this turning point is a minimum

When $x = \dfrac{3}{2}$, $\quad y = \dfrac{2}{3}\left(\dfrac{3}{2}\right)^3 + \dfrac{1}{2}\left(\dfrac{3}{2}\right)^2 - 6\left(\dfrac{3}{2}\right)$

$$= \frac{9}{4} + \frac{9}{8} - 9 = \underline{-5\tfrac{5}{8}} \text{ ANS}$$

Substituting $x = -2$ in $\dfrac{d^2y}{dx^2} = 4x + 1$ gives $4(-2) + 1 = -7$ (negative)

$\therefore$ this turning point is a maximum

When $x = -2$, $y = \dfrac{2}{3}(-2)^3 + \dfrac{1}{2}(-2)^2 - 6(-2)$

$$= \frac{-16}{3} + 2 + 12 = \underline{8\tfrac{2}{3}} \quad \text{ANS}$$

2. Find the maximum rectangular area of ground that can be enclosed by fencing of overall length 100 metres.

Let 'a' be the length of one side of the rectangle in metres and 'b' be the length of the other (adjacent) side in metres. The perimeter is 100 metres.

$$\therefore\ 2a + 2b = 100$$
$$\text{i.e. } a + b = 50$$
$$\text{i.e. } a = 50 - b$$
$$\text{Area} = ab$$

Let A represent the area of the rectangle, then $A = ab$.

Substituting $a = (50 - b)$ gives:

$$A = (50 - b)b$$
$$\text{i.e.} \quad A = 50b - b^2$$
$$\frac{dA}{db} = 50 - 2b,$$

$$\text{and} \quad \frac{d^2A}{db^2} = -2 \quad \text{(negative)}$$

The negative result of $\dfrac{d^2A}{db^2}$ means that the area is a maximum in this case.

Putting $\dfrac{dA}{db} = 0$ gives $\quad 0 = 50 - 2b$

$$\therefore\ 2b = 50$$
$$\therefore\ b = 25 \quad \text{metres}$$

Substituting $b = 25$ in the earlier equation $a = 50 - b$ gives

$$a = 50 - 25 = 25 \text{ metres}$$

The maximum rectangular area that can be enclosed by 100 metres of fencing is therefore formed by a square of side 25 metres.

$$\text{Area} = 25 \times 25$$
$$= \underline{625\,\text{m}^2} \quad \text{ANS}$$

Exercise 54 Find the maximum and minimum values of the following curves, and distinguish between them:

1. $y = 3x^2 - 2x$ **2.** $y = \frac{1}{2}x^2 - x + 1$

3. $y = 4 + 3x - 3x^2$ **4.** $y = \dfrac{2x^3}{3} - x^2 - 4x$

5. $y = x^3 + 3x^2$ **6.** $y = 3x^3 - 6x^2 + 3x$

7. Given that $f(x) = x^3 - 2x^2 - 4x - 3$, find the maxima and minima of $f(x)$.

8. Find the maximum rectangular area of field that a farmer can enclose using portable fencing of overall length 80 metres.

9. If the farmer in question **8** uses the same 80 metres of fencing for three sides of the rectangle, and a straight hedge for the fourth side, what is the maximum rectangular area he can then enclose? [*Hint:* If the width of the enclosure is 'a' metres, then the length will be $(80 - 2a)$ metres.]

10. The sum of two numbers is 12. What is the maximum possible product of two such numbers?

Linear Kinematics

Let x be the distanced travelled (metres) or (m)
t be the time taken (seconds) or (s)
v be the velocity (metres/second) or (m/s)
a be the acceleration (metres/second/second) or (m/s^2)

then $v = \dfrac{dx}{dt}$,

and $a = \dfrac{dv}{dt} = \dfrac{d^2x}{dt^2}$

Note

$\dfrac{dx}{dt}$ compares the rate of change of distance with that of time = **velocity**.

$\dfrac{dv}{dt}$ compares the rate of change of velocity with that of time = **acceleration**.

Examples

1. A particle moves along the x-axis in such a way that its distance x metres from the origin after t seconds is given by $x = 3t^2 - t^3$.
(a) Find the distance travelled after 2 seconds
(b) Find the speed of the particle at this instant
(c) Find the acceleration of the particle at this instant.

(a) Given $x = 3t^2 - t^3$

Substituting $t = 2$ gives

$$x = 3(2)^2 - (2)^3$$
$$= 12 - 8 = \underline{4\text{m}} \quad \text{ANS}$$

(b) Given $x = 3t^2 - t^3$

$$v = \frac{dx}{dt} = 6t - 3t^2$$

Substituting $t = 2$ gives

$$v = 6(2) - 3(2)^2 = 12 - 12 = \underline{0\text{m/s}} \quad \text{ANS}$$

(i.e. the particle is momentarily at rest)

(c) From $v = 6t - 3t^2$

$$a = \frac{dv}{dt} = 6 - 6t$$

Substituting $t = 2$ gives

$$a = 6 - 6(2) = 6 - 12 = \underline{-6\text{m/s}^2} \quad \text{ANS}$$

(i.e. there is an acceleration of 6m/s^2 back towards the origin).

2. A particle is projected vertically upwards and after t seconds its height h metres reached is given by the formula $h = 50t - 5t^2$.
(a) Find the time taken to reach its maximum height.
(b) Find the acceleration of the particle on its downward path.

(a) The particle's height will be a maximum when $\frac{dh}{dt} = 0$ and $\frac{d^2h}{dt^2}$ is negative.

Given $h = 50t - 5t^2$

$$v = \frac{dh}{dt} = 50 - 10t$$

and $\frac{d^2h}{dt^2} = -10$ (negative)

Putting $\frac{dh}{dt} = 0$ gives $0 = 50 - 10t$

$$\text{i.e. } 10t = 50$$
$$\therefore t = \underline{5 \text{ seconds}} \quad \text{ANS}$$

Note — At its maximum height the particle will be momentarily at rest, and its velocity at that instant will be zero as in the previous example

$$\text{i.e.} \quad v = \frac{dh}{dt} = 0$$

(b) From $v = 50 - 10t$

$$a = \frac{dv}{dt} = -10$$

The negative value of 'a' indicates acceleration in the opposite direction to the original motion. This downwards acceleration is constant, there being no variable 't' in the answer.

i.e. $a = \underline{10 \text{m/s}^2 \text{ vertically downwards}}$ ANS

Exercise 55

1. A particle moves along the x-axis in such a way that its distance x metres from the origin after t seconds is given by $x = 5t - 3t^2$.
   (a) Find the distance travelled after 1 second.
   (b) Find the distance travelled after 2 seconds and explain your answer.
   (c) Find the speed of the particle after 2 seconds.
   (d) Find the acceleration of the particle after 2 seconds.
   (e) Is this acceleration constant?
2. A particle is projected vertically upwards and after t seconds its height h metres reached is given by the formula $h = 80t - 5t^2$.
   (a) Find the time taken to reach its maximum height.
   (b) Find the acceleration of the particle after this time.
3. A stone is thrown vertically downwards from a cliff top. After t seconds the distance fallen, x metres, is given by the formula $x = 4t + 5t^2$. Find (a) the velocity of the stone after 3 seconds, (b) the constant acceleration during this time, (c) the velocity of the stone when thrown downwards.
4. A particle moves along the x-axis in such a way that, t seconds after it leaves the origin, its displacement $x = 4t^2 - t^3$ metres.
   (a) Show that the particle returns to the origin when $t = 4$.
   (b) Find the value of x when $t = 2$.
   (c) Find the speed at which the particle returns to the origin.
   (d) Find the particle's acceleration when it returns to the origin.
5. The distance x metres moved by a particle along a line in t seconds is given by the formula $x = 2t^2 + t$. Find (a) its velocity after s seconds, (b) its acceleration after $2s$ seconds.

Integration

We know that if $y = x^2$, then $\dfrac{dy}{dx} = 2x$,

but if $y = x^2 + 2$, then $\dfrac{dy}{dx} = 2x$ also.

If, therefore, we are asked to find the equation of a curve whose gradient is $2x$, the answer is $y = x^2 +$ (a number). This number is a constant, its value does not change as y and x vary. We need more information if we are to find the precise equation of the curve.

Example Find the equation of the curve with gradient $3x^2$ that passes through the point (1, 2).

Given $\frac{dy}{dx} = 3x^2$,

then $y = x^3 + \text{constant}$ [Check this yourself by differentiating x^3]

The curve passes through (1, 2), i.e. when $x = 1$, $y = 2$

$\therefore y = x^3 + \text{constant}$

becomes $2 = 1^3 + \text{constant}$

$\therefore 2 - 1 = \text{constant} = \underline{1}$

$\therefore$ The equation of the curve is $\underline{y = x^3 + 1}$ ANS

This process of reversing differentiation is called integration. The sign used to denote integration is $\int$, (an elongated 's' showing summation). The constant term is known as the **constant of integration** and is abbreviated to 'c'.

e.g. $\int 2x\,dx$ *means* the integral of $2x$ with respect to x.

and $\int 2x\,dx = x^2 + c$

The rule for integrating is $\int ax^n\,dx = \frac{ax^{n+1}}{n+1} + c$

where a and n are ordinary numbers.

e.g. $\int 2x^1\,dx = \frac{2x^{1+1}}{1+1} + c$

[Cancelling] $= \frac{{}^1\not{2}x^2}{\not{2}_1} + c$

$= \underline{x^2 + c}$

Note $\int 3\,dx = 3x + c$, because 3 can be written as $3x^{\circ}$.

Examples

1. Integrate $2x^3$

[Here $a = 2$, $n = 3$.]

Substituting in $\frac{ax^{n+1}}{n+1} + c$

gives $\frac{2x^4}{4} + c$

$= \underline{\frac{x^4}{2} + c}$ ANS

2. Find $\int (3x^2 - x + 2)dx$.

[Integrate each term separately] $= \frac{3x^3}{3} - \frac{x^2}{2} + 2x + c$

$= \underline{x^3 - \frac{x^2}{2} + 2x + c}$ ANS

Exercise 56 Find:

1. $\int (x^2+1)\,dx$ 2. $\int x^4\,dx$

3. $\int 4x^3\,dx$ 4. $\int \frac{x^2}{2}\,dx$

5. $\int (5x^4-3x^2)\,dx$ 6. $\int \left(\frac{3x^2}{2}+x-3\right)dx$

7. $\int \left(\frac{3}{4}x^2-2x+1\right)dx$ 8. $\int t^2\,dt$

9. $\int 4y^3\,dy$ 10. $\int \left(v-\frac{v^2}{3}\right)dv$

Earlier we stated that the velocity $v=\frac{dx}{dt}$, where $\frac{dx}{dt}$ was the rate of change of distance (x) with respect to time (t).

Therefore as $v = \frac{dx}{dt}$, and $a=\frac{dv}{dt}$

then $\int v\,dt = x+c$, and $\int a\,dt = v+c$

Example The acceleration of a particle is given in terms of t by $a=t+2$. Find formulae for the velocity (v) and distance (x) given that when $t=0$; $x=0$ and $v=0$.

Given $a=t+2$

$$\therefore v=\int (t+2)\,dt=\frac{t^2}{2}+2t+c$$

When $t=0$, $v=0$

Substituting in $v=\frac{t^2}{2}+2t+c$ gives

$$0=0+0+c$$

$$\therefore c=\underline{0}$$

$$\therefore v=\underline{\frac{t^2}{2}+2t} \quad \text{ANS}$$

From $v=\frac{t^2}{2}+2t$

$$x=\int\left(\frac{t^2}{2}+2t\right)dt=\frac{t^3}{6}+\frac{{}^1\cancel{2}t^2}{\cancel{2}_1}+c$$

When $t=0$, $x=0$

Substituting in $x = \frac{t^3}{6} + t^2 + c$, gives

$$0 = 0 + 0 + c,$$

$$\therefore\ c = \underline{0}$$

$$\therefore\ x = \frac{t^3}{6} + t^2 \quad \text{ANS}$$

Exercise 57

1. Given that the velocity (v m/sec) in terms of time (t sec) is given by $v = t^2 + 2t$, find the distance (x metres) after 3 seconds if $x = 0$ when $t = 0$.
2. Given that the acceleration (a m/s^2) in terms of time (t s) is given by $a = 2t^3 + t^2$, find the velocity (v m/s) after 2 seconds if $v = 5$ when $t = 0$.
3. Given that the acceleration (a m/s^2) in terms of time (t sec) is given by $a = t^4 + \frac{1}{t^3}$, find the distance (x metres) after 2 seconds if when $t = 0$, $v = 0$ and $x = 0$.
4. Given that $\frac{dx}{dt} = 3t^4 + t^3 + 2t$, find x in terms of t.
5. If the velocity of a particle starting from rest is 8 m/sec after 2 seconds; find the distance travelled after 4 seconds.
6. A stone is dropped from the edge of a cliff and falls with an acceleration of 10 m/s^2. Calculate the velocity of the stone 3 seconds after being dropped.
7. Calculate the distance fallen by the stone in question **6** after 3 seconds.
8. A particle is projected in a straight line with an initial velocity of 20 m/sec. If its acceleration is -4 m/s^2 (i.e. a retardation), find the distance of the particle from its starting place after 2 seconds.
9. Find the velocity of the particle in question **8** after 5 seconds.
10. Find the furthest distance the particle in question **8** reaches from its starting point. [*Hint*: the particle is momentarily at rest before returning.]

Area Under a Curve

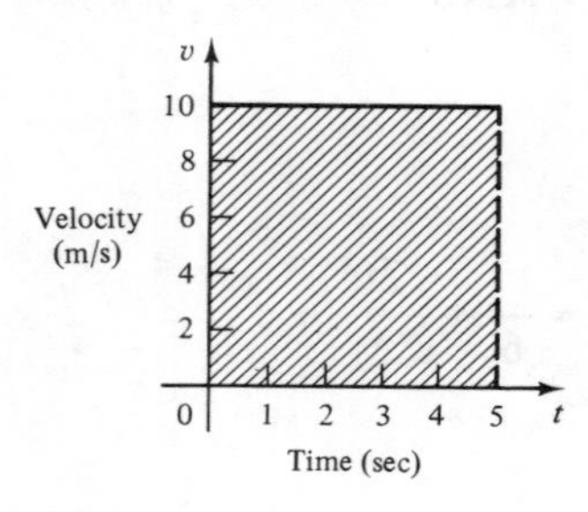

A cyclist is travelling at a constant speed of 10 m/sec. A stopwatch is started and stopped again after 5 seconds as shown on the graph. He will then have travelled $10 \times 5 = 50$ metres. (i.e. Distance = Velocity × Time). This is the same as finding the area under the graph between the times 0s and 5s.

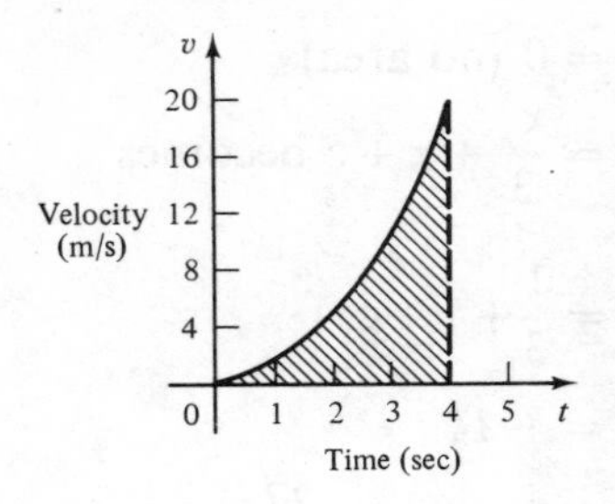

The velocity (v) of a particle is given in terms of the time (t) by the formula $v = t^2 + t$. This graph shows the particle's motion for the first 4 seconds. The distance travelled during the 4 seconds can be found by drawing the graph on graph paper and finding the area beneath the curve by counting the small squares. The area could also be found by dividing the area into small strips.

A quicker and more accurate solution may be found using integration. We know that $\int v\,dt = x$ (distance), therefore the area beneath the curve may be found by integrating the equation of the curve.

$$\text{i.e. } A(\text{area}) = \int (t^2 + t)\,dt = \frac{t^3}{3} + \frac{t^2}{2} + c$$

When $t = 0$, $A = 0$ (there is no area at that point)

Substituting in $A = \frac{t^3}{3} + \frac{t^2}{2} + c$, gives

$$0 = 0 + 0 + c$$

$$\therefore\ c = 0$$

$$\therefore\ A = \frac{t^3}{3} + \frac{t^2}{2}$$

When $t = 4$, $A = \frac{64}{3} + \frac{16}{2} = 21\frac{1}{3} + 8 = 29\frac{1}{3}\text{units}^2$

(i.e. the particle travelled a distance of $29\frac{1}{3}$ metres in the first 4 seconds.)

We have found the area under the graph between the values $t = 0$ and $t = 4$. These are called the limits of the integration performed.

The integral $A = \int (t^2 + t)\,dt$ is referred to as an **indefinite integral** as no limits are set.

The integral $A = \int_0^4 (t^2 + t)\,dt$ is referred to as a **definite integral** as definite limits have been set.

Example

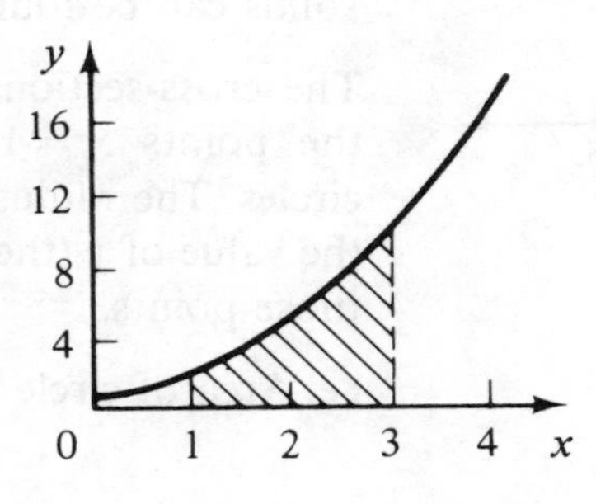

Find the area beneath the graph of $y = x^2 + 1$ between the limits $x = 1$ and 3.

$$A = \int (x^2 + 1)\,dx$$

$$= \frac{x^3}{3} + x + c$$

When $x = 1$ $\quad A = 0$ (no area)

$$\therefore \; A = \frac{x^3}{3} + x + c \text{ becomes}$$

[Substituting for x and A] $\quad 0 = \frac{1}{3} + 1 + c$

$$\therefore \; c = -1\tfrac{1}{3}$$

When $x = 3$ $\quad A(\text{the shaded area}) = \frac{27}{3} + 3 - 1\tfrac{1}{3}$

[Substituting for x, A and c] $\quad \therefore \; A = 12 - 1\tfrac{1}{3} = \underline{10\tfrac{2}{3}\text{units}^2}$ ANS

The problem can be solved more easily by using the definite integral

$$A = \int_1^3 (x^2 + 1)\,dx$$

which is found by subtracting the value of the lower limit from that of the upper limit. Therefore the definite integral has no constant of integration 'c'.

i.e. $$A = \int_1^3 (x^2 + 1)dx$$

[Integrating] $$= \left[\frac{x^3}{3} + x + c\right]_1^3$$

[Substituting] $$= \left[\left(\frac{27}{3} + 3 + c\right) - \left(\frac{1}{3} + 1 + c\right)\right]$$

$$\therefore \; A = [12 - 1\tfrac{1}{3}] = \underline{10\tfrac{2}{3}\text{units}^2} \quad \text{ANS}$$

Volumes of Revolution

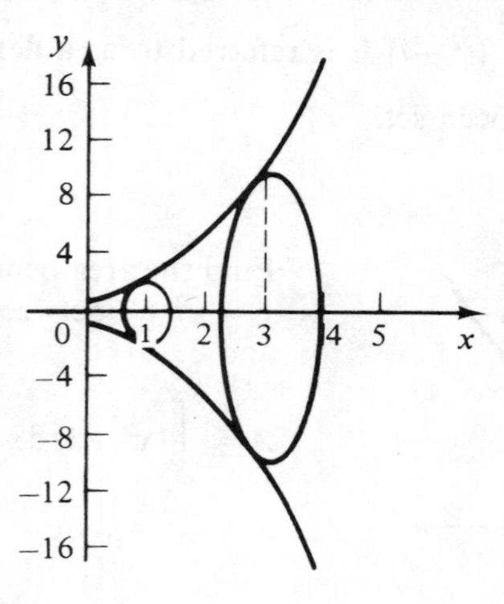

If the graph of $y = x^2 + 1$ discussed above is revolved around the x-axis, the shape generated is called a **solid of revolution**. The volume of such solids can be found.

The cross-sections of the solid at the points $x = 1$ and $x = 3$ are circles. The radius of each circle is the value of y (the y-coordinate) at those points.

i.e. Area of circle $= \pi y^2$

If we consider the solid of revolution as being made up of an infinite number of such circles between the limits $x = 1$ and $x = 3$, the volume of the solid could be found by finding the sum of these circles. As mentioned earlier, integration means summation and can therefore be used in this case.

i.e. $$V = \pi \int_1^3 y^2 \,dx \text{ units}^3$$

[Substituting for $y = x^2 + 1$] $$= \pi \int_1^3 (x^2+1)^2 \,dx$$

[Squaring] $$= \pi \int_1^3 (x^4 + 2x^2 + 1)\,dx$$

[Integrating] $$= \pi \left[\frac{x^5}{5} + \frac{2x^3}{3} + x\right]_1^3$$

[Substituting for $x = 3$, then $x = 1$] $$= \pi\left[\left(\frac{243}{5} + \frac{54}{3} + 3\right) - \left(\frac{1}{5} + \frac{2}{3} + 1\right)\right]$$

[Common denominator is 15] $$= \pi[(48\tfrac{3}{5} + 18 + 3) - (1\tfrac{13}{15})]$$

$$\therefore\ V = \pi[69\tfrac{3}{5} - 1\tfrac{13}{15}] = \underline{67\tfrac{11}{15}\pi \text{ units}^3} \quad \text{ANS}$$

Note The volume of revolution between the limits $x = a$ and b is

$$V = \pi \int_a^b y^2 \,dx \text{ units}^3$$

Exercise 58 Make sketches of each problem before calculation:

1. Find the area beneath the graph of $y = x^2$ between the limits $x = 1$ and $x = 3$.
2. Find the area beneath the curve $y = 3x^2 - 2$ between the limits $x = 0$ and $x = 2$.
3. Find the area between the curve $y = 3x - x^2$ and the x-axis.
4. Find the area between the curve $y = x^2 - 4x + 3$ and the x-axis.
5. Find the volume of the solid formed between the limits $x = 1$ and $x = 3$, by revolving the curve $y = x^2$ about the x-axis.
6. Find the volume of the solid cone formed between the limits $x = 0$ and $x = 3$, by revolving the line $y = 2x$ about the x-axis.
7. Find the volume of the solid cylinder formed, between the limits $x = 0$ and $x = 3$, by revolving the line $y = 3$ about the x-axis.
8. Find the volume of the solid of revolution, between the limits $x = 1$ and $x = 3$, formed by rotating the curve $y = x^2 - 4x + 3$ around the axis of x.

Examination Standard Questions

Worked examples

1. If $y = 18x^2 - 6x^3$, find $\frac{dy}{dx}$. Show by calculation that y has both a minimum and a maximum value and find these values.

Given $y = 18x^2 - 6x^3$

$$\frac{dy}{dx} = 36x - 18x^2 = 18x(2 - x)$$

Putting $\frac{dy}{dx} = 0$ gives $\quad 0 = 18x(2 - x)$

i.e. $\quad 18x = 0 \quad$ **or** $\quad 2 - x = 0$

i.e. $\quad x = 0 \quad$ **or** $\quad x = +2$

$$\frac{d^2y}{dx^2} = 36 - 36x$$

When $x = 0 \quad \frac{d^2y}{dx^2} = 36$ (positive)

When $x = 2 \quad \frac{d^2y}{dx^2} = 36 - (36 \times 2) = -36$ (negative)

$\therefore$ y has both a minimum and a maximum value

When $x = 0 \quad y = 18(0)^2 - 6(0)^3 = \underline{0 \text{ (minimum)}}$ ANS

When $x = 2 \quad y = 18(2)^2 - 6(2)^3 = 72 - 48$

$= \underline{24 \text{ (maximum)}}$ ANS

2. A particle moves in a straight line such that its distance, x metres, from a fixed point 0 is given after time t seconds by the equation

$$x = 4t + 2t^2 - t^3$$

Calculate (a) the initial velocity, (b) the acceleration after 1 second.

(a) Given $x = 4t + 2t^2 - t^3$

$$v = \frac{dx}{dt} = 4 + 4t - 3t^2$$

Substituting $t = 0$ to find the initial velocity gives

$$v = 4 + 4(0) - 3(0)^2$$

$= \underline{4 \text{ m/s}}$ ANS

(b) From $\quad v = 4 + 4t - 3t^2$

$$a = \frac{dv}{dt} = 4 - 6t$$

Substituting $t = 1$ gives

$a = 4 - 6(1) = \underline{-2 \text{ m/s}^2}$ ANS

3. A solid cylinder has a radius 'r' centimetres and height 7 centimetres. If A represents the total surface area of the cylinder in square centimetres, find a formula for A in terms of r. Take $\pi = \frac{22}{7}$.

Calculate the rate of change of A with respect to r when $r = 3\frac{1}{2}$.

[From chapter 3] Total surface area of a cylinder $= 2\pi r(h + r)$, where h is the height and r the radius.

Substituting $\pi = \frac{22}{7} h = 7$ gives

$$A = 2 \times \frac{22}{7} r(7 + r) = 44r + \frac{44}{7} r^2 \text{ units}^2 \quad \text{ANS}$$

[Rate of change is represented by $\frac{dA}{dr}$]

$$\frac{dA}{dr} = 44 + \frac{88}{7} r$$

Substituting $r = 3\frac{1}{2}$ gives

$$\frac{dA}{dr} = 44 + \left(\frac{\cancel{88}^{44}}{\cancel{7}_1} \times \frac{\cancel{7}^1}{\cancel{2}_1}\right) = 88 \quad \text{ANS}$$

4. P and Q are two points on a straight line 18cm apart. A particle moving along the line passes through P with a velocity of 6cm/s towards Q. If the displacement of the particle from P after t seconds is represented by $x$cm, its velocity by $v$cm/s and its acceleration by $(2t - 5)$ cm/s^2, find (a) Formulae for v and x in terms of t. (b) The times at which the particle is momentarily at rest.

Show also that (c) the particle passes through Q 6 seconds after passing through P.

(a) Given $a = (2t - 5)$ cm/s^2

$$v = \int a \, dt = \int (2t - 5) \, dt$$

$$= \frac{{}^1\cancel{2}t^2}{\cancel{2}_1} - 5t + c$$

When $t = 0$, $v = 6$,

[Substituting] $\therefore v = t^2 - 5t + c$, becomes

$$6 = 0 - 0 + c$$

$$\therefore c = 6$$

[Substituting for 'c'] $\therefore v = t^2 - 5t + 6$cm/s ANS

From $v = t^2 - 5t + 6$

$$x = \int v\,dt = \int (t^2 - 5t + 6)\,dt$$

$$= \frac{t^3}{3} - \frac{5t^2}{2} + 6t + c$$

When $t = 0$, $x = 0$,

[Substituting] $\therefore\ x = \frac{t^3}{3} - \frac{5t^2}{2} + 6t + c$, becomes

$$0 = 0 - 0 + 0 + c$$

$$\therefore\ c = 0$$

[Substituting for 'c'] $\therefore\ x = \underline{\frac{t^3}{3} - \frac{5t^2}{2} + 6t\,\text{cm}}$ ANS

(b) The particle is momentarily at rest when $v = 0$.

$\therefore\ v = t^2 - 5t + 6$

becomes

$$0 = t^2 - 5t + 6$$

[Factorising the quadratic] $\therefore\ (t-3)(t-2) = 0$

i.e. $t - 3 = 0$ **or** $t - 2 = 0$

$\therefore\ \underline{t = 3\text{s}}$ **or** $\underline{2\text{s}}$ ANS

(c) When $t = 6$

$$x = \frac{t^3}{3} - \frac{5t^2}{2} + 6t$$

becomes

$$x = \frac{6^3}{3} - \frac{5(6^2)}{2} + 6(6)$$

$$= 72 - 90 + 36$$

$= \underline{18\text{cm}}$, which is the distance between P and Q. ANS

5. Prove that the curve $y = 1 - x^2$ is intercepted by the straight line $y = x + 1$ at the points A, $(-1, 0)$ and B, $(0, 1)$. Find the area between the curve and the line.

The two graphs cross when $1 - x^2 = x + 1$

[Rearrange into a quadratic] i.e. $0 = x^2 + x$

$\therefore\ x(x+1) = 0$

$\therefore\ x = 0$ **or** $x + 1 = 0$

$\therefore\ x = \underline{0}$ **or** $\underline{-1}$

[Substituting in $y = x + 1$ or in $y = 1 - x^2$]

When $x = 0$, $y = 1$

When $x = -1$, $y = 0$

$\therefore$ The points of interception are $(0, 1)$ and $(-1, 0)$.

[Sketching the graphs]

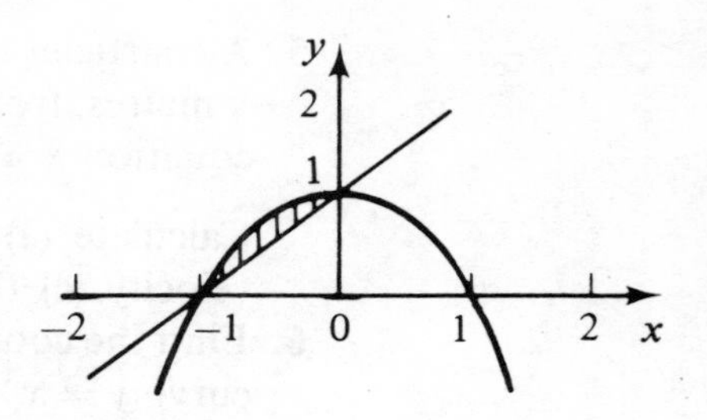

The area beneath the curve between the limits $x = -1$ and 0 is found by:

$$\int_{-1}^{0} (1-x^2)\,dx$$

$$= \left[x - \frac{x^3}{3}\right]_{-1}^{0}$$

$$= [(0) - (-1 + \tfrac{1}{3})] = \tfrac{2}{3}\text{units}^2 \quad \text{ANS}$$

The area of the triangle AOB is found by:

$\frac{1}{2}$ base × perpendicular height

[We ignore the negative sign of the x-coordinate (-1)] $= \frac{1}{2} \times 1 \times 1 = \frac{1}{2}$units^2

∴ The area between the curve and the line (shaded)

$$= \frac{2}{3} - \frac{1}{2} = \frac{4-3}{6} = \frac{1}{6}\text{units}^2 \quad \text{ANS}$$

Exercise 59

1. A cylinder is to be constructed so that $V = \pi(6r^2 - r^3)$, where V cm^3 is the volume and r cm the radius. By differentiation show that the value of r which makes V a maximum is 4, and find this maximum value. (O)
2. (a) Use differentiation to find the maximum value of $9 + 6x - 2x^2$
   (b) Find the coordinates of the two points on the curve whose equation is $y = 6x^2 - x^3$ where the gradient is 9. (O)
3. A rectangular block has a height of x centimetres, breadth of x centimetres and length $2x$ centimetres. If the total surface area is A square centimetres, find a formula for A in terms of x. Hence calculate the rate of change of A with respect to x when $x = 3$. (WJEC)
4. A particle moves in a straight line such that its velocity, v metres per second, is given after time t seconds by the equation $v = 6 + 5t - t^2$.

   Calculate (a) the time when the particle is at rest instantaneously, (b) the initial velocity, (c) the acceleration after 1.5 seconds, (d) the maximum velocity. (WJEC)

5. A particle moves in a straight line such that its distance, x metres, from a fixed point 0 is given after time t seconds by the equation $x = 4t + 2t^2 - t^3$.

Calculate (a) the distance from 0 after 2 seconds, (b) the initial velocity, (c) the distance travelled in the second second. (WJEC)

6. Find the coordinates of the maximum and minimum points of the curve $y = x^3 - 6x^2 + 9x + 2$. (O)

Indicate clearly how you distinguish the maximum from the minimum.

7. A particle moves in a straight line so that, t seconds after passing a fixed point in the line, its velocity v metres per second is given by $v = \frac{t^2}{4} - 3t + 5$.

Calculate (a) the velocity after 10 seconds, (b) the acceleration when $t = 0$, (c) the minimum velocity. (L)

8. A particle moves from rest in a straight line. Its velocity, v cm per second, after t seconds, is given by $v = 5t - 2t^2$.

Calculate (a) the velocity after 2 seconds, (b) the time taken for the particle to come to rest again, (c) the acceleration after 3 seconds, (d) the velocity when the acceleration is zero. (L)

9. Calculate the coordinates of the points at which the curve $y = 15 + 4x - 4x^2$ cuts the x-axis and the y-axis, and hence sketch the curve.

Calculate (a) the value of x which gives the maximum value of y, (b) the maximum value of y. (L)

10. Find the coordinates of the maximum and minimum points on the graph of $y = 2x^3 - 9x^2 + 12x - 5$.

State the range of values of x for which the gradient of the graph is negative. (O)

11. At which two of the marked points is the gradient of the graph equal to zero? (L)

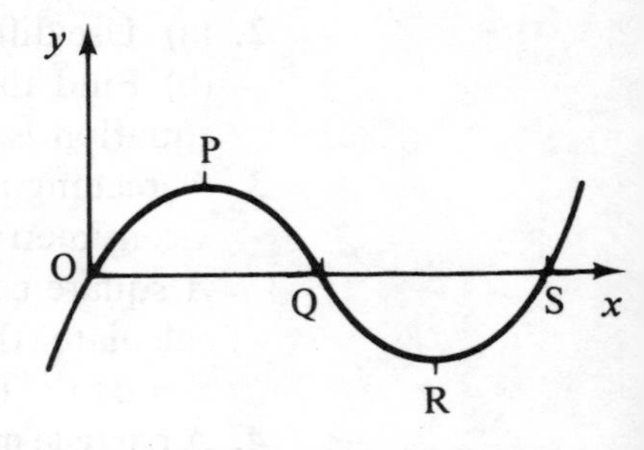

12. Given that $y = 25 + 4x - 3x^2$, copy and complete the following table.

x	-3	-2	-1	0	1	2	3	4	5
y	-14	5				21		-7	-30

Taking 2cm to represent 1 unit of x and 2cm to represent 10units of y, draw the graph of $y = 25+4x-3x^2$ from $x = -3$ to $x = 5$.
(a) Use your graph to solve the equation $25+4x-3x^2 = 16$,
(b) Draw on the same axes the straight line graph of $y = -5x$.
(c) Use your graphs to solve the equation $25+4x-3x^2 = -5x$.
(i.e. find the values of x at their points of intersection.) (C)

13. Draw the graph of $y = 4x-x^2-1$ for values of x from -2 to $+6$. Use scales of 2cm for 1 unit on the x-axis and 2cm for 4 units on the y-axis.
Use your graph to solve the equations
(a) $4x-x^2-1 = 0$; (b) $x^2-4x = 3$.
On the same axes draw the graph of $y = \frac{6}{x}$ plotting the points for which $x = -2, -1, -\frac{1}{2}, \frac{1}{2}, 1, 2, 3, 4, 5, 6$. Determine from your graphs the ranges of values of x for which $\frac{6}{x} < 4x-x^2-1$. (O)

14. A and B are two points 6cm apart on a straight line. At time $t = 0$ a particle moving along the line passes through A with velocity 8cm/s towards B. At time t seconds the particle's displacement from A is $x$cm, its velocity is $v$cm/s and its acceleration is $(2t-6)$cm/s^2.
(a) Obtain formulae for v and x in terms of t.
(b) Find the times at which the particle is momentarily at rest.
(c) Show that the particle passes through B three seconds after passing through A.
(d) Sketch the graph of $t \to v$ for $0 \leqslant t \leqslant 6$. (JBM)

15. Show that the line whose equation is $y = x+2$ meets the curve $y = 4-x^2$ at the points whose coordinates are $(-2, 0)$ and $(1, 3)$. Find, by integration, the finite area between the line and the arc of the curve intercepted by the line. (O)

16. The sketch shows part of the curve $y = x^2+1$. P is the point (2, 5) and PM, PN are perpendiculars from P onto the axes.

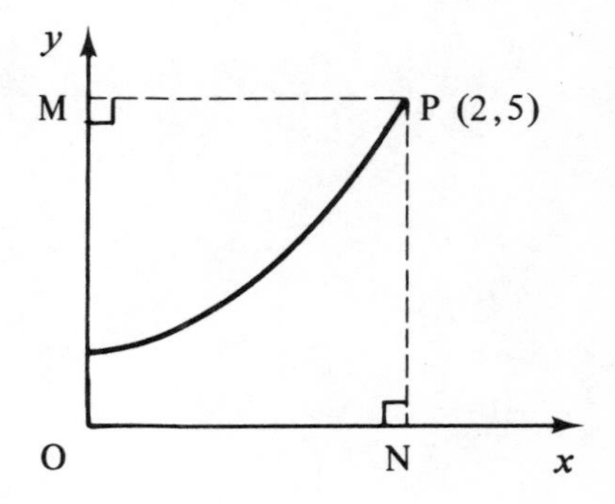

(a) Calculate the area bounded by the curve, the line PN and the axes.
(b) Hence, or otherwise, find the area bounded by the curve, the line PM and the y-axis.

(c) Calculate the volume obtained by rotating the area in (a) completely about the x-axis.

(d) Calculate the volume of the cylinder obtained by rotating the rectangle ONPM about the x-axis. (L)

17. Sketch the curve $y = x^2 - 2x$ between $x = 0$ and $x = 3$. Show that the area bounded by the curve and the x-axis between 0 and 2 is equal to the area bounded by the curve, the x-axis from 2 to 3 and the line $x = 3$. (O)

18. A particle moves in a straight line so that, t seconds after it passes through a point 0, its displacement from 0 is. x metres and its velocity v metres per second where $v = t^2 - 5t + 6$.

(a) Find the velocity when $t = 0$.

(b) Find the values of t when $v = 0$.

(c) Sketch the time → velocity graph for $0 \leqslant t \leqslant 4$.

(d) Find an expression for x as a function of t and sketch the time → displacement graph for $0 \leqslant t \leqslant 4$.

(e) Find the distance moved in the third second. (JMB)

19. The graph of $y = \frac{1}{2}x^2$ intersects the graph of $y = x$ at the origin and the point P. Calculate the coordinates of P. The region included between these two graphs is revolved completely about the x-axis. Calculate the volume of revolution so formed, leaving your answer as a multiple of π. (JMB)

20. A spot of light moves along a straight line so that its acceleration t seconds after passing a fixed point O on the line is $(2 - 2t)\text{cm/s}^2$. Three seconds after passing O the spot has a velocity of 5cm/s. Find, in terms of t, expressions for (a) the velocity of the spot of light after t seconds, (b) the distance of the spot from O after t seconds.

Calculate (c) the time when the spot comes to rest, (d) the distance of the spot from O when it comes to rest. (L)

8
Further Work on Graphs

We have seen how linear equations may be represented graphically. Inequations may also be shown graphically.

Example Shade the regions represented by the inequation $y \geqslant x-2$.

[Draw the graph of $y = x-2$]
Substituting $x = 0$
gives $y = \underline{-2}$

Substituting $x = -2$
gives $y = \underline{-4}$

Substituting $x = 2$
gives $y = \underline{0}$

Plot the points $(0, -2)$, $(-2, -4)$ $(2, 0)$ and draw a line through them.

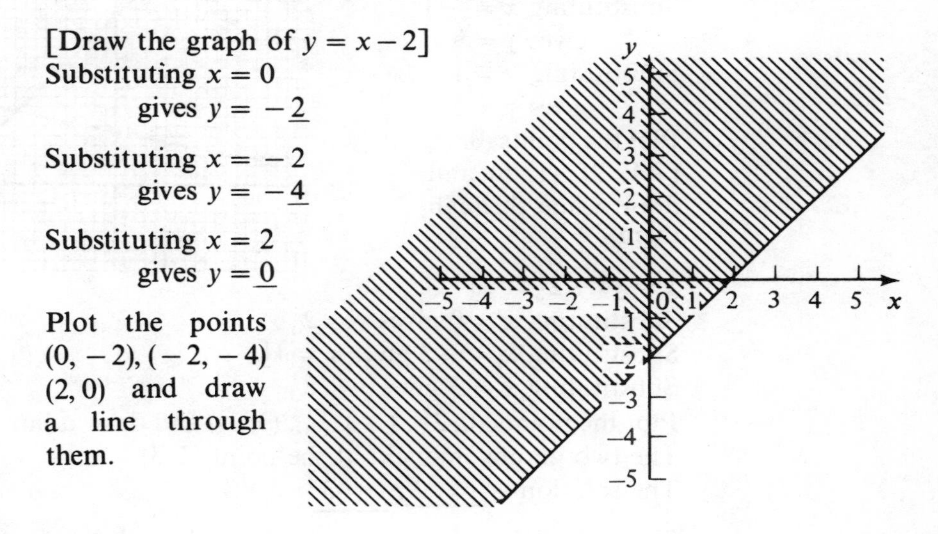

$\left[\begin{array}{l} y = x-2 \text{ is indicated by the straight line graph, to test if } y > x-2 \\ \text{above or below this line substitute } x = 0,\ y = 0 \text{ in the inequation.} \end{array}\right]$

$\therefore$ When $x = 0$, $y = 0$; $y > x-2$ becomes $0 > 0-2$ i.e. $0 > -2$,

which is correct since -2 is below $y = 0$.
Therefore the required region contains the origin.

Note If the region $y > x-2$ only had been required, the graph line $y = x-2$ would have been drawn as a dotted-line.

Exercise 60 Represent the following conditions graphically:

1. $y \geqslant x$ **2.** $y \geqslant x+1$ **3.** $y \leqslant x$
4. $y \leqslant x-1$ **5.** $y \leqslant 3x-1$ **6.** $y < 2x$
7. $y+x > 3$ **8.** $2y+3x > 4$ **9.** $y+2x-3 < 0$
10. $3 \leqslant y+2x$

Graphical Solutions of Simultaneous Equations

In Chapter 5 a method for the solution of simultaneous equations was given. Another (sometimes less accurate) method is now given.

Examples

1. Solve graphically the equations $2x+y=7$, $x=3y-7$.
[Rearranging both equations in the form $y=mx+c$:]

$y=7-2x$
Substituting $x=0$
gives $y=\underline{7}$
Substituting $x=-1$
gives $y=\underline{9}$
Substituting $x=1$
gives $y=\underline{5}$
Plot the points (0, 7), (−1, 9), (1, 5) and draw a line through them.

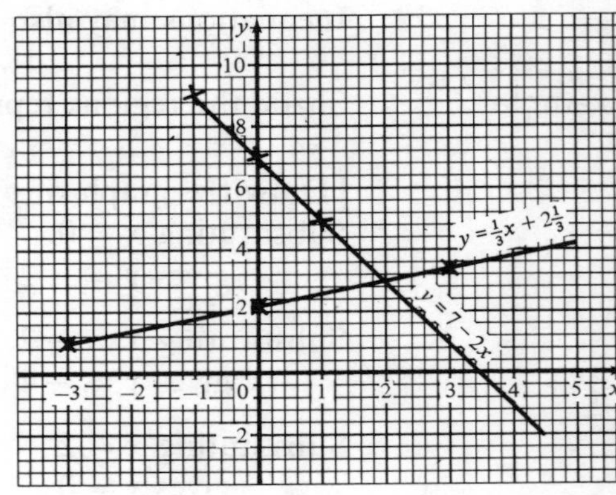

$y=\frac{1}{3}x+2\frac{1}{3}$
Substituting $x=0$ gives $y=2\frac{1}{3}$
Substituting $x=-3$ gives $y=1\frac{1}{3}$
Substituting $x=3$ gives $y=3\frac{1}{3}$
Plot the points $(0, 2\frac{1}{3})$, $(-3, 1\frac{1}{3})$, $(3, 3\frac{1}{3})$ and draw a line through them.
The two graphs intersect at the point (2, 3)
The solution is $\underline{x=2}$, $\underline{y=3}$ ANS

2. Shade the region represented by the solutions of $x \geqslant 0$, $y \geqslant 0$, $2x+y \leqslant 5$.

[Draw the graph of $2x+y=5$, i.e. $y=5-2x$.]
In $y=5-2x$,
when $x=0$, $y=\underline{5}$
when $x=-1$, $y=\underline{7}$
when $x=1$, $y=\underline{3}$
Plot the points (0, 5), (−1, 7) (1, 3) and draw in the straight line. The inequation $x \geqslant 0$ represents all values of x on or to the right of the y-axis.

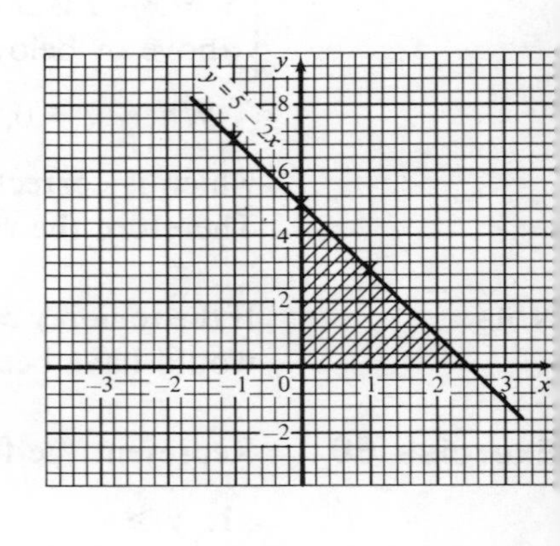

The inequation $y \geqslant 0$ represents all values of y on or above the x-axis. The inequation $2x+y \leqslant 5$ represents all values of x and y on or below the graph line. The region common to all these conditions is shown by shading.

Exercise 61 Show graphically the solutions of:

1. $x+y=3,\ 3x=y+1$ **2.** $y=4+x,\ x=y-4$
3. $2x-3y=-1,\ x+2y+4=0$ **4.** $x \geqslant 0,\ y \geqslant 0,\ x+y \geqslant 1$
5. $x \leqslant 0,\ y \leqslant 0,\ 2x-y \geqslant 4$ **6.** $x > 0,\ y > 0,\ 2x+y > 3$
7. $x \leqslant 0,\ y \leqslant 0,\ x+y \geqslant -4$ **8.** $x > 0,\ y < 0,\ x+2y > 3$

Cubic functions

In chapter 7 we noted that the expression ax^3+bx^2+cx+d, where a, b, c and d are constants was called a **cubic function.**

Example

1. Draw the graph of $y = x^3 - 4x^2 - 3$ for the range $-2 \leqslant x \leqslant 5$.
2. From the graph determine as accurately as you can the coordinates of the points where the gradient is zero.
3. The gradient at the point $x = 4$ by drawing.

1. [Set up a table to calculate y.]

x	-2	-1	0	1	2	3	4	5
x^3	-8	-1	0	1	8	27	64	125
$-4x^2$	-16	-4	0	-4	-16	-36	-64	-100
-3	-3	-3	-3	-3	-3	-3	-3	-3
y	-27	-8	-3	-6	-11	-12	-3	22

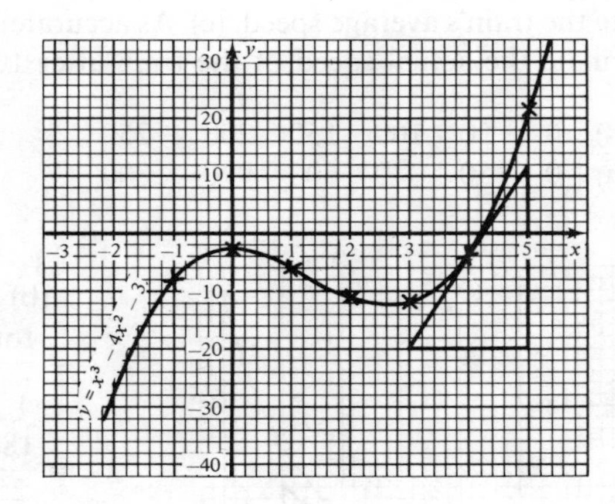

2. The gradient is zero where the graph turns, (i.e. at the turning points).

From the graph, these points are $\underline{(0, -3)}$ and $\underline{(2\tfrac{2}{3}, -12\tfrac{1}{2})}$ ANS

3. Draw a tangent to the curve at the point $x = 4$, and form a right-angled triangle ensuring that its base is an exact number of units.

Height of triangle = 32 units, base length = 2 units

$$\therefore \text{Gradient} = \frac{\text{height}}{\text{base}} = \frac{32}{2} = \underline{16} \quad \text{ANS}$$

Exercise 62

1. Draw the graph of $y = 2x^3 + 3x^2 - 4$ for the range $-3 \leqslant x \leqslant 3$ and from your graph determine (a) the coordinates of the points where the gradient is zero, (b) the gradient at the point $x = 1$.
2. Draw the graph of $y = \frac{1}{3}x^3 + x^2 + x$, for the range $-4 \leqslant x \leqslant 4$ and from your graph deduce (a) if there are any turning points, and if so state their coordinates, (b) the value of the gradient at the point where $x = 2$.
3. Draw the graph of $y = x^3 - 2\frac{1}{4}x$ for the range $-3 \leqslant x \leqslant 3$. What are the gradients of the curve when $x = \frac{1}{2}$ and when $x = 2$, stating if either gradient is negative?

Linear Kinematics

Problems involving **time**, **speed** and **distance** were met in Chapter 7. Here are some further examples requiring a purely graphical approach to their solution.

Examples

1. A train leaves Waterloo Station for Bournemouth, and the distance from its starting point is recorded every 50 seconds. (a) Draw a distance-time graph from the tabulated results which follow. (b) Calculate the train's average speed. (c) As accurately as your graph permits calculate the actual speed of the train after 100 seconds.

Time (s)	0	50	100	150	200	250
Distance (m)	0	150	340	600	990	1800

(a)

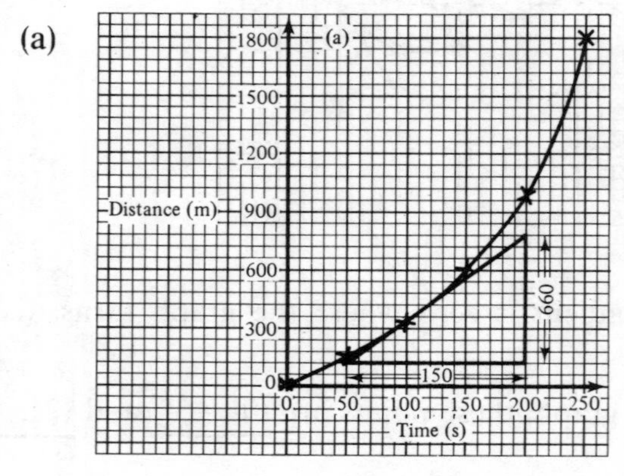

(b) Average speed

$$= \frac{\text{total distance}}{\text{total time}}$$

$$= \frac{1800}{250}$$

$$= \underline{7.2\,\text{m/s}} \quad \text{ANS}$$

(c) Draw a tangent at $t = 100$ and construct a triangle. The calculation is made easier if the base length is an exact number of units.

Gradient of tangent (i.e. speed) $= \dfrac{\text{distance}}{\text{time}}$

$$= \frac{660}{150}$$

$$= \underline{4.4\,\text{m/s}} \quad \text{ANS}$$

2. The speed of a car is recorded every 5 seconds as it accelerates from rest.
(a) Draw a speed-time graph from the tabulated results which follow.
(b) Calculate the car's average acceleration.
(c) As accurately as your graph allows calculate the car's actual acceleration after 10 seconds.

Time (s)	0	5	10	15	20
Speed (m/s)	0	3	7	14	42

(a)

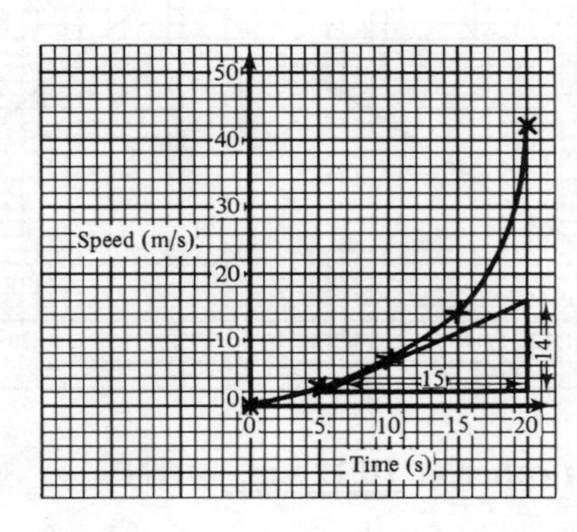

(b) $$\text{Average acceleration} = \frac{\text{Speed increase}}{\text{Time taken}}$$

$$= \frac{42}{20}$$

$$= \underline{2.1 \text{ m/s}^2} \quad \text{ANS}$$

(c) Actual acceleration

= gradient of tangent

$$= \frac{\text{speed}}{\text{time}} = \frac{14}{15}$$

$$= \underline{0.93 \text{ m/s}^2} \quad \text{ANS}$$

(correct to 2 decimal places)

3. A train leaves Waterloo Station for Portsmouth at 9 a.m. It travels at an average speed of 40 km/h until it arrives at Guildford 50 km away. It waits at Guildford for 5 minutes and then travels the remaining 60 km at an average speed of 75 km/h. Also at 9 a.m. the same day another train leaves Portsmouth for Waterloo. The second train travels non-stop to Waterloo taking the same time for the journey as the first train. At what time do the trains pass each other?

The first train travels the 50 km to Guildford at 40 km/h.

It therefore takes $\frac{50 \times 60}{40}$ minutes

$= \underline{75 \text{ minutes}}$

This point can be plotted on the graph, followed by a horizontal line representing the 5 minute wait.

The remainder of the journey takes $\frac{60 \times 60}{75}$ minutes $= \underline{48 \text{ minutes}}$

The journey of the second train can be shown on the graph as a straight line reaching Waterloo at the same time as the first train reaches Portsmouth.

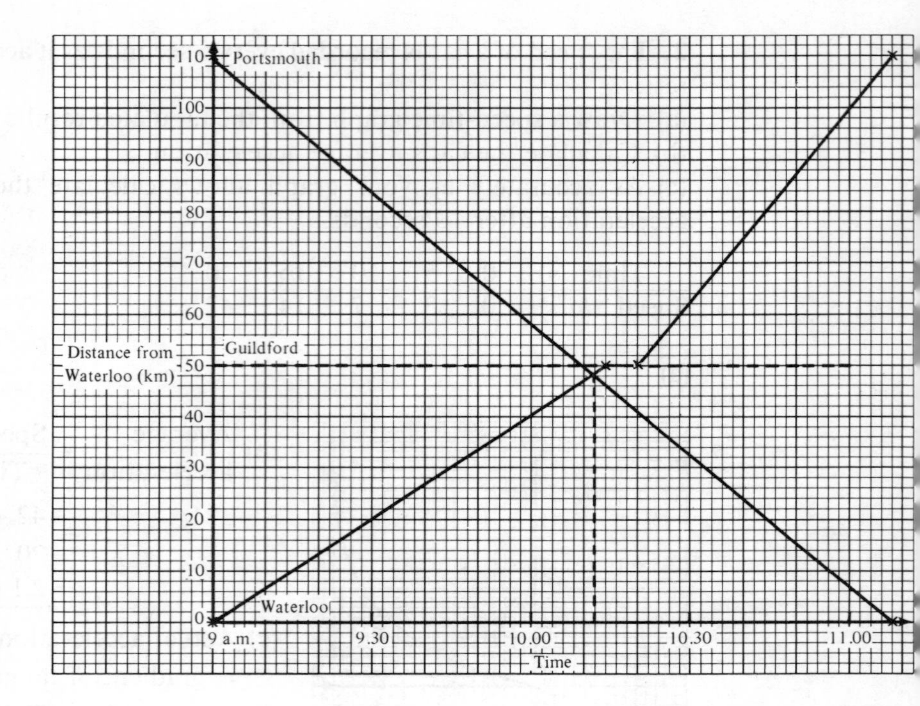

The trains pass each other at 10.12a.m. ANS

4. The velocity of a particle for the first 5 seconds of motion is given by the equation $v = 100t^2$, where v is its velocity in m/s and t the time in seconds. Draw a velocity-time graph for the first 5 seconds of motion, and from your graph calculate (a) the average acceleration of the particle, (b) the actual acceleration after 4 seconds.

Constructing a table:

t(s)	0	1	2	3	4	5
v(m/s)	0	100	400	900	1600	2500

[When $t = 0$, $v = 0$; when $t = 1$, $v = 100 \times 1^2 = 100 \ldots$]

(a) Average acceleration

$$= \frac{\text{speed increase}}{\text{time taken}}$$

$$= \frac{2500}{5} = \underline{500\,\text{m/s}^2} \quad \text{ANS}$$

(b) Actual acceleration at $t = 4$

$$= \frac{\text{speed}}{\text{time}} = \frac{800}{1}$$

$$= \underline{800\,\text{m/s}^2} \quad \text{ANS}$$

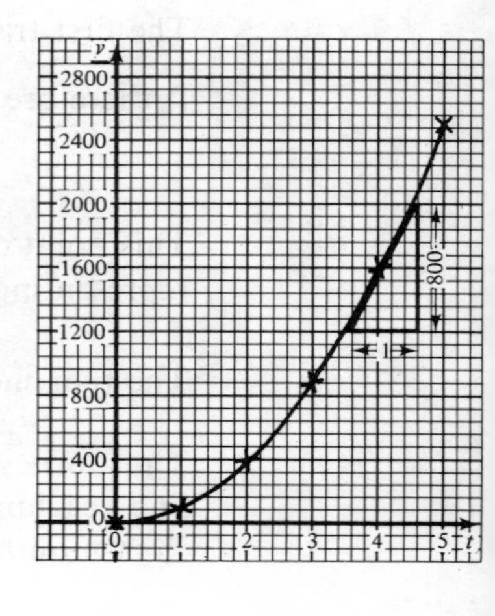

Exercise 63

1. A coach joins a motorway and accelerates. Its distance from the junction is recorded every 5 seconds. Draw a distance-time graph from the tabulated results which follow and use your graph to calculate (a) the coach's average speed, (b) the actual speed of the coach after 15 seconds.

Time (s)	0	5	10	15	20	25
Distance (m)	0	14	55	110	185	320

2. The speed of a train as it leaves the station is recorded every 10 seconds. From the following observations draw a speed-time graph and use the graph to calculate (a) the train's average acceleration, (b) the train's actual acceleration after 30 seconds.

Time (s)	0	10	20	30	40	50
Speed (m/s)	0	9	23	42	65	100

3. A train leaves London at mid-day and travels the 160km non-stop to Dover at an average speed of 60km/h. A second train leaves Dover also at mid-day that day and travels at an average speed of 40km/h until it reaches Ashford 45km away where it stops for 5 minutes. If it arrives, without further stops, in London at 3 p.m., find
   (a) at what time the trains pass each other,
   (b) the time of arrival in Dover of the first train mentioned,
   (c) the average speed of the second train between Ashford and London.
4. The distance fallen by a stone dropped from a cliff edge is given by the equation $x = 5t^2$,
   where x is the distance fallen in metres and t is the time elapsed in seconds. Draw a distance-time graph for the first 5 seconds of falling and from this graph calculate (a) the average speed of the stone, (b) the actual speed of the stone after 2 seconds.
5. The velocity of a particle during the first 5 seconds of motion is given by

$$v = 20t^2 + 5,$$

   where v is the velocity in m/s and t is the time elapsed in seconds. Draw a velocity-time graph for the first 5 seconds of flight and use this graph to calculate (a) the average acceleration of the particle, (b) the actual acceleration of the particle after 3 seconds.
6. A cyclist starts from home at noon and rides towards his friend's house which is 50km away at an average speed of 20km/h. After one hour a car leaves the friend's house and heads towards the cyclist at an average speed of 45 km/h. Draw a suitable graph to find (a) at what time they pass each other, (b) the distance of this point from the cyclist's home.

Examination Standard Questions

Exercise 64

1. 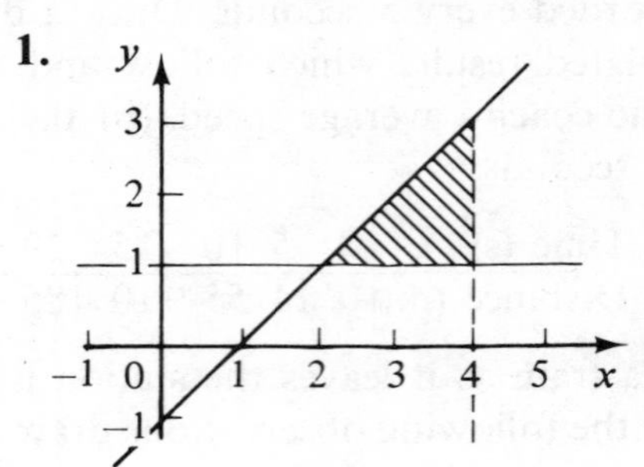

Write down the inequations of the three intersecting lines in the diagram illustrated by the shaded area. The full line indicates that the line itself is included in the shading while the broken line is not included in the shaded area. (O)

2. 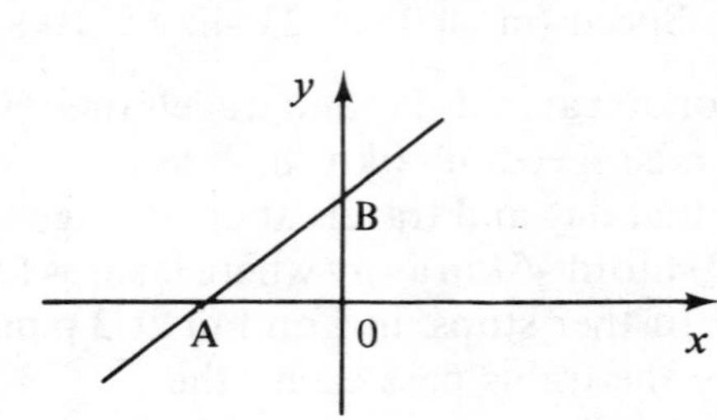

The diagram shows the graph of the line $y = x+3$.
(a) Write down the coordinates of the points A and B at which the line crosses the x- and y-axes.
(b) Indicate clearly in the diagram the region which represents the set $\{(x, y): y \leqslant 0 \text{ and } y \geqslant x+3\}$.
(This notation means 'the set of all values of x and y such that $y \leqslant 0$ and $y \geqslant x+3$). (JMB)

3. (a) Express the inequalities $8 > 15-2t > 3$ in the form $p > t > q$.
(b) Draw, on squared paper, the straight line graphs $x = 4$, $x-4y+28 = 0$ and $5x+3y-66 = 0$. Shade in the region in which $y \geqslant 0$, $x \geqslant 4$, $x-4y+28 \geqslant 0$, $5x+3y-66 \leqslant 0$.
    (i) Find the greatest value of $x+y$ subject to these inequalities.
    (ii) If x and y are also restricted to be integers, find the greatest and least values of $x+y$ subject to these inequalities. (O)

4. Draw the graph of $y = 2x^3-3x^2-12x$ using a scale of 2cm to 1 unit on the x-axis and 2cm to 10 units on the y-axis. Draw this graph for values of x from -3 to $+4$ at unit intervals. (i.e. $-3 \leqslant x \leqslant +4$) Use your graph to solve the equations
(a) $2x^3-3x^2-12x = 0$, (b) $2x^3-3x^2-12x = 3$,
(c) $2x^3-3x^2-12x-10 = 0$. (JMB)

5. Copy and complete the table of values of the function $y = \dfrac{x}{3}+\dfrac{3}{x}$, giving values of y correct to two decimal places.

x	$\frac{1}{2}$	1	2	3	4	5	6	7
$\frac{x}{3}$	0.17	0.33	0.67		1.33	1.67		2.33
$\frac{3}{x}$	6.00	3.00						0.43
y	6.17	3.33						2.76

Draw the graph of the function for values of x from $\frac{1}{2}$ to 7, using a scale of 2cm to 1 unit on each axis.

From your graph find the range of values of x for which $\frac{x}{3}+\frac{3}{x}$ is less than 2.6.

Using the same scale and axes, draw the graph of the function $y = 5 - x$. Find the values of x at the points of intersection of the two graphs and write down, but do not simplify, the equation which is satisfied by these two values of x. (L)

6. Copy and complete the table of values of the function $y = x - 4 + \frac{11}{x}$ giving values of x correct to two decimal places.

x	1	2	3	4	5	6	7
-4	-4	-4	-4	-4	-4	-4	-4
$\frac{11}{x}$	11			2.75			1.57
y	8			2.75			4.57

Using a scale of 2cm to represent 1 unit on each axis, draw the graph of the function for these values. From your graph find (a) the minimum value of $x - 4 + \frac{11}{x}$ and the value of x at which it occurs, (b) the roots of the equation $x + \frac{11}{x} = 8$.

From the point (1, 1) draw a tangent to the curve and find its gradient. (L)

7.

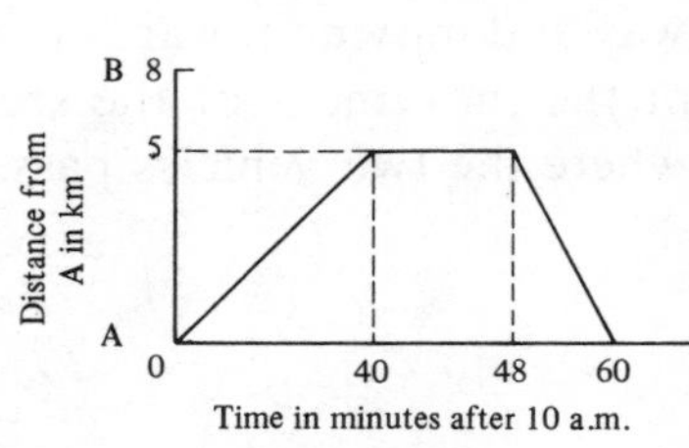

The graph, which is not drawn to scale, represents the journey of a cyclist who left A at 10.00a.m. and rode towards B, 8km away, at a steady speed. At 10.40 a.m. he had a puncture which he mended in 8 minutes before returning to A, which he reached at 11.00 a.m.

Using this information draw an accurate graph with a scale of 2cm for 10 minutes on the time axis and 2 cm for 1 km on the distance axis. Determine the average speeds in km per hour on the outward and return stages of the journey.

At 10.30 a.m. a second cyclist left B, riding at a steady speed to A. If he passed the first cyclist whilst he was repairing the puncture, find, by drawing suitable lines, the earliest and latest times he could have reached A. (O)

8. A container of uniform cross-section was filled with oil which leaked through a small hole in the base so that, t hours after the container was filled, the depth of oil was h cm where

$$h = 10(3 - \tfrac{1}{2}t)^2.$$

Prepare a table of values of h for $t = 0, 1, 2, 3, 4, 5, 6$ and draw the graph of h against t.
From your graph estimate the time that elapsed before the container was half full. (O)

9. An empty storage tank is filled with water so that t hours later it contains V gallons where $V = 1080t - 10t^3$.
The tank is completely full when $t = 6$. What is its capacity? Plot a graph of V against t for values of t from 0 to 6. As scales use 2cm to represent 1 hour on the axis of t and 500 gallons on the axis of V.
From your graph determine the value of t at which the tank is half full. (O)

10. A monorail vehicle accelerates from rest. After time t seconds, its distance, x metres, from its starting place (ignoring the vehicle's length) is given in the following table:

t	0	5	10	15	20	25	30
x	0	60	210	460	800	1250	1800

Taking 2cm as 5 seconds and 1cm as 100metres, draw a graph of the movement. From your graph obtain
(a) the time, to the nearest second, when the vehicle is 1500 metres from the starting point,
(b) the speed of the vehicle after ten seconds, clearly stating your units.
When the vehicle starts to move, another vehicle, on a parallel rail, is 1200 metres away and moving towards it at a constant speed of 30m/s. Represent the movement of the second vehicle on your graph and find where the two vehicles pass. (WJEC)

11.
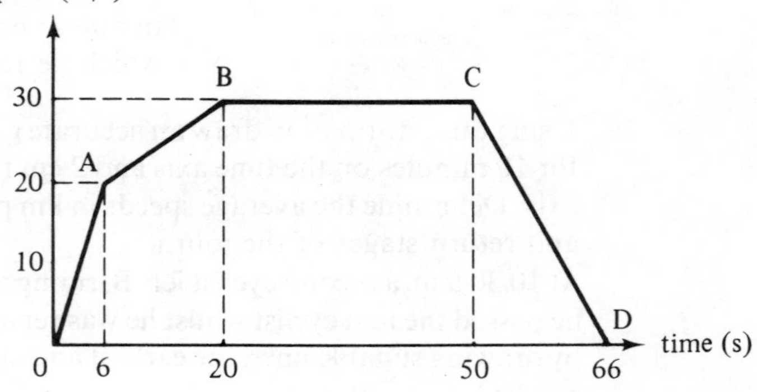

The speed-time graph of a car consists of the straight lines shown in the diagram. Calculate the distance travelled by the car during the 66 seconds of motion. (C)

12. A lorry starts from rest at a point P and after t seconds its speed, v metres per second, is given by the following table:

t(s)	0	1	2	3	4	5	6
v(m/s)	0	2.4	4.2	5.6	6.6	7.2	7.6

Using a horizontal scale of 2 cm to represent 1 second and a vertical scale of 1 cm to represent 1 m/s, plot the speed against time.
(a) Estimate the distance the lorry travels in the six seconds.
(b) Two cars A and B start from rest at P when $t = 2$ and move with constant accelerations in the same direction as the lorry. When $t = 6$ car A has the same speed as the lorry and car B has travelled the same distance as the lorry.
Add to your graph the two straight lines which represent the movements of the cars, labelling them A and B respectively. (C)

13. The following table gives the speed of a vehicle, v metres per second, at time t seconds after it passes a given point.

t	0	10	20	30	40
v	11	20	24	23	17

Taking 2 cm to represent 5 seconds on the t-axis and 2 cm to represent 5 metres/second on the v-axis, plot these values on graph paper and join them with a smooth curve.
(a) By drawing a tangent to the curve, find the acceleration in metres/second/second when $t = 10$.
(b) For what value of t is the acceleration zero? (WJEC)

14. In a laboratory experiment a heavy spring was suspended vertically from a horizontal beam. A mass of m kilograms was hung on the lower end of the spring which was stretched and then released. The time of oscillation, t seconds, of the mass was measured and the experiment was repeated for various values of m. The following results were obtained:

m	0.13	0.25	0.37	0.50	0.63	0.71	0.85
t	1.8	2.0	2.2	2.4	2.6	2.7	2.9

Construct a new table showing m against t^2 and by plotting these new values on a graph show that m and t^2 are connected by a law of the form $t^2 = am + b$ where a and b are constants. (Scales: take 2 cm to represent 0.1 kg on the m-axis and take 2 cm to represent 1 unit on the t^2-axis)
Use your graph to estimate (a) values of a and b, (b) the value of m for which $t = 2.5$, (c) the percentage increase in t as m increases from 0.4 to 0.8. (AEB)

9

Modern Algebra

Set Notation

A **set** is a collection of distinct items. These items are called **members** o **elements** of the set.

e.g. A = {1, 2, 3, 4} means that A represents the set of natural number less than 5. The set has four members. The symbol used to denot membership is $\in$, and means 'is a member of'.

e.g. $2 \in A$, means that 2 is a member of the set A,
and $5 \notin A$, means that 5 is *not* a member of the set A.

Consider the set B = {1, 2}. It can be seen that the members of set B ar also members of set A. It is therefore said that B is a **subset** of A, and i written as $B \subset A$. If C = {3, 4} it can also be said that $C \subset A$, whicl reads 'C is a subset of A'.

Exercise 65

Let P = {1, 2, 3, 4, 5}; Q = {2, 4, 6, 8, 10}; R = {2, 3, 5, 7, 11}:

1. Describe each of the sets P, Q, R in words.
2. List the members of set P.
3. Copy the following statements, writing after each, the letter 'T' i you consider the statement is true, or the letter 'F' if you consider th statement is false.

(a) $4 \in P$, (b) $6 \in Q$, (c) $6 \in R$, (d) $13 \in R$,
(e) $6 \notin P$, (f) $7 \notin R$, (g) $10 \notin Q$, (h) $P \subset R$.

4. If S = {2, 4} is (a) $S \subset Q$, (b) $S \subset R$, (c) $S \subset P$?
5. Write down all the possible subsets of Q having four members.

Consider N = {prime numbers between 8 and 10}. This set has n members, and is denoted by either of the symbols { } or $\varnothing$. Such sets ar called **empty** or **null** sets.

All of the examples met so far in this chapter have been sets containin various types of number. It can be said that all of the sets discussed s far have been subsets of a much larger 'infinite' set; the set of all rea numbers. This larger set of all items we wish to consider in a specifi situation is called the **universal** set and is denoted by the symbol $\mathscr{E}$ ('ensemble' or 'entirety').

i.e. If $\mathscr{E}$ = {all real numbers}
then $A \subset \mathscr{E}$, $B \subset \mathscr{E}$, $P \subset \mathscr{E}$, $Q \subset \mathscr{E}$ and $R \subset \mathscr{E}$

considering the sets used earlier in this chapter. Suppose our universa set $\mathscr{E} = \{1,2,3,4,5,6,7,8,9\}$ and a set $D = \{1,3,5,7,9\}$. Then $\{2,4,6,8\}$ is the set of all members of $\mathscr{E}$ which do not belong to D, and is referred to as the **complement** of D, denoted by D′.

i.e. Given $\mathscr{E} = \{1, 2, 3, 4, 5, 6, 7, 8, 9\}$; $D = \{1, 3, 5, 7, 9\}$; then $D' = \{2, 4, 6, 8\}$.

Exercise 66 Let $\mathscr{E} = \{1, 2, 3, 4, 5, 6, 7, 8, 9\}$; $E = \{2, 4\}$; $F = \{6, 7, 8, 9\}$; $G = \{6, 7\}$:

1. List the set E′.
2. Is it correct to write that $E \subset \mathscr{E}$?
3. List the set F′.
4. Is $G \subset E$?
5. Is $G \subset F$?
6. Is $E \subset F'$?
7. Given that $\mathscr{E} \supset E$ means that $\mathscr{E}$ **contains** E, is it correct to say that $\mathscr{E} \supset E'$?
8. Is it correct to say that $F \supset G$?
9. Is it correct to say that $4 \in F'$?
10. Is it correct to say that $10 \in E'$?

Consider $P = \{1, 2, 3, 4, 5\}$ and $Q = \{2, 4, 6, 8, 10\}$, then the **union** of the two sets P and Q is $\{1, 2\ 3, 4, 5, 6, 8, 10\}$; i.e. the set of members which belong to P **or** to Q, (we do not write the common members 2 and 4 twice). The union of two sets is denoted by the symbol $\cup$.

i.e. $P \cup Q = \{1, 2, 3, 4, 5, 6, 8, 10\}$

Considering th same sets P and Q, then the **intersection** of the two sets is $\{2, 4\}$; i.e. the set of members which belong to both sets. The intersection of two sets is denoted by the symbol $\cap$.

i.e. $P \cap Q = \{2, 4\}$

Sets which have no members in common are called **disjoint** sets.

Examples Let $\mathscr{E} = \{\text{all real numbers}\}$; $P = \{1, 2, 3, 4, 5\}$; $R = \{2, 3, 5, 7, 11\}$; $T = \{7, 9, 11, 13\}$.

1. Find the union of P and R listing the resulting set.

[Members belonging to P or R.] i.e. $P \cup R = \{1, 2, 3, 4, 5, 7, 11\}$

2. List the set representing the intersection of P and R.

[Members common to P and R.] i.e. $P \cap R = \{2, 3, 5\}$

3. List the set $P \cup T$.

[Members belonging to P or T.] i.e. $P \cup T = \{1, 2, 3, 4, 5, 7, 9, 11, 13\}$

4. List the set $P \cap T$

i.e. $P \cap T = \varnothing$.

Note

There are no members common to P and T, and therefore their intersecting set is empty. P and T are disjoint sets.

Exercise 67

Let $\mathscr{E} = \{1, 2, 3, 4, 5, 6, 7, 8, 9, 10\}$; $X = \{2, 4, 6, 8, 10\}$; $Y = \{1, 3, 5, 7, 9\}$; $Z = \{5, 6, 7, 8, 9\}$.

1. List the set $X \cup Y$.
2. List the set $X \cup Z$.
3. List the set $Y \cup Z$.
4. List the set $X \cap Y$.
5. List the set $X \cap Z$.
6. List the set $Y \cap Z$.
7. Name one pair of disjoint sets.
8. List the set $X \cup Y \cup Z$.
9. List the set $X \cap Y \cap Z$.
10. If $V = X \cup Y$ and $W = Y \cup Z$, list the set $V \cup W$.
11. List the set X'.
12. List the set Y'.
13. List the set Z'.
14. List the set $X' \cup Y'$.
15. List the set $Y' \cup Z'$.
16. What can you say about the sets X and Y'?
17. Considering the set V in question **10**, is $Z \subset V$?
18. Considering the set W in question **10**, is $X \subset W$?

Venn Diagrams

Diagrams are often used to illustrate sets. Such diagrams are called **Venn diagrams**, named after the English mathematician who was responsible for their introduction. A rectangle is usually used to represent the universal set $\mathscr{E}$, and circles or ovals are drawn within this rectangle to represent subsets of $\mathscr{E}$.

Examples

1. Let $\mathscr{E} = \{1, 2, 3, 4, 5, 6, 7, 8\}$; $A = \{1, 2, 3, 4\}$; $B = \{3, 4, 5, 6\}$.

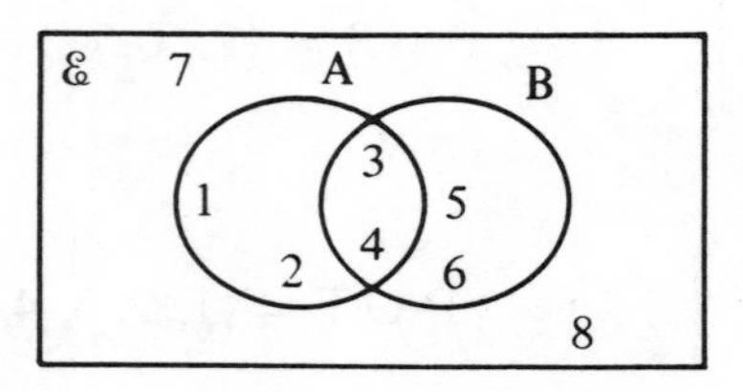

The sets could be illustrated by this Venn diagram, where $A \cap B = \{3, 4\}$, and $A \cup B = \{1, 2, 3, 4, 5, 6\}$.

2. Let $\mathscr{E} = \{1, 2, 3, 4, 5, 6, 7, 8\}$; $C = \{1, 2, 3\}$; $D = \{4, 5, 6\}$.

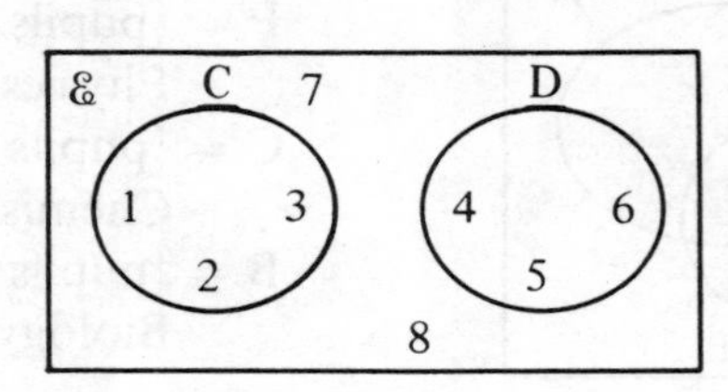

The Venn diagram can be drawn as shown, where C and D are disjoint sets. From this diagram it can be deduced that $C \cap D = \varnothing$

$C \cup D = \{1, 2, 3, 4, 5, 6\}$

$C' = \{4, 5, 6, 7, 8\}$

$D' = \{1, 2, 3, 7, 8\}$

3. Let $\mathscr{E} = \{1, 2, 3, 4, 5, 6, 7, 8\}$; $A = \{1, 2, 3, 4\}$; $C = \{1, 2, 3\}$.

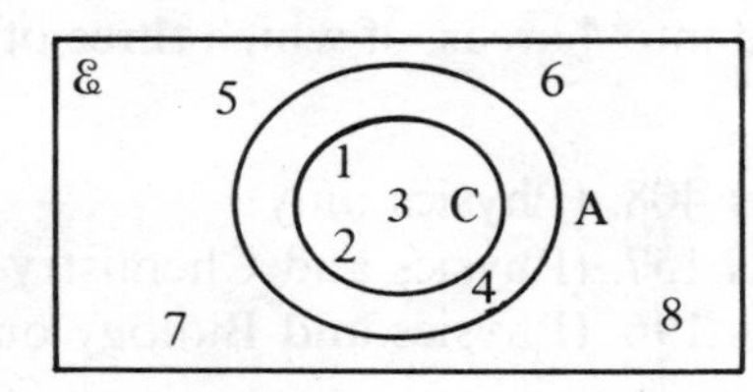

From the Venn diagram it can be deduced that $C \subset A$,

$A' = \{5, 6, 7, 8\}$

$C' = \{4, 5, 6, 7, 8\}$

therefore that $A' \subset C'$

Exercise 68 Draw Venn diagrams to illustrate the following:

1. $\mathscr{E} = \{1, 2, 3, 4, 5, 6\}$; $A = \{1, 2\}$; $B = \{3, 4\}$.
2. $\mathscr{E} = \{2, 4, 6, 8, 10\}$; $C = \{2, 4, 6\}$; $D = \{6, 8\}$.
3. $\mathscr{E} = \{0, 1, 2, 3, 4, 5\}$; $F = \{2, 3, 4\}$; $G = \{3\}$.
4. $\mathscr{E} = \{\text{natural numbers less than } 12\}$; $H = \{1, 2, 3, 4\}$; $J = \{4, 5, 6, 7\}$; $K = \{3, 4, 5, 8\}$.
List $H \cup J$; $H \cup K$; $J \cup K$; $H \cap J$; $H \cap K$; $J \cap K$; $H \cap J \cap K$.
5. $\mathscr{E} = \{\text{natural numbers less than } 12\}$; $L = \{3, 4, 5, 6, 7\}$; $M = \{5, 6\}$.
List $L \cap M$; $M \cup M'$; $L \cup L'$; $M \cap M'$; $L \cap L'$.

Problem Solving

Examples **1.** During their third year 1029 pupils were asked to choose which of the following three optional science subjects they wished to study. They could choose one, two or all three and their timetables would be organised to meet their wishes if possible.
The total numbers of pupils opting for each subject for at least part of the time were:

Physics 761 Chemistry 409 Biology 346

Of these, 408 pupils wanted Physics only, 118 pupils wanted Chemistry only, 157 pupils wanted Physics and Chemistry only, and 196 pupils wanted Physics and Biology only.
Find (a) The number of pupils taking all three subjects, (b) The number of pupils opting for Biology only.

The Venn diagram illustrating the problem is:

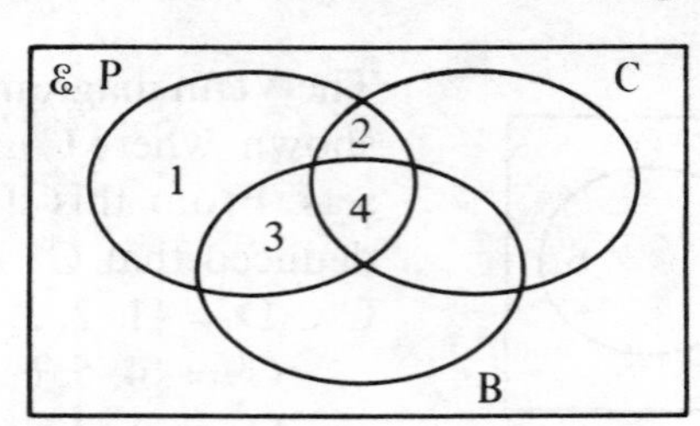

where $\mathscr{E}$ = {pupils in 3rd year}
P = {pupils requiring Physics}
C = {pupils requiring Chemistry}
B = {pupils requiring Biology}

and $n(\mathscr{E}) = 1029$, $n(\text{P}) = 761$, $n(\text{C}) = 409$, $n(\text{B}) = 346$.

Note $n(\mathscr{E})$ means the number of members of the set $\mathscr{E}$.

(a) The set P is divided into 4 areas, of which three of the population are known.

Subdivision **1** contains 408, (Physics only)
Subdivision **2** contains 157, (Physics and Chemistry only)
Subdivision **3** contains 196, (Physics and Biology only)

Total 761

The total requiring *any* Physics was 761, therefore the population o subdivision **4** is zero.

∴ The number of pupils opting for all three subjects [subdivision **4**] $= n(\text{P} \cap \text{C} \cap \text{B}) = \underline{0}$ ANS

(b) The Venn diagram is now:

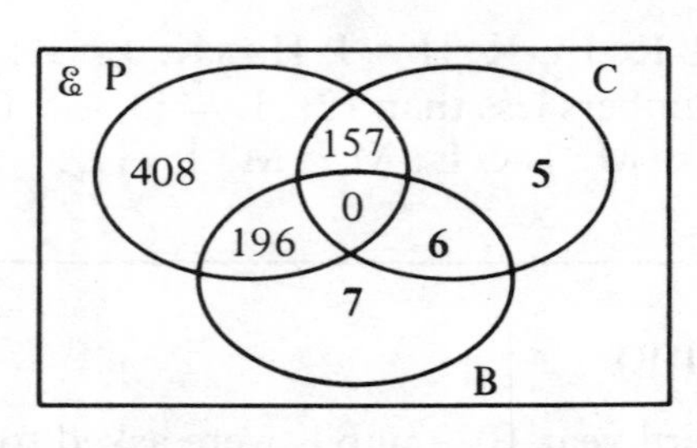

The population of subdivision **5** i known to be 118, (Chemistry only)

∴ Subdivision **6**
$= 409 - (157 + 118 + 0) = 13$
Subdivision **7**
$= 346 - (196 + 0 + 134) = 16$

∴ The number of pupils opting fo Biology only $= \underline{16}$ ANS

2. 74 pupils saving for a school visit to Europe were asked to vote fo which of three countries they would like to visit. They could vote to sta in one country only, or to divide their stay between two or all thre countries.

The total numbers of pupils wishing to visit the following countries fo at least part of the time were:

France 55 Holland 46 Belgium 39

Of these, 29 wished to visit France and Holland of which some als wanted to see Belgium, 24 wished to visit France and Belgium of whic some also asked to see Holland, 21 wished to visit Holland and Belgiu of which some also wanted to visit all three countries.

Find (a) the number of pupils wishing to visit France *only*, (b) the total number of pupils wishing to visit only one country.

Let $\mathscr{E}$ = {pupils saving for the visit} where $n(\mathscr{E}) = 74$,
F = {those wishing to see France} where $n(F) = 55$,
H = {those wishing to see Holland} where $n(H) = 46$,
B = {those wishing to see Belgium} where $n(B) = 39$.

From the information given $n(F \cap H) = 29$
$n(F \cap B) = 24$
$n(H \cap B) = 21$
$n(F \cap H \cap B) = 8$

The Venn diagram is therefore:

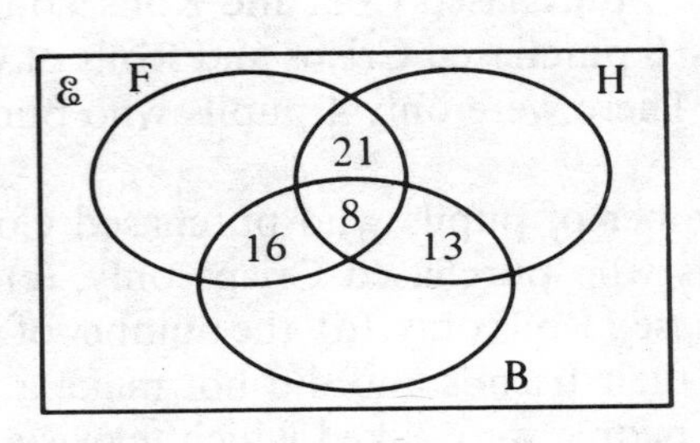

(a) Since $F \cap H$ contains 29 members, of which 8 chose Belgium; there must be $29 - 8 = 21$ pupils who chose to visit France and Holland but not Belgium.
Similarly, those wishing to visit France and Belgium but not Holland must be $24 - 8 = 16$ pupils.
Those wishing to visit Belgium and Holland but not France must be $21 - 8 = 13$ pupils.
Further, as 55 pupils asked to see France, of which $(16 + 8 + 21) = 45$ pupils are already placed on the Venn diagram, there must be $55 - 45 =$ 10 pupils who asked to stay only in France. ANS

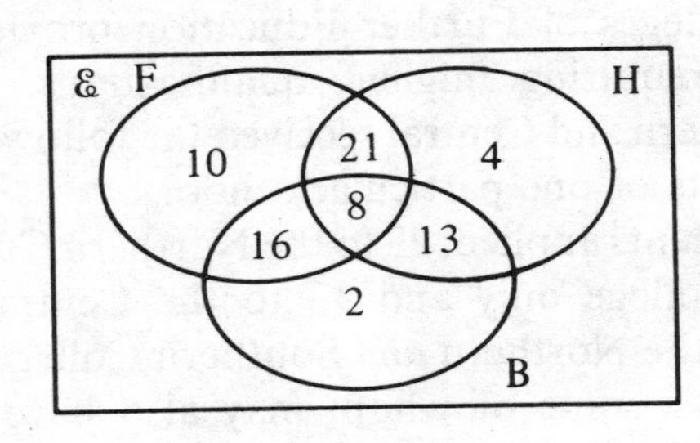

(b) There were 46 pupils who asked to see Holland, of which $(21 + 8 + 13) = 42$ pupils are already placed on the diagram. Therefore, $46 - 42 = 4$ pupils wished to visit only Holland. Similarly, $39 - (16 + 8 + 13) = 2$ pupils asked to see only Belgium. $10 + 4 + 2 =$ 16 pupils asked to see one country only. ANS

Exercise 69 Use Venn diagrams to assist in the solution of these problems:

1. Of 280 pupils in a sixth form it was found that the following subjects contained these numbers of pupils:

   Mathematics 147 English Language 157 Physics 121.

Of these, 57 pupils studied Mathematics only, 47 pupils studied English Language only, 43 pupils studied Mathematics and English Language only, and 40 pupils studied Mathematics and Physics only.

Find (a) the number of pupils who studied all three subjects, (b) the number of pupils who studied Physics only, (c) the number of pupils who studied none of these subjects, i.e. $n(M \cup E \cup P)'$.

2. 170 pupils visited the school Tuck Shop one break time. Three items were on sale; Cola, Crisps and Rolls. The total numbers of pupils who purchased each item were:

Cola 101 Crisps 78 Rolls 31

Of these, 35 purchased Cola and Crisps, some of whom also purchased Rolls; 27 purchased Cola and Rolls, some of whom also purchased Crisps; 8 purchased Crisps and Rolls of which some also purchased Cola. There were only 4 pupils who purchased all three items.

Find (a) the number of pupils who purchased Cola only, (b) the number of pupils who purchased Crisps only, (c) the number of pupils who purchased Rolls only, (d) the number of pupils who were only waiting for their friends and did not make a purchase.

3. A sample of 200 people were asked which television channels they had watched the previous day:

118 watched BBC 1, 137 watched ITV, 77 watched BBC 2

Of these, 89 watched at least BBC 1 and ITV, 26 watched at least BBC 1 and BBC 2, 36 watched at least ITV and BBC 2, 5 watched all three channels that day.

Find (a) the number of people who watched BBC 1 and ITV but *not* BBC 2, (b) the number of people who watched BBC 2 only, (c) the number of people who watched no television that day.

4. Three local Colleges of Further Education provided courses for school-leavers requiring higher qualifications. The Colleges; Northern, Southern and Central received the following applications from the students of one particular school.

A total of 99 students applied, 23 to the Northern College only, 27 to the Southern College only and 13 to the Central College only. Applications to the Northern *and* Southern Colleges were received from 12 students, some of whom may also have applied to the Central College. Similarly, 11 students applied to the Southern *and* Central, some of whom may also have applied to the Northern; 15 students applied to the Central *and* Northern, some of whom may have applied to the Southern.

How many students applied to all three colleges?

Hint: Let x represent the number of students in $N \cap C \cap S$, then $N \cap S = 12 - x$, etc.

The Algebra of Sets

Exercise 70 Let $\mathscr{E} = \{a, b, c, d, e, f, g, h, i, j, k\}$; $A = \{a, b, c, d, e\}$; $B = \{c, e, f, g, h\}$; $C = \{a, e, h, i, j, k\}$.

1. (a) Evaluate (i) $3+4$, (ii) $4+3$. Are (i) and (ii) equal?
(b) Does $x+y = y+x$?
(c) (i) List the set $A \cup B$, (ii) list the set $B \cup A$. Are (i) and (ii) equal?
(d) Evaluate (i) 3×4, (ii) 4×3. Are (i) and (ii) equal?
(e) Does $xy = yx$?
(f) List the sets (i) $A \cap B$, (ii) $B \cap A$. Are (i) and (ii) equal?

Note: The law investigated here is called the **Commutative Law.**

2. (a) Evaluate (i) $3+(4+5)$, (ii) $(3+4)+5$. Are (i) and (ii) equal?
(b) Does $x+(y+z) = (x+y)+z = x+y+z$?
(c) List the sets (i) $B \cup C$, (ii) $A \cup (B \cup C)$, (iii) $(A \cup B) \cup C$, (iv) $A \cup B \cup C$. Are (ii), (iii) and (iv) equal?
(d) Evaluate (i) $3(4 \times 5)$, (ii) $(3 \times 4)5$. Are (i) and (ii) equal?
(e) Does $x(yz) = (xy)z = xyz$?
(f) List the sets (i) $B \cap C$, (ii) $A \cap (B \cap C)$, (iii) $(A \cap B) \cap C$, (iv) $A \cap B \cap C$. Are (ii), (iii) and (iv) equal?

Note: The law investigated here is called the **Associative Law.**

3. (a) Evaluate (i) $3(4+5)$, (ii) $(3 \times 4)+(3 \times 5)$. Are (i) and (ii) equal?
(b) Does $x(y+z) = xy+xz$?
(c) List the sets (i) $A \cap (B \cup C)$, (ii) $(A \cap B) \cup (A \cap C)$. Are (i) and (ii) equal?
(d) Evaluate (i) $3+(4 \times 5)$, (ii) $(3+4) \times (3+5)$. Are (i) and (ii) equal?
(e) Does $x+yz = (x+y)(x+z)$?
(f) List the sets (i) $A \cup (B \cap C)$, (ii) $(A \cup B) \cap (A \cup C)$. Are (i) and (ii) equal?

Note: The law investigated here is called the **Distributive Law.**

Draw Venn diagrams to illustrate that:

4. (a) $A \cap (A \cup B) = A$, (b) $A \cup (A \cap B) = A$.
5. (a) $A \cup A = A$, (b) $A \cap A = A$.
6. (a) $A \cup A' = \mathscr{E}$, (b) $A \cap A' = \varnothing$.
7. (a) $\mathscr{E} \cap A = A$, (b) $\varnothing \cup A = A$.
8. (a) $\mathscr{E} \cup A = \mathscr{E}$, (b) $\varnothing \cap A = \varnothing$.
9. (a) $\mathscr{E}' = \varnothing$, (b) $\varnothing' = \mathscr{E}$.
10. (a) $(A \cap B)' = A' \cup B'$, (b) $(A \cup B)' = A' \cap B'$.
11. $(A')' = A$.

These facts can now be formulated into a set of rules. The laws of the Algebra thus formed derive initially from the work of the English mathematician Boole and is thus called a **Boolean Algebra**. It has important uses in the fields of logic and computing.

It is only necessary to remember part (a) of each rule, part (b) can be formed easily by interchanging the symbols $\cup$ and $\cap$, $\mathscr{E}$ and $\varnothing$. This is known as the **principle of duality**.

Laws

	a	b
1	$A \cup B = B \cup A$	$A \cap B = B \cap A$
2	$A \cup (B \cup C) = (A \cup B) \cup C$	$A \cap (B \cap C) = (A \cap B) \cap C$
3	$A \cap (B \cup C) = (A \cap B) \cup (A \cap C)$	$A \cup (B \cap C) = (A \cup B) \cap (A \cup C)$
4	$A \cap (A \cup B) = A$	$A \cup (A \cap B) = A$
5	$A \cup A = A$	$A \cap A = A$
6	$A \cup A' = \mathscr{E}$	$A \cap A' = \varnothing$
7	$\mathscr{E} \cap A = A$	$\varnothing \cup A = A$
8	$\mathscr{E} \cup A = \mathscr{E}$	$\varnothing \cap A = \varnothing$
9	$\mathscr{E}' = \varnothing$	$\varnothing' = \mathscr{E}$
10	$(A \cap B)' = A' \cup B'$	$(A \cup B)' = A' \cap B'$
11	$(A')' = A$	

Example

Simplify $A \cup (A \cap B) \cap \mathscr{E}$.

From law 4b $\quad A \cup (A \cap B) \cap \mathscr{E} = A \cap$

From law 7a $\quad A \cap \mathscr{E} = \underline{A}$ ANS

Exercise 71

Simplify:

1. $A \cup (A \cup A')$ **2.** $(A \cap A')'$

3. $A' \cup (A \cup B)$ **4.** $A' \cup (A \cap B)$

5. $(\varnothing \cap A)' \cup [A \cap (A \cup B)]$

Vectors

In earlier chapters we have met the quantities Length, Area and Volume; all such quantities are referred to as **Scalars.** We have also met the quantities Displacement, Velocity and Acceleration; these are referred to as **Vector** quantities.

A scalar quantity has magnitude only, a vector has both magnitude and direction.

A vector is usually denoted in print by a small letter in bold type, e.g. **u**, **v**. When writing such vectors yourself place a curled line beneath the letter, e.g. $\underset{\sim}{u}$, $\underset{\sim}{v}$.

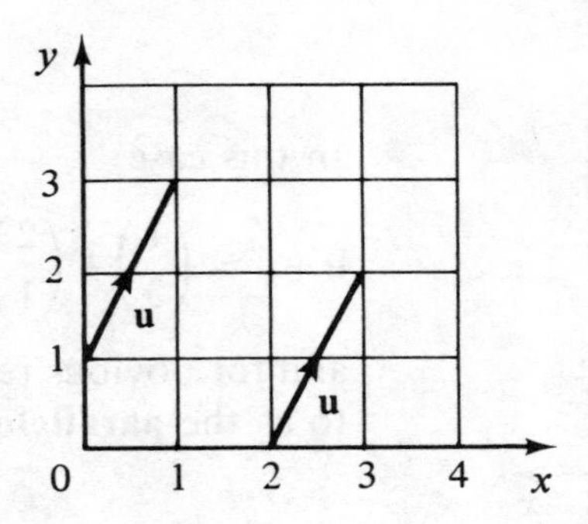

In this diagram the arrowed lines indicate a move, or translation; 'one unit parallel to the x-axis followed by two units parallel to the y-axis.'

The moves have direction as well as magnitude and are therefore vectors. Let **u** denote this vector. Such vectors are usually shown in the form of column vectors, in this case $\begin{pmatrix}1\\2\end{pmatrix}$. The 1 shows the increase in the x value and the 2 the increase in the y value.

i.e. $\mathbf{u} = \begin{pmatrix}1\\2\end{pmatrix}$. 1 and 2 are the **components** of **u**.

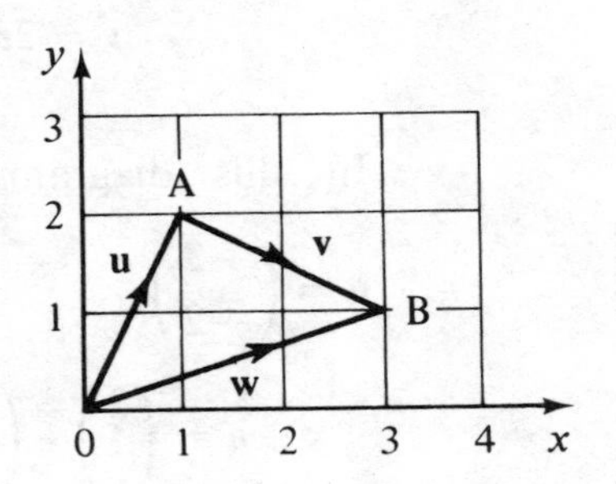

In this diagram the vector **u** is represented by the line segment OA. To show direction an arrow can be placed above $\overrightarrow{OA}$ and this can also represent the vector.

i.e. $\mathbf{u}$ or $\overrightarrow{OA} = \begin{pmatrix}1\\2\end{pmatrix}$

$\mathbf{v}$ or $\overrightarrow{AB} = \begin{pmatrix}2\\-1\end{pmatrix}$

and $\mathbf{w}$ or $\overrightarrow{OB} = \begin{pmatrix}3\\1\end{pmatrix}$

It can be seen that if corresponding components of the vectors are added we get:

$$\mathbf{u}+\mathbf{v} = \begin{pmatrix}1\\2\end{pmatrix} + \begin{pmatrix}2\\-1\end{pmatrix} = \begin{pmatrix}1+2\\2+(-1)\end{pmatrix} = \begin{pmatrix}3\\1\end{pmatrix} = \mathbf{w}$$

$\therefore$ $\mathbf{u}+\mathbf{v} = \mathbf{w}$, which is apparent from the diagram, and for obvious reasons is referred to as the **triangle rule**. We now know that vector[s] may be added.

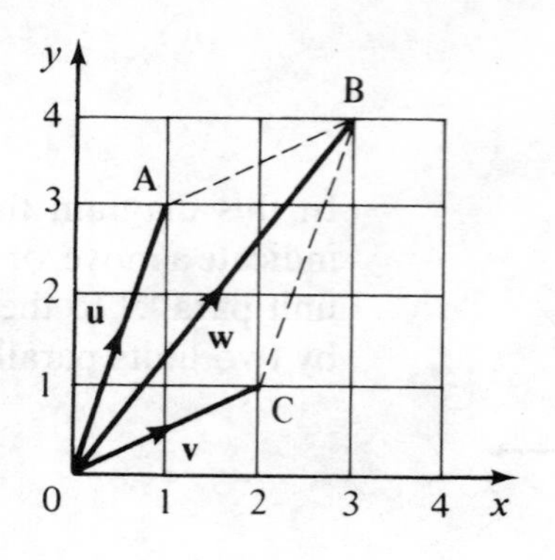

In this case

$$\mathbf{u}+\mathbf{v}=\begin{pmatrix}1\\3\end{pmatrix}+\begin{pmatrix}2\\1\end{pmatrix}=\begin{pmatrix}3\\4\end{pmatrix}=\mathbf{w},$$

and for obvious reasons is referre[d] to as the **parallelogram rule**.

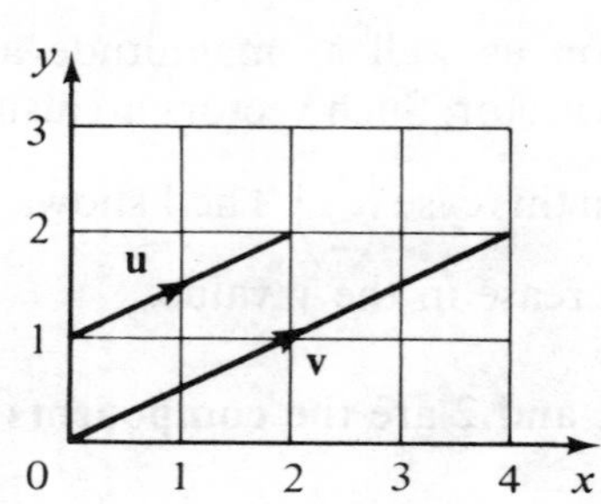

In this diagram $\mathbf{u}=\begin{pmatrix}2\\1\end{pmatrix}$ an[d] $\mathbf{v}=\begin{pmatrix}4\\2\end{pmatrix}$.

It can be seen that if the vector **u** i[s] multiplied by the scalar 2 we get

$$2\mathbf{u}=2\begin{pmatrix}2\\1\end{pmatrix}=\begin{pmatrix}4\\2\end{pmatrix}=\mathbf{v}$$

$$\therefore\ \mathbf{v}=2\mathbf{u}$$

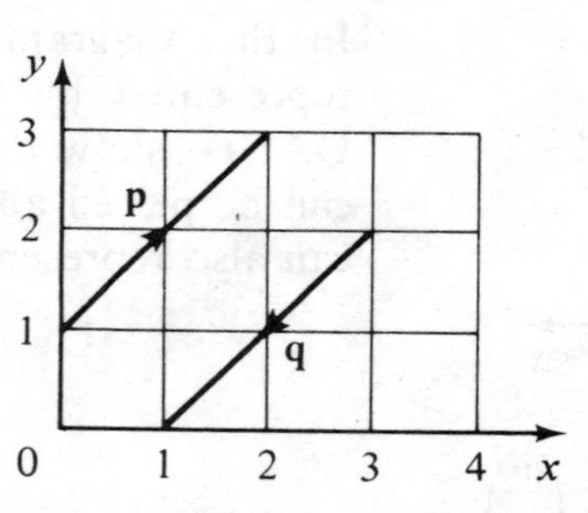

In this diagram, $\mathbf{p}=\begin{pmatrix}2\\2\end{pmatrix}$ and $\mathbf{q}=\begin{pmatrix}-2\\-2\end{pmatrix}$.

$$\mathbf{p}+\mathbf{q}=\begin{pmatrix}2\\2\end{pmatrix}+\begin{pmatrix}-2\\-2\end{pmatrix}=\begin{pmatrix}0\\0\end{pmatrix}$$

termed the **zero vector**.

Since $\mathbf{p}+\mathbf{q}=0$, $\mathbf{q}$ is said to be the negative (or additive inverse) of **p**

i.e. $\mathbf{q}=-\mathbf{p}$.

The **magnitude** of a vector **u** is denoted by |**u**|, read as the 'modulus of **u**'.

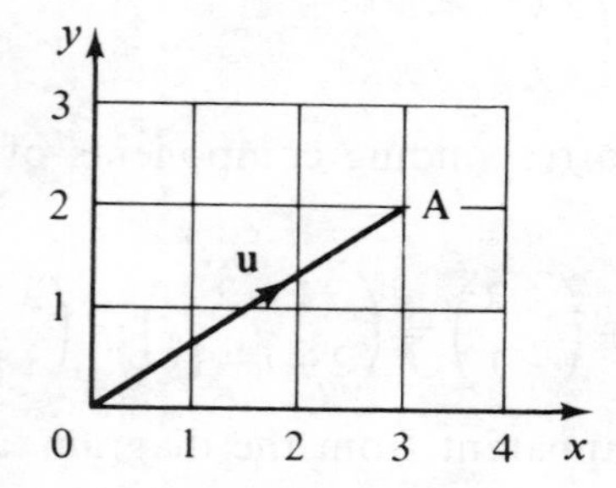

In this case the vector $\mathbf{u}=\begin{pmatrix}3\\2\end{pmatrix}$, and |**u**| is found by measurement of the line segment $\overrightarrow{OA}$.

i.e. $|\overrightarrow{OA}|=|\mathbf{u}|=$ <u>3.6 units</u>

Alternatively, the magnitude of **u** can be found by using Pythagoras Theorem, dealt with in a later chapter. For those familiar with its use:

A, 2, O, 3, B

$$(|\overrightarrow{OA}|)^2 = 3^2 + 2^2 = 9 + 4 = 13$$
$$\therefore |\overrightarrow{OA}| = \sqrt{13} = \underline{3.606 \text{ units}}$$

Note

1. In the same way that $3\mathbf{u} = 3\begin{pmatrix}3\\2\end{pmatrix} = \begin{pmatrix}9\\6\end{pmatrix}$;

$$|3\mathbf{u}| = 3|\mathbf{u}| = 3(3.6) = \underline{10.8 \text{ units}}$$

2. The modulus cannot be negative. The magnitude of a vector can be measured in either direction along the line segment.

Exercise 72

Illustrate questions **1** to **6** with diagrams: Given $\mathbf{u} = \begin{pmatrix}2\\1\end{pmatrix}$, $\mathbf{v} = \begin{pmatrix}1\\3\end{pmatrix}$ find

1. $\mathbf{u} + \mathbf{v}$, **2.** $\mathbf{u} - \mathbf{v}$,
3. $2\mathbf{u}$, **4.** $3\mathbf{v}$,
5. $|\mathbf{u}|$, **6.** $|\mathbf{v}|$,
7. $|3\mathbf{u}|$, **8.** Does $|\mathbf{u}| + |\mathbf{v}| = |\mathbf{u} + \mathbf{v}|$?
9. Does $\mathbf{u} + \mathbf{v} = \mathbf{v} + \mathbf{u}$? **10.** Does $2\mathbf{u} + 2\mathbf{v} = 2(\mathbf{u} + \mathbf{v})$?
11. Find $|\mathbf{v}| - |\mathbf{u}|$, **12.** Find $|\mathbf{v} - \mathbf{u}|$.

Matrices

The need to store information in an easy to read form often arises. Consider the following table that shows the situation in an inter-House Football Competition after the first week.

	Played	Won	Drawn	Points
Yellows	1	1	0	2
Blues	2	1	0	2
Reds	3	0	1	1
Greens	2	0	1	1

If we omit the row and column headings, and enclose the array of numbers in brackets, the result is called a **matrix**.

$$\text{i.e.} \quad \begin{pmatrix}1 & 1 & 0 & 2\\2 & 1 & 0 & 2\\3 & 0 & 1 & 1\\2 & 0 & 1 & 1\end{pmatrix}$$

This matrix has 4 rows and 4 columns and is therefore referred to as a '4 by 4' matrix. This is known as the **order** of the matrix.

e.g. $\begin{pmatrix} 2 & 4 & 7 \\ 3 & 1 & 6 \end{pmatrix}$ is an example of a '2 by 3' matrix, (2 rows and 3 columns).

$\begin{pmatrix} 1 & 7 \\ 3 & 2 \\ 4 & 8 \end{pmatrix}$ is an example of a '3 by 2' matrix, (3 rows and 2 columns).

At the conclusion of the House competition another table showing the results of the concluding matches read:

	P	W	D	P
Yellows	2	1	1	3
Blues	1	0	0	0
Reds	0	0	0	0
Greens	1	0	1	1

The combined results from the two tables are found easily by the addition of the two matrices:

$$\begin{pmatrix} 1 & 1 & 0 & 2 \\ 2 & 1 & 0 & 2 \\ 3 & 0 & 1 & 1 \\ 2 & 0 & 1 & 1 \end{pmatrix} + \begin{pmatrix} 2 & 1 & 1 & 3 \\ 1 & 0 & 0 & 0 \\ 0 & 0 & 0 & 0 \\ 1 & 0 & 1 & 1 \end{pmatrix} = \begin{pmatrix} 3 & 2 & 1 & 5 \\ 3 & 1 & 0 & 2 \\ 3 & 0 & 1 & 1 \\ 3 & 0 & 2 & 2 \end{pmatrix}$$

This means that all the teams have played three matches, that Yellow have achieved 5 points, etc.

Note

Matrices may be added or subtracted in the same way as vectors. Provided their order is the same, matrices may be added by adding their corresponding elements, or subtracted by subtracting their corresponding elements.

We found earlier in this chapter that a vector could be multiplied by a scalar. The same rule applies to matrices.

Consider the matrix $A = \begin{pmatrix} 2 & 1 \\ 1 & 3 \end{pmatrix}$, we know that we can find $A + A + A$ by adding the corresponding elements.

i.e. $$\begin{pmatrix} 2 & 1 \\ 1 & 3 \end{pmatrix} + \begin{pmatrix} 2 & 1 \\ 1 & 3 \end{pmatrix} + \begin{pmatrix} 2 & 1 \\ 1 & 3 \end{pmatrix} = \begin{pmatrix} 6 & 3 \\ 3 & 9 \end{pmatrix}$$

It follows that the same result is achieved by multiplying the matrix A by the scalar 3.

i.e. $$3A = 3\begin{pmatrix} 2 & 1 \\ 1 & 3 \end{pmatrix} = \begin{pmatrix} 6 & 3 \\ 3 & 9 \end{pmatrix}$$

Consider the milk and bread deliveries to two neighbours on a specific day:

	Pints of Milk	Loaves of Bread
Mrs. Brown	2	1
Mrs. Green	3	2

The price of each item was: Milk 14p per pint, Bread 28p per loaf. The day's bill for both neighbours could be calculated by using the matrices:

$$\begin{pmatrix} 2 & 1 \\ 3 & 2 \end{pmatrix} \times \begin{pmatrix} 14 \\ 28 \end{pmatrix}$$

This means that Mrs. Brown's bill for milk was $2 \times 14 = 28\text{p}$
and her bill for bread was $1 \times 28 = \underline{28\text{p}}$
Mrs. Brown owed a total of $\underline{\underline{56\text{p}}}$

Mrs. Green's bill for milk was $3 \times 14 = 42\text{p}$
and her bill for bread was $2 \times 28 = \underline{56\text{p}}$
Mrs. Green owed a total of $\underline{\underline{98\text{p}}}$

Consider now the sales at a school tuck shop on the five days of a particular week:

	Cola (Bottles)	Crisps (Packets)	Rolls
Monday	20	15	30
Tuesday	15	21	24
Wednesday	18	28	14
Thursday	22	18	21
Friday	9	8	10

The price-list displayed in the shop is:

Cola 10p per bottle
Crisps 7p per packet
Rolls 5p each

The master in charge of the shop is a mathematics teacher and uses matrices to calculate his total takings at the end of the week.

i.e. $$\begin{pmatrix} 20 & 15 & 30 \\ 15 & 21 & 24 \\ 18 & 28 & 14 \\ 22 & 18 & 21 \\ 9 & 8 & 10 \end{pmatrix} \times \begin{pmatrix} 10 \\ 7 \\ 5 \end{pmatrix}$$

This means that the total for Monday was:

$$\begin{aligned}
&20 \text{ Colas @ } 10\text{p} = 20 \times 10 = 200\text{p}\\
&15 \text{ Crisps @ } 7\text{p} = 15 \times 7 = 105\text{p}\\
&30 \text{ Rolls @ } 5\text{p} = 30 \times 5 = \underline{150\text{p}}\\
&\qquad\qquad\qquad\qquad\qquad \underline{455\text{p}} \quad = \underline{£4.55}
\end{aligned}$$

It therefore follows that when multiplying matrices we must multiply each *row* of the first matrix by each *column* of the second matrix and find their sum.

$$\text{i.e.} \begin{pmatrix} 20 & 15 & 30 \\ 15 & 21 & 24 \\ 18 & 28 & 14 \\ 22 & 18 & 21 \\ 9 & 8 & 10 \end{pmatrix} \begin{pmatrix} 10 \\ 7 \\ 5 \end{pmatrix} = \begin{pmatrix} (20\times 10)+(15\times 7)+(30\times 5) \\ (15\times 10)+(21\times 7)+(24\times 5) \\ (18\times 10)+(28\times 7)+(14\times 5) \\ (22\times 10)+(18\times 7)+(21\times 5) \\ (9\times 10)+(8\times 7)+(10\times 5) \end{pmatrix}$$

$$= \begin{pmatrix} 455 \\ 417 \\ 446 \\ 451 \\ 196 \end{pmatrix}$$

$$\begin{aligned}
\text{Total takings} &= 455 + 417 + 446 + 451 + 196 \\
&= 1965\text{p} \quad = \underline{£19.65}
\end{aligned}$$

Note

It is only possible to multiply two matrices if the number of **columns** in the first matrix is the same as the number of **rows** in the second matrix.

Examples

1. Find the sums of the following matrices if possible:

(a) $$\begin{pmatrix} 1 & 4 & -6 \\ 2 & 7 & 3 \end{pmatrix} + \begin{pmatrix} 2 & 9 & 8 \\ 1 & -4 & 7 \end{pmatrix} = \begin{pmatrix} 1+2 & 4+9 & -6+8 \\ 2+1 & 7-4 & 3+7 \end{pmatrix}$$

$$= \underline{\begin{pmatrix} 3 & 13 & 2 \\ 3 & 3 & 10 \end{pmatrix}} \quad \text{ANS}$$

(b) $\begin{pmatrix} 2 & 1 \\ 7 & -2 \end{pmatrix} + \begin{pmatrix} 4 \\ 6 \end{pmatrix}$ This is not possible as the two matrices are of different order.

2. Find the products of the following matrices if possible:

(a) $\begin{pmatrix} a & b \\ c & d \end{pmatrix} \begin{pmatrix} e & f \\ g & h \end{pmatrix}$

[Each row 'onto' each column in turn]

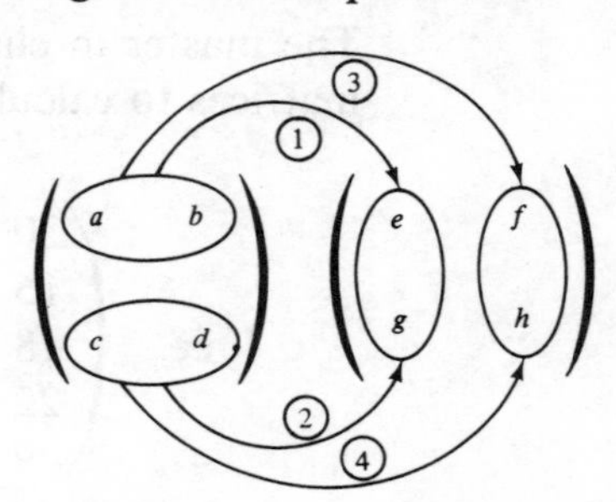

$$= \begin{pmatrix} ae+bg & af+bh \\ ce+dg & cf+dh \end{pmatrix} \quad \text{ANS}$$

(b) $\begin{pmatrix} 2 & 1 \\ -1 & 3 \end{pmatrix}\begin{pmatrix} 1 & 4 \\ 2 & -2 \end{pmatrix} = \begin{pmatrix} 2+2 & 8-2 \\ -1+6 & -4-6 \end{pmatrix} = \begin{pmatrix} 4 & 6 \\ 5 & -10 \end{pmatrix}$ ANS

(c) $\begin{pmatrix} 3 & 1 \\ -2 & 4 \end{pmatrix}\begin{pmatrix} 1 & 6 \\ 3 & -9 \\ 4 & -3 \end{pmatrix}$ This is not possible as there are only two elements in each row of the first matrix, compared with three elements in each column of the second matrix.

Exercise 73 Given $A = \begin{pmatrix} 1 \\ 3 \end{pmatrix}$, $B = \begin{pmatrix} 4 & 3 \\ -1 & 2 \end{pmatrix}$, $C = \begin{pmatrix} 2 & 9 \\ 3 & -1 \end{pmatrix}$,

$D = \begin{pmatrix} 4 & 6 & 1 & -3 \\ 2 & 1 & 3 & 9 \end{pmatrix}$, $E = \begin{pmatrix} 3 & 1 \\ 7 & 2 \end{pmatrix}$:

1. State the order of each of the matrices A, B, C, D, E.

2. Calculate where possible

(a) $A+B$, (b) $B+C$, (c) $D+E$, (d) $C+E$,
(e) $C+B$, (f) $E+D$, (g) $B-C$, (h) $C-B$,
(i) AB, (j) BC, (k) CD, (l) DE,
(m) CB, (n) DC, (o) 3A, (p) 3C,
(q) $\frac{1}{2}E$, (r) B(CE), (s) (BC)E, (t) $B(C+E)$,
(u) $BC+BE$.

3. (a) Does $B+C = C+B$? (b) Does $BC = CB$? (c) Does $B(CE) = (BC)E$? (d) Does $B(C+E) = BC+BE$?

4. Given that $B^2 = B \times B$, $B^3 = B \times (B \times B) = B \times B^2$. Calculate (a) B^2, (b) C^2, (c) B^3.

5. Given that $I = \begin{pmatrix} 1 & 0 \\ 0 & 1 \end{pmatrix}$ and $O = \begin{pmatrix} 0 & 0 \\ 0 & 0 \end{pmatrix}$

calculate (a) BI, (b) IB, (c) BO, (d) OB, (e) $O+C$, (f) $C+O$.

6. (a) Is it true that $BI = IB = B$?
(b) Is it true that $BO = OB = O$?
(c) Is it true that $O+C = C+O = C$?

Note

In question **5** above; I is referred to as the **unit** or **identity matrix**, and O as the **zero** or **null matrix** of matrices of order 2 by 2.

Consider the matrices $A = \begin{pmatrix} 3 & 2 \\ 1 & 1 \end{pmatrix}$, $B = \begin{pmatrix} 1 & -2 \\ -1 & 3 \end{pmatrix}$

then $AB = \begin{pmatrix} 3 & 2 \\ 1 & 1 \end{pmatrix}\begin{pmatrix} 1 & -2 \\ -1 & 3 \end{pmatrix} = \begin{pmatrix} 3-2 & -6+6 \\ 1-1 & -2+3 \end{pmatrix} = \begin{pmatrix} 1 & 0 \\ 0 & 1 \end{pmatrix} = I$

and $BA = \begin{pmatrix} 1 & -2 \\ -1 & 3 \end{pmatrix}\begin{pmatrix} 3 & 2 \\ 1 & 1 \end{pmatrix} = \begin{pmatrix} 3-2 & 2-2 \\ -3+3 & -2+3 \end{pmatrix} = \begin{pmatrix} 1 & 0 \\ 0 & 1 \end{pmatrix} = I$

It is therefore said that B is the multiplicative inverse of A, and is denoted by A^{-1}. (i.e. $A \times A^{-1} = A^{-1} \times A = I$).

It appears that the inverse matrix can be found by interchanging the elements of the main diagonal, the 3 and the 1, and changing the signs of the other diagonal.

Consider the matrix $C = \begin{pmatrix} 3 & 1 \\ 1 & 2 \end{pmatrix}$, then try $D = \begin{pmatrix} 2 & -1 \\ -1 & 3 \end{pmatrix}$ by the above 'rule'.

$$\therefore CD = \begin{pmatrix} 3 & 1 \\ 1 & 2 \end{pmatrix}\begin{pmatrix} 2 & -1 \\ -1 & 3 \end{pmatrix} = \begin{pmatrix} 6-1 & -3+3 \\ 2-2 & -1+6 \end{pmatrix}$$

$$= \begin{pmatrix} 5 & 0 \\ 0 & 5 \end{pmatrix}$$

$$= 5\begin{pmatrix} 1 & 0 \\ 0 & 1 \end{pmatrix} = 5I$$

i.e. $D = 5 \times C^{-1}$

$\therefore C^{-1} = D \div 5 = \begin{pmatrix} \frac{2}{5} & -\frac{1}{5} \\ -\frac{1}{5} & \frac{3}{5} \end{pmatrix}$

(Prove this for yourself by finding $C \times C^{-1}$ and $C^{-1} \times C$.)

Consider the matrices $X = \begin{pmatrix} a & b \\ c & d \end{pmatrix}$, and $Y = \begin{pmatrix} d & -b \\ -c & a \end{pmatrix}$

$$\therefore XY = \begin{pmatrix} a & b \\ c & d \end{pmatrix}\begin{pmatrix} d & -b \\ -c & a \end{pmatrix} = \begin{pmatrix} ad-bc & -ab+ab \\ cd-cd & -bc+ad \end{pmatrix}$$

$$= \begin{pmatrix} ad-bc & 0 \\ 0 & ad-bc \end{pmatrix}$$

$$= (ad-bc)\begin{pmatrix} 1 & 0 \\ 0 & 1 \end{pmatrix}$$

$(ad-bc)$ is called the **determinant** of X. If the determinant is equal to 1, the inverse matrix can be found simply by using the 'interchange main elements and change other signs rule'.

e.g. Considering the earlier matrix $A = \begin{pmatrix} 3 & 2 \\ 1 & 1 \end{pmatrix}$,

and comparing with $\begin{pmatrix} a & b \\ c & d \end{pmatrix}$

the determinant $= (3 \times 1) - (2 \times 1) = \underline{1}$.

Considering the matrix C above $= \begin{pmatrix} 3 & 1 \\ 1 & 2 \end{pmatrix}$,

the determinant $= (3 \times 2) - (1 \times 1) = \underline{5}$

Therefore each element of the rearranged matrix needed division by 5.

If the determinant is equal to zero, the matrix has no inverse since we cannot divide by zero. Such a matrix is called a **singular matrix.**

Examples

Find the inverses of the following matrices if they exist:

1. $A = \begin{pmatrix} 2 & 1 \\ 3 & 6 \end{pmatrix}$

The determinant of $A = (2 \times 6) - (1 \times 3) = 9$. Hence A^{-1} exists.

$$\therefore A^{-1} = \frac{1}{9}\begin{pmatrix} 6 & -1 \\ -3 & 2 \end{pmatrix} = \begin{pmatrix} \frac{2}{3} & -\frac{1}{9} \\ -\frac{1}{3} & \frac{2}{9} \end{pmatrix} \quad \text{ANS}$$

(Prove that this is so yourself by finding $A \times A^{-1}$ and $A^{-1} \times A$.)

2. $B = \begin{pmatrix} 2 & 1 \\ 6 & 3 \end{pmatrix}$

The determinant of $B = (2 \times 3) - (1 \times 6) = 0$.

$\therefore B^{-1}$ does not exist and B is a singular matrix

Exercise 74

State which of the following are singular matrices. If an inverse exists, find it:

1. $\begin{pmatrix} 1 & 2 \\ 3 & 7 \end{pmatrix}$ **2.** $\begin{pmatrix} 2 & 1 \\ 5 & 2 \end{pmatrix}$ **3.** $\begin{pmatrix} 3 & 6 \\ 1 & 2 \end{pmatrix}$ **4.** $\begin{pmatrix} 3 & 4 \\ 2 & 6 \end{pmatrix}$ **5.** $\begin{pmatrix} 4 & 6 \\ 2 & 3 \end{pmatrix}$

Transformations

Vectors and matrices have geometrical applications.

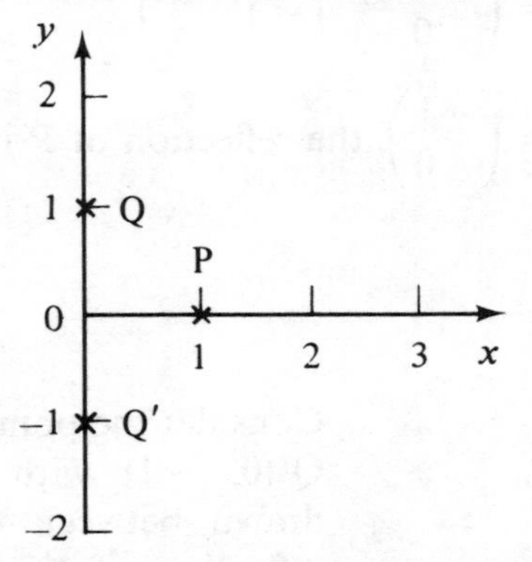

Consider the points P (1, 0) and Q (0, 1). These may be shown as the column vectors

$$\begin{pmatrix} 1 \\ 0 \end{pmatrix} \text{ and } \begin{pmatrix} 0 \\ 1 \end{pmatrix}.$$

$$\text{IP gives } \begin{pmatrix} 1 & 0 \\ 0 & 1 \end{pmatrix}\begin{pmatrix} 1 \\ 0 \end{pmatrix} = \begin{pmatrix} 1+0 \\ 0+0 \end{pmatrix} = \begin{pmatrix} 1 \\ 0 \end{pmatrix}.$$

Therefore pre-multiplication of P by the unit matrix causes no change in the coordinates.

Consider the point Q′(0, −1). This is the reflection of point Q in the x-axis, called the **image** of Q under this transformation. Point P is unaltered by such a reflection.

The only change in Q is one of sign in the y-coordinate. We therefore attempt to find a matrix that will effect such a change.

i.e. $\begin{pmatrix}1\\0\end{pmatrix}$ is unchanged and $\begin{pmatrix}0\\1\end{pmatrix}$ becomes $\begin{pmatrix}0\\-1\end{pmatrix}$.

$\therefore$ Try $\begin{pmatrix}1 & 0\\0 & -1\end{pmatrix}\begin{pmatrix}x\\y\end{pmatrix} = \begin{pmatrix}x+0\\0+y\end{pmatrix} = \begin{pmatrix}x\\-y\end{pmatrix}$, which is the required result.

$\therefore \begin{pmatrix}1 & 0\\0 & -1\end{pmatrix}\begin{pmatrix}0\\1\end{pmatrix} = \begin{pmatrix}0\\-1\end{pmatrix}$, the reflection of Q in the x-axis.

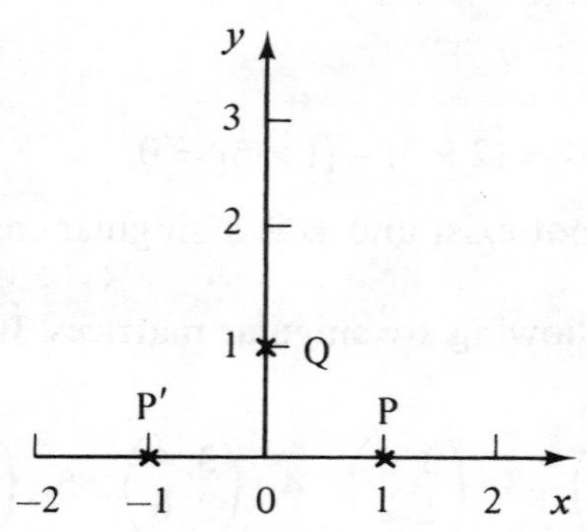

Consider the point P′ $(-1, 0)$. This is the reflection of point P in the y-axis. Point Q is unaltered by such a reflection.

The only change in P is one of sign in the x-coordinate.

i.e. $\begin{pmatrix}1\\0\end{pmatrix}$ becomes $\begin{pmatrix}-1\\0\end{pmatrix}$ and $\begin{pmatrix}0\\1\end{pmatrix}$ is unchanged.

$\therefore$ Try $\begin{pmatrix}-1 & 0\\0 & 1\end{pmatrix}\begin{pmatrix}x\\y\end{pmatrix} = \begin{pmatrix}-x+0\\0+y\end{pmatrix} = \begin{pmatrix}-x\\y\end{pmatrix}$

$\therefore \begin{pmatrix}-1 & 0\\0 & 1\end{pmatrix}\begin{pmatrix}1\\0\end{pmatrix} = \begin{pmatrix}-1\\0\end{pmatrix}$, the reflection of P in the y-axis.

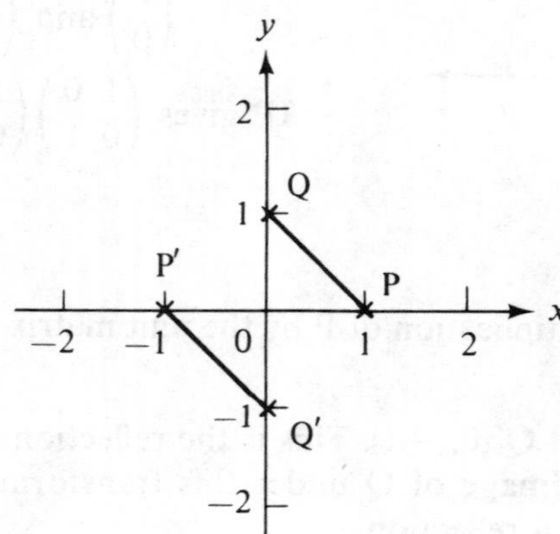

Consider the points P′ $(-1, 0)$ and Q′$(0, \ -1)$ with a straight line drawn between them. This is a reflection of the line PQ in the origin, or a rotation of 180° about the origin.

The changes are ones of sign in the coordinates.

i.e. $\begin{pmatrix}1\\0\end{pmatrix}$ becomes $\begin{pmatrix}-1\\0\end{pmatrix}$ and $\begin{pmatrix}0\\1\end{pmatrix}$ becomes $\begin{pmatrix}0\\-1\end{pmatrix}$

$\therefore$ Try $\begin{pmatrix}-1 & 0\\0 & -1\end{pmatrix}\begin{pmatrix}x\\y\end{pmatrix}=\begin{pmatrix}-x\\-y\end{pmatrix}$, which is the required result

$$\therefore \begin{pmatrix}-1 & 0\\0 & -1\end{pmatrix}\begin{pmatrix}1\\0\end{pmatrix}=\begin{pmatrix}-1\\0\end{pmatrix}, \text{ and } \begin{pmatrix}-1 & 0\\0 & -1\end{pmatrix}\begin{pmatrix}0\\1\end{pmatrix}=\begin{pmatrix}0\\-1\end{pmatrix}$$

the reflections of P and Q in the origin.

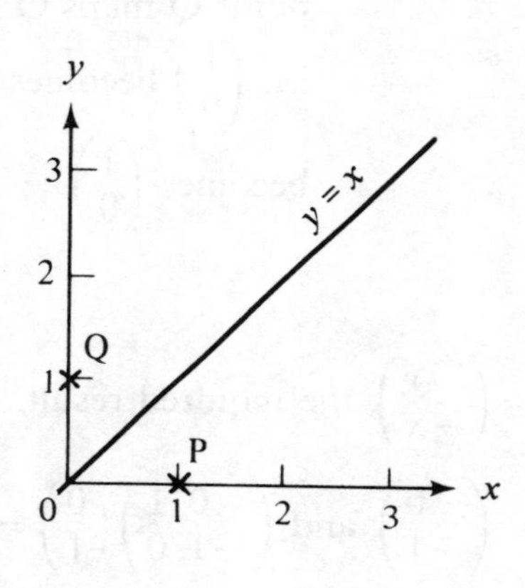

Consider the reflection of point P in the line $y = x$, such a translation maps P onto Q. Similarly, the reflection of point Q in the line $y = x$ maps Q onto P. This is to say that the values of the coordinates interchange.

i.e. $\begin{pmatrix}1\\0\end{pmatrix}$ becomes $\begin{pmatrix}0\\1\end{pmatrix}$

and $\begin{pmatrix}0\\1\end{pmatrix}$ becomes $\begin{pmatrix}1\\0\end{pmatrix}$.

$\therefore$ Try $\begin{pmatrix}0 & 1\\1 & 0\end{pmatrix}\begin{pmatrix}x\\y\end{pmatrix}=\begin{pmatrix}y\\x\end{pmatrix}$, which is the required result

$$\therefore \begin{pmatrix}0 & 1\\1 & 0\end{pmatrix}\begin{pmatrix}1\\0\end{pmatrix}=\begin{pmatrix}0\\1\end{pmatrix}, \text{ and } \begin{pmatrix}0 & 1\\1 & 0\end{pmatrix}\begin{pmatrix}0\\1\end{pmatrix}=\begin{pmatrix}1\\0\end{pmatrix}$$

the reflections of P and Q in the line $y = x$.

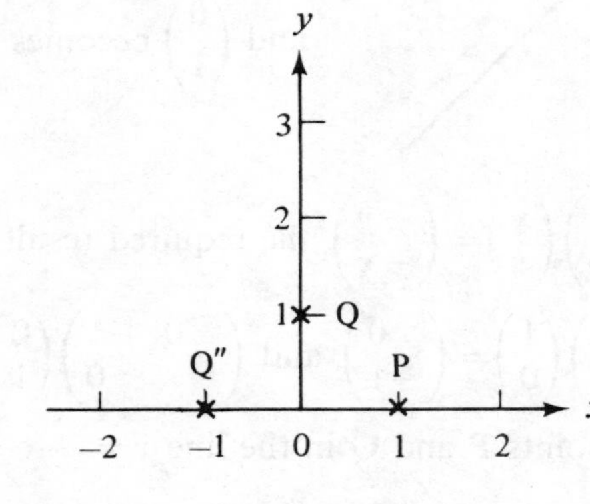

Consider the point Q″ $(-1, 0)$. This is a rotation of point Q through 90° anticlockwise, a similar rotation of point P maps P onto Q.

i.e. $\begin{pmatrix}1\\0\end{pmatrix}$ becomes $\begin{pmatrix}0\\1\end{pmatrix}$

and $\begin{pmatrix}0\\1\end{pmatrix}$ becomes $\begin{pmatrix}-1\\0\end{pmatrix}$.

$\therefore$ Try $\begin{pmatrix} 0 & -1 \\ 1 & 0 \end{pmatrix}\begin{pmatrix} x \\ y \end{pmatrix} = \begin{pmatrix} -y \\ x \end{pmatrix}$, the required result

$\therefore$ $\begin{pmatrix} 0 & -1 \\ 1 & 0 \end{pmatrix}\begin{pmatrix} 1 \\ 0 \end{pmatrix} = \begin{pmatrix} 0 \\ 1 \end{pmatrix}$, and $\begin{pmatrix} 0 & -1 \\ 1 & 0 \end{pmatrix}\begin{pmatrix} 0 \\ 1 \end{pmatrix} = \begin{pmatrix} -1 \\ 0 \end{pmatrix}$

the rotation of points P and Q through 90° anticlockwise.

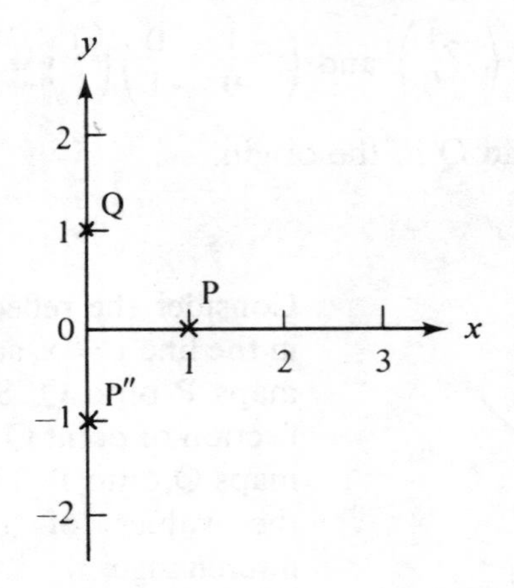

Consider the point P″ (0, −1). This is a rotation of point P through 90° clockwise, a similar rotation of point Q maps Q onto P.

i.e. $\begin{pmatrix} 1 \\ 0 \end{pmatrix}$ becomes $\begin{pmatrix} 0 \\ -1 \end{pmatrix}$, and $\begin{pmatrix} 0 \\ 1 \end{pmatrix}$ becomes $\begin{pmatrix} 1 \\ 0 \end{pmatrix}$.

$\therefore$ Try $\begin{pmatrix} 0 & 1 \\ -1 & 0 \end{pmatrix}\begin{pmatrix} x \\ y \end{pmatrix} = \begin{pmatrix} y \\ -x \end{pmatrix}$, the required result.

$\therefore$ $\begin{pmatrix} 0 & 1 \\ -1 & 0 \end{pmatrix}\begin{pmatrix} 1 \\ 0 \end{pmatrix} = \begin{pmatrix} 0 \\ -1 \end{pmatrix}$, and $\begin{pmatrix} 0 & 1 \\ -1 & 0 \end{pmatrix}\begin{pmatrix} 0 \\ 1 \end{pmatrix} = \begin{pmatrix} 1 \\ 0 \end{pmatrix}$,

the rotation of points P and Q through 90° clockwise.

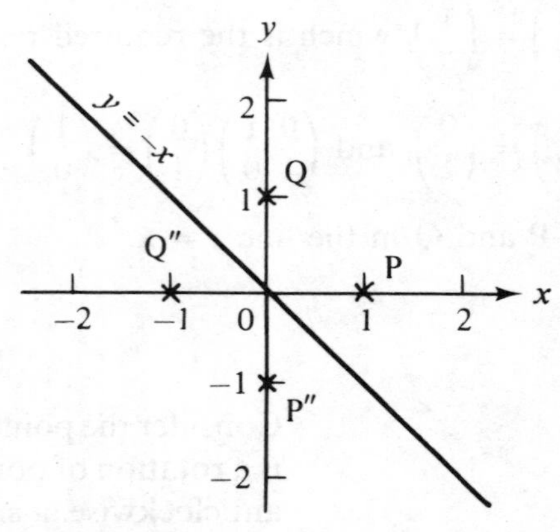

Consider the points P″ (0, −1) and Q″ (−1, 0). These are reflections of points P and Q in the line $y = -x$.

i.e. $\begin{pmatrix} 1 \\ 0 \end{pmatrix}$ becomes $\begin{pmatrix} 0 \\ -1 \end{pmatrix}$

and $\begin{pmatrix} 0 \\ 1 \end{pmatrix}$ becomes $\begin{pmatrix} -1 \\ 0 \end{pmatrix}$.

$\therefore$ Try $\begin{pmatrix} 0 & -1 \\ -1 & 0 \end{pmatrix}\begin{pmatrix} x \\ y \end{pmatrix} = \begin{pmatrix} -y \\ -x \end{pmatrix}$, the required result

$\therefore$ $\begin{pmatrix} 0 & -1 \\ -1 & 0 \end{pmatrix}\begin{pmatrix} 1 \\ 0 \end{pmatrix} = \begin{pmatrix} 0 \\ -1 \end{pmatrix}$, and $\begin{pmatrix} 0 & -1 \\ -1 & 0 \end{pmatrix}\begin{pmatrix} 0 \\ 1 \end{pmatrix} = \begin{pmatrix} -1 \\ 0 \end{pmatrix}$

the reflection of points P and Q in the line $y = -x$.

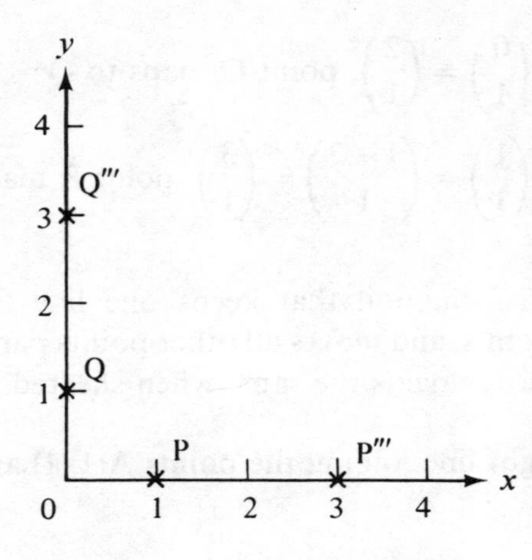

Consider the points P''' (3, 0) and Q''' (0, 3). These are **enlargements** or magnifications of the points P and Q, by a scale factor 3 with the origin as centre.

i.e. $3P = 3\begin{pmatrix}1\\0\end{pmatrix} = \begin{pmatrix}3\\0\end{pmatrix} = P'''$

and $3Q = 3\begin{pmatrix}0\\1\end{pmatrix} = \begin{pmatrix}0\\3\end{pmatrix} = Q'''$

$\therefore$ Try $\begin{pmatrix}3 & 0\\0 & 3\end{pmatrix}\begin{pmatrix}x\\y\end{pmatrix} = \begin{pmatrix}3x\\3y\end{pmatrix}$, the required result.

$\therefore \quad \begin{pmatrix}3 & 0\\0 & 3\end{pmatrix}\begin{pmatrix}1\\0\end{pmatrix} = \begin{pmatrix}3\\0\end{pmatrix}$, and $\begin{pmatrix}3 & 0\\0 & 3\end{pmatrix}\begin{pmatrix}0\\1\end{pmatrix} = \begin{pmatrix}0\\3\end{pmatrix}$,

the enlargements of P and Q, by a scale factor 3 with the origin as centre.

Note The vectors $\begin{pmatrix}1\\0\end{pmatrix}$ and $\begin{pmatrix}0\\1\end{pmatrix}$ are called **base vectors**. They form a basis for the construction of all 2 by 1 vectors.

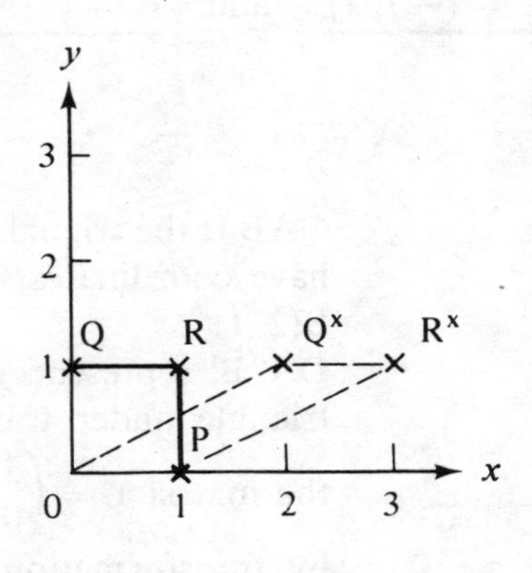

Consider the square OQRP. The points OQ^xR^xP are the vertices of a parallelogram formed by shearing the square OQRP.

i.e. $\begin{pmatrix}1\\0\end{pmatrix}$ remains unchanged and $\begin{pmatrix}0\\1\end{pmatrix}$ becomes $\begin{pmatrix}2\\1\end{pmatrix}$,

the points Q and R have moved 2 units parallel to the line OP in the positive direction.

$\therefore$ Try $\begin{pmatrix}1 & 2\\0 & 1\end{pmatrix}\begin{pmatrix}x\\y\end{pmatrix} = \begin{pmatrix}x+2y\\y\end{pmatrix}$.

$\therefore \quad \begin{pmatrix}1 & 2\\0 & 1\end{pmatrix}\begin{pmatrix}0\\0\end{pmatrix} = \begin{pmatrix}0\\0\end{pmatrix}$, point O remains the same

$\begin{pmatrix}1 & 2\\0 & 1\end{pmatrix}\begin{pmatrix}1\\0\end{pmatrix} = \begin{pmatrix}1\\0\end{pmatrix}$, point P remains the same

$$\begin{pmatrix}1 & 2\\0 & 1\end{pmatrix}\begin{pmatrix}0\\1\end{pmatrix} = \begin{pmatrix}2\\1\end{pmatrix}, \text{ point Q maps to } Q^x$$

$$\begin{pmatrix}1 & 2\\0 & 1\end{pmatrix}\begin{pmatrix}1\\1\end{pmatrix} = \begin{pmatrix}1+2\\1\end{pmatrix} = \begin{pmatrix}3\\1\end{pmatrix}, \text{ point R maps to } R^x$$

Note A **shear** is a transformation that keeps one line fixed, called the **invariant** line of points, and moves all other points parallel to this line. The area of a figure remains the same when sheared.

Examples **1.** AB is the straight line joining the points A(1, 1) and B(3, 3).

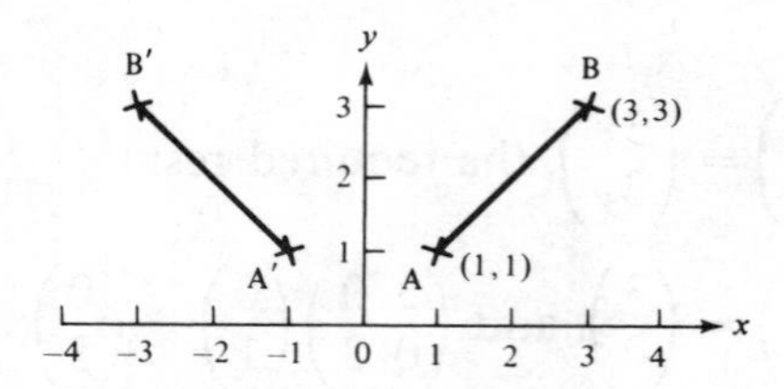

A′B′ represents the image of AB under reflection in the *y*-axis. Find the coordinates of A′ and B′.

The matrix representing the transformation under reflection in the *y*-axis is $\begin{pmatrix}-1 & 0\\0 & 1\end{pmatrix}$.

$$\therefore \begin{pmatrix}-1 & 0\\0 & 1\end{pmatrix}\begin{pmatrix}1\\1\end{pmatrix} = \begin{pmatrix}-1\\1\end{pmatrix}, \text{ and } \begin{pmatrix}-1 & 0\\0 & 1\end{pmatrix}\begin{pmatrix}3\\3\end{pmatrix} = \begin{pmatrix}-3\\3\end{pmatrix}$$

$\therefore$ $\underline{A' = (-1, 1)}$ and $\underline{B' = (-3, 3)}$. ANS

2.

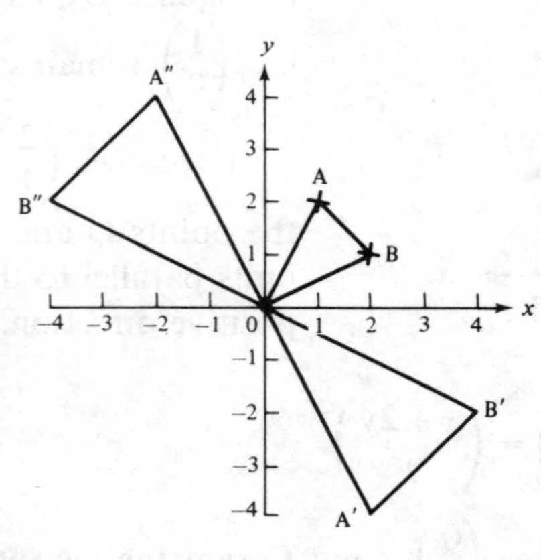

OAB is the triangle whose vertices have coordinates O(0, 0), A(1, 2); B(2, 1).
OA″B″ represents the image of this triangle under transformation by the matrix $L = \begin{pmatrix}2 & 0\\0 & -2\end{pmatrix}$ followed by transformation by the matrix $M = \begin{pmatrix}-1 & 0\\0 & -1\end{pmatrix}$. Find the coordinates of A″ and B″.

First transformation $= \begin{pmatrix}2 & 0\\0 & -2\end{pmatrix}\begin{pmatrix}1\\2\end{pmatrix} = \begin{pmatrix}2\\-4\end{pmatrix}$, i.e. A maps to A′(2, −4)

$$\begin{pmatrix} 2 & 0 \\ 0 & -2 \end{pmatrix}\begin{pmatrix} 2 \\ 1 \end{pmatrix} = \begin{pmatrix} 4 \\ -2 \end{pmatrix}$$ i.e. B maps to B′(4, −2)

$$\begin{pmatrix} 2 & 0 \\ 0 & -2 \end{pmatrix}\begin{pmatrix} 0 \\ 0 \end{pmatrix} = \begin{pmatrix} 0 \\ 0 \end{pmatrix}$$ i.e. O is unaltered

Second transformation $= \begin{pmatrix} -1 & 0 \\ 0 & -1 \end{pmatrix}\begin{pmatrix} 2 \\ -4 \end{pmatrix} = \begin{pmatrix} -2 \\ 4 \end{pmatrix}$ i.e. A′ maps to A″ (−2, 4)

$$\begin{pmatrix} -1 & 0 \\ 0 & -1 \end{pmatrix}\begin{pmatrix} 4 \\ -2 \end{pmatrix} = \begin{pmatrix} -4 \\ 2 \end{pmatrix}$$ i.e. B′ maps to B″ (−4, 2)

$$\begin{pmatrix} -1 & 0 \\ 0 & -1 \end{pmatrix}\begin{pmatrix} 0 \\ 0 \end{pmatrix} = \begin{pmatrix} 0 \\ 0 \end{pmatrix},$$ i.e. O is unaltered.

∴ The coordinates of A″ and B″ are (−2, 4) and (−4, 2).

Note

The **combined transformation** $\text{ML} = \begin{pmatrix} 2 & 0 \\ 0 & -2 \end{pmatrix}\begin{pmatrix} -1 & 0 \\ 0 & -1 \end{pmatrix}$

$$= \begin{pmatrix} -2 & 0 \\ 0 & 2 \end{pmatrix}$$

and $\overset{\text{ML}}{\begin{pmatrix} -2 & 0 \\ 0 & 2 \end{pmatrix}}\overset{\text{AB}}{\begin{pmatrix} 1 & 2 \\ 2 & 1 \end{pmatrix}} = \begin{pmatrix} -2 & -4 \\ 4 & 2 \end{pmatrix} = \text{A}''\text{B}''$

Therefore, a transformation by matrix L followed by matrix M can be represented by the combined transformation ML.

Exercise 75

1. Find the images A′B′ of the straight line AB, given that A is the point (1, 2) and B is the point (3, 1), under the following transformations. State the coordinates of A′ and of B′ in each case:
(a) reflection in the x-axis, (b) a rotation of 180° about the origin, (c) reflection in the line $y = x$, (d) enlargement, centre the origin, by scale factor 2, (e) reflection in the y-axis followed by a rotation of 180° about the origin.
2. A triangle OAB has vertices with coordinates O(0, 0); A(0, 3); B(2, 0). A point A′ has coordinates (3, 3). A shear maps A onto A′ leaving OB invariant. Draw sketches of the triangle OAB and its image OA′B under this shear. Calculate the area of triangle OA′B.
3. A rectangle OABC has vertices with coordinates O(0, 0); A(0, 3); B(2, 3); C(2, 0).

OC is the invariant line under a shear represented by the matrix $\begin{pmatrix} 1 & -2 \\ 0 & 1 \end{pmatrix}$. If this shear maps A onto A′ and B onto B′, find the coordinates of A′ and of B′.

Examination Standard Questions

Worked examples

1. In a group of 20 people, 8 are wearing hats, 10 are wearing overcoats and 6 are wearing neither hats nor overcoats. How many are wearing both hats and overcoats?

Let $\mathscr{E}$ = {the group of people}, then $n(\mathscr{E}) = 20$
H = {all people wearing hats}, then $n(\text{H}) = 8$
O = {all people wearing overcoats}, then $n(\text{O}) = 10$
Those wearing hats and overcoats are represented by $\{\text{H} \cup \text{O}\}$, therefore those wearing neither = $\{(\text{H} \cup \text{O})'\}$ and $n[(\text{H} \cup \text{O})'] = 6$.
$n[\text{H} \cup \text{O}] = n[\mathscr{E}] - n[(\text{L} \cup \text{S})'] = 20 - 6 = 14$
But $\quad n[\text{H}] + n[\text{O}] = 8 + 10 = 18$
Therefore $\underline{4}$ are wearing both hats and overcoats ANS

2. If $\mathbf{p} = \begin{pmatrix} 2 \\ 4 \end{pmatrix}$, $\mathbf{q} = \begin{pmatrix} 3 \\ 5 \end{pmatrix}$, $\mathbf{r} = \begin{pmatrix} 8 \\ 14 \end{pmatrix}$ find the value of n if $\mathbf{p} + n\mathbf{q} = \mathbf{r}$.

i.e. $\begin{pmatrix} 2 \\ 4 \end{pmatrix} + n\begin{pmatrix} 3 \\ 5 \end{pmatrix} = \begin{pmatrix} 8 \\ 14 \end{pmatrix}$

$\therefore \quad n\begin{pmatrix} 3 \\ 5 \end{pmatrix} = \begin{pmatrix} 8 \\ 14 \end{pmatrix} - \begin{pmatrix} 2 \\ 4 \end{pmatrix}$

$\therefore \quad n\begin{pmatrix} 3 \\ 5 \end{pmatrix} = \begin{pmatrix} 6 \\ 10 \end{pmatrix}$

$\therefore \quad n = \underline{2}$ ANS

3.

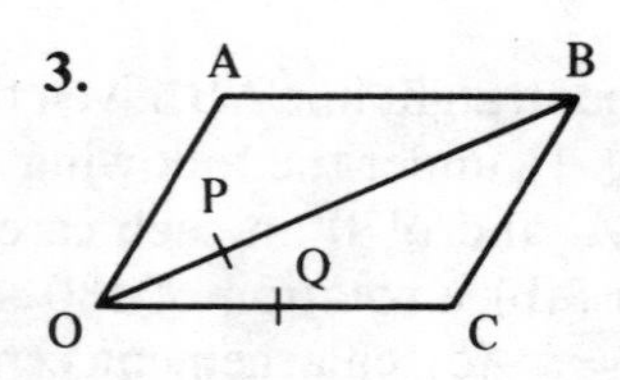

OABC is a parellelogram with $\overrightarrow{OA} = \mathbf{a}$ and $\overrightarrow{OC} = \mathbf{c}$. $\overrightarrow{OP} = \frac{1}{4}\overrightarrow{OB}$. $\overrightarrow{OQ} = \frac{1}{2}\overrightarrow{OC}$
(a) Express the vectors $\overrightarrow{OB}$ and $\overrightarrow{AC}$ in terms of **a** and **c**. (b) Show that $\overrightarrow{PQ} = \frac{1}{4}\overrightarrow{AC}$

(a)

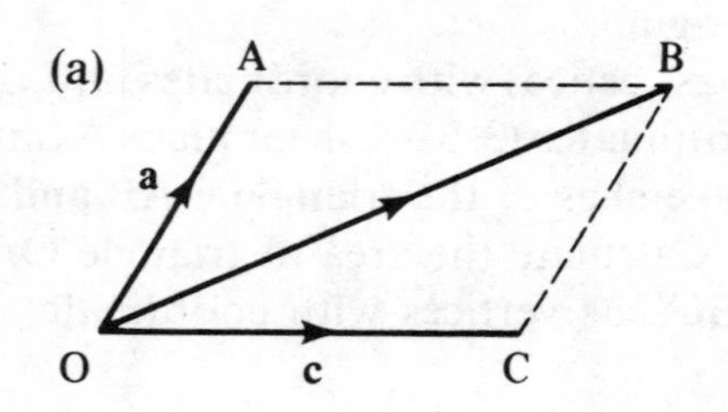

Using the parallelogram law:
$\overrightarrow{OA} + \overrightarrow{OC} = \overrightarrow{OB}$
i.e. $\underline{\mathbf{a} + \mathbf{c}} = \overrightarrow{OB}$ ANS

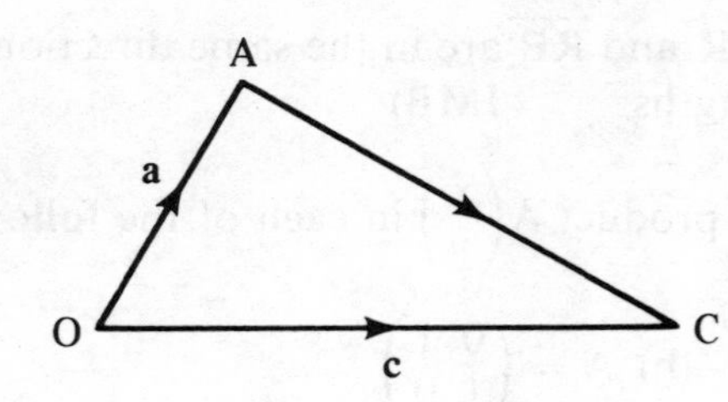

Using the triangle law:
$\overrightarrow{OA} + \overrightarrow{AC} = OC$
i.e. $\mathbf{a} + \overrightarrow{AC} = \mathbf{c}$
$\therefore \overrightarrow{AC} = \underline{\mathbf{c} - \mathbf{a}}$ ANS

(b)

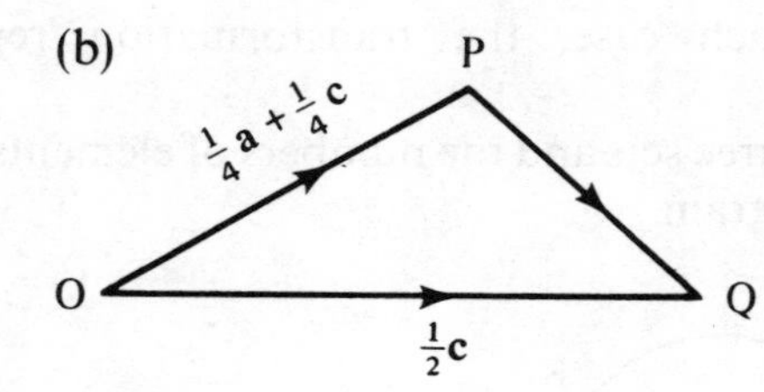

By the triangle law:
$\overrightarrow{OP} + \overrightarrow{PQ} = \overrightarrow{OQ}$
i.e. $\frac{1}{4}\mathbf{a} + \frac{1}{4}\mathbf{c} + \overrightarrow{PQ} = \frac{1}{2}\mathbf{c}$
$\overrightarrow{PQ} = \frac{1}{2}\mathbf{c} - \frac{1}{4}\mathbf{a} - \frac{1}{4}\mathbf{c}$
$= \frac{1}{4}\mathbf{c} - \frac{1}{4}\mathbf{a}.$

From part (a) $\overrightarrow{AC} = \mathbf{c} - \mathbf{a}$
$\therefore \underline{\overrightarrow{PQ} = \frac{1}{4}\overrightarrow{AC}}$

4. Find the value of $x + y$ where $\begin{pmatrix} x & 3 \\ -2 & y \end{pmatrix}\begin{pmatrix} 4 \\ 3 \end{pmatrix} = \begin{pmatrix} 13 \\ 1 \end{pmatrix}$

[Multiplying the matrices gives:] $\begin{pmatrix} 4x+9 \\ -8 \;+3y \end{pmatrix} = \begin{pmatrix} 13 \\ 1 \end{pmatrix}$

[Comparing corresponding components] i.e. $4x + 9 = 13$ and also $3y - 8 = 1$
$\therefore x = \underline{1}$ $\qquad \therefore y = \underline{3}$
$\therefore x + y = \underline{4}$ ANS

Exercise 76

1. In a class of 30 pupils, 12 have logarithm tables, 16 have slide rules and 8 have neither logarithm tables nor a slide rule. How many have both logarithm tables and a slide rule? (JMB)

2. If $\mathbf{a} = \begin{pmatrix} 1 \\ 3 \end{pmatrix}$, $\mathbf{b} = \begin{pmatrix} 2 \\ 5 \end{pmatrix}$, $\mathbf{c} = \begin{pmatrix} 7 \\ 18 \end{pmatrix}$ find the value of n if $\mathbf{a} + n\mathbf{b} = \mathbf{c}$. (AEB)

3.

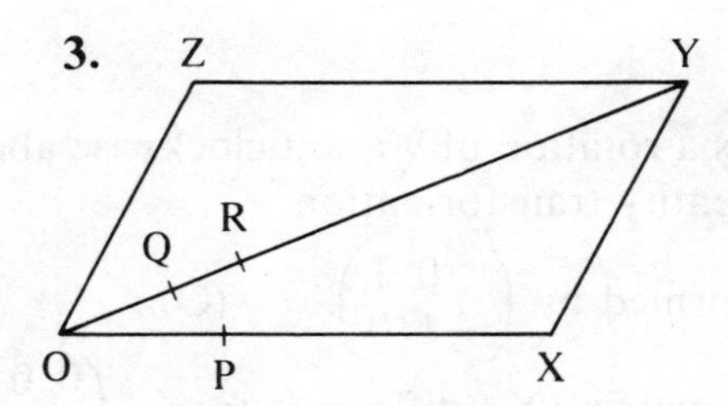

In the diagram, OXYZ is a parallelogram and $\overrightarrow{OX} = \mathbf{x}$, $\overrightarrow{OY} = \mathbf{y}$, $\overrightarrow{OP} = \frac{1}{3}\mathbf{x}$, $\overrightarrow{OQ} = \frac{1}{6}\mathbf{y}$, $\overrightarrow{OR} = \frac{1}{4}\mathbf{y}$.

(a) Express the vectors $\overrightarrow{QP}$, $\overrightarrow{OZ}$ and $\overrightarrow{ZR}$ in terms of $\mathbf{x}$ and $\mathbf{y}$.
(b) Show that $\overrightarrow{QP} = \frac{1}{6}\overrightarrow{ZX}$.

(c) Show that $\overrightarrow{ZR}$ and $\overrightarrow{RP}$ are in the same direction and find the ratio of their lengths. (JMB)

4. Find the matrix product $A\begin{pmatrix} x \\ y \end{pmatrix}$ in each of the following cases:

(a) $A = \begin{pmatrix} 3 & 0 \\ 0 & 3 \end{pmatrix}$, (b) $A = \begin{pmatrix} 0 & 1 \\ 1 & 0 \end{pmatrix}$

Describe, in each case, the transformation represented by A. (C)

5. A, B and C are three sets and the numbers of elements are as shown in the Venn diagram.

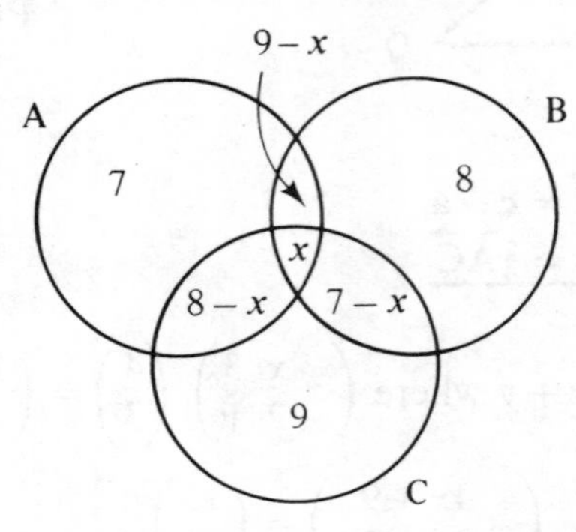

Given that $\mathscr{E} = A \cup B \cup C$ and that $n(\mathscr{E}) = 34$, find (a) the value of x (b) $n(A \cap B \cap C')$ (C)

6. Find the value of $|x+y| - |x| - |y|$ (a) when $x = 3$, $y = -3$, (b) when $x = 4$, $y = -5$. (JMB)

7. Given that O is the point (0, 0), A is (3, 0), B is (0, 4), C is (5, 5) and that A′, B′ and C′ are the images of A, B and C respectively under the enlargement with scale factor 3 and centre O, find (a) the length A′B′, (b) the ratio of the areas of the triangles ABC and A′B′C′. (JMB)

8. The transformation A transforms the vector $\begin{pmatrix} a \\ b \end{pmatrix}$ into $\begin{pmatrix} -b \\ a \end{pmatrix}$ for all numbers a and b.

Which of the following statements are true and which are false?

(a) The transformation A can be carried out by the matrix $\begin{pmatrix} 0 & -1 \\ 1 & 0 \end{pmatrix}$.

(b) A represents a rotation of 90° anticlockwise about the origin.

(c) A^2 is the identity transformation.

(d) A^{-1} is performed by $\begin{pmatrix} 0 & 1 \\ -1 & 0 \end{pmatrix}$. (O)

9. Find a 2×2 matrix X, different from $\begin{pmatrix} 0 & 0 \\ 0 & 0 \end{pmatrix}$, such that $X\begin{pmatrix} 0 & 1 \\ 1 & 0 \end{pmatrix} = X$. (O)

10. Let A(n) be the set of factors of the positive whole number n. Thus A(6) = {1, 2, 3, 6}.
(a) Write down the sets (i) A(30); (ii) A(45)
(b) Write down the sets (i) A(30) $\cap$ A(45); (ii) A(30) $\cup$ A(45).
(c) Write down the number r which has the property that A(30) $\cap$ A(45) = A(r).
(d) What is the smallest number s such that A(30) $\cup$ A(45) $\subset$ A(s)? (O)

11. Quadrilateral ABCD is such that, with origin O, A, C and D have coordinates (0, 2), (4, 2) and (-2, -2) respectively and $\overrightarrow{AB} = \begin{pmatrix} 3 \\ 2 \end{pmatrix}$.
Calculate (a) the coordinates of B,
(b) the vector $\overrightarrow{DC}$ in the form $\begin{pmatrix} x \\ y \end{pmatrix}$.
Write down a relation between vectors $\overrightarrow{AB}$ and $\overrightarrow{DC}$, and interpret this geometrically. (WJEC)

12. In an office where 43 people work, none use all three of bus, car and train; 5 use car and train; 2 use car and bus; 3 use bus and train; 19 use a car; 12 use a bus; and 8 use a train but neither bus nor car. By drawing a Venn diagram and by letting

A = {people who use a train}
B = {people who use a bus}
C = {people who use a car},

find (a) the number who use a bus and neither car nor train, (b) the number who use a train, (c) the number who use none of these methods of transport. (WJEC)

13.

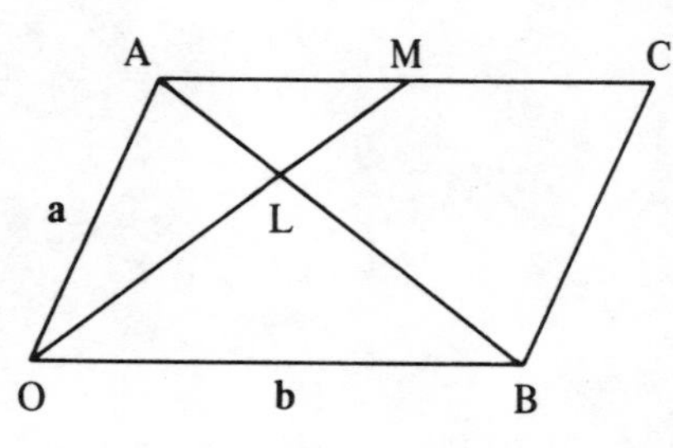

OACB is a parallelogram. M is the mid-point of the side AC, and OM intersects AB in L. Given that $\overrightarrow{OA} = \mathbf{a}$ and $\overrightarrow{OB} = \mathbf{b}$, express the vectors $\overrightarrow{AB}$ and $\overrightarrow{OM}$ in terms of **a** and **b**. Use your answers to find constants p and q such that $\overrightarrow{OL} = \overrightarrow{OA} + p\overrightarrow{AB}$ and $\overrightarrow{OL} = q\overrightarrow{OM}$. (C)

14. Matrices, A and B, are defined by $A = \begin{pmatrix} 4 & 2 \\ 3 & 2 \end{pmatrix}$, $B = \begin{pmatrix} 1 & 0 \\ 1 & 2 \end{pmatrix}$.
(a) Express as single matrices (i) A + B, (ii) A − B, (iii) A^2 (iv) B^2. Hence, determine whether or not, in this case, $A^2 - B^2 = (A + B)(A - B)$.
(b) Express as single matrices (i) A^{-1}, the inverse of A, (ii) $(A^2)^{-1}$, the inverse of A^2. Hence determine whether or not, in this case, $(A^{-1})^2 = (A^2)^{-1}$. (O)

15. The sets A and B are defined as follows:

$A = \{x: x^2 + px + 3q = 0\}$,

$B = \{x: x^2 + 2px + 4q = 0\}$.

(a) If $p = -1$ and $q = -2$, find the members of set A and of set B. Name the element of $A \cap B$ and list the three elements of $A \cup B$ in this case.

(b) If, instead, $A \cap B = \{2\}$, write down two equations to find p and q, and solve them for p and q. (L)

10
Angles & Symmetry

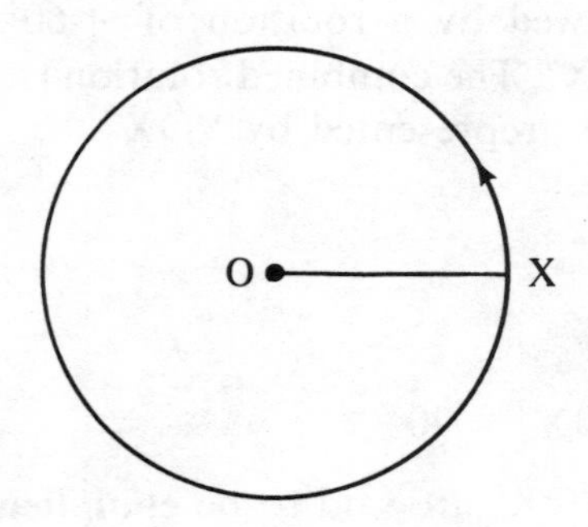

Consider the line OX, if we rotate this line about the fixed point O the path described by X is a circle. The Babylonians were the first people to divide such a turn into 360 parts, called **degrees**. Therefore, during a complete revolution the line OX turns through an angle of 360°.

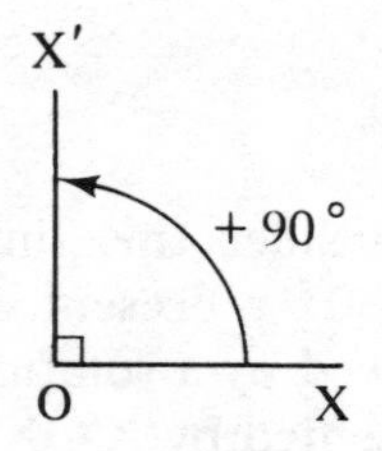

Let OX′ be the position of OX after a quarter turn anti-clockwise; that is, a rotation of $+90°$. (A turn anti-clockwise is taken as positive, clockwise as negative.) The angle described by this rotation is known as a **right angle**. (The sign ⊾ is used to denote a right angle.)

The lines OX′ and OX are said to be **perpendicular** to each other. Two symbols usually used to denote angles of any size are $\angle$ and $\hat{}$.

i.e. $\angle X'OX = 90°$ **or** $X'\hat{O}X = 90°$

If $X'\hat{O}X < 90°$, the angle is said to be an **acute angle**,
if $90° < X'\hat{O}X < 180°$, the angle is said to be an **obtuse angle**,
and if $180° < X'\hat{O}X < 360°$, the angle is called a **reflex angle**.

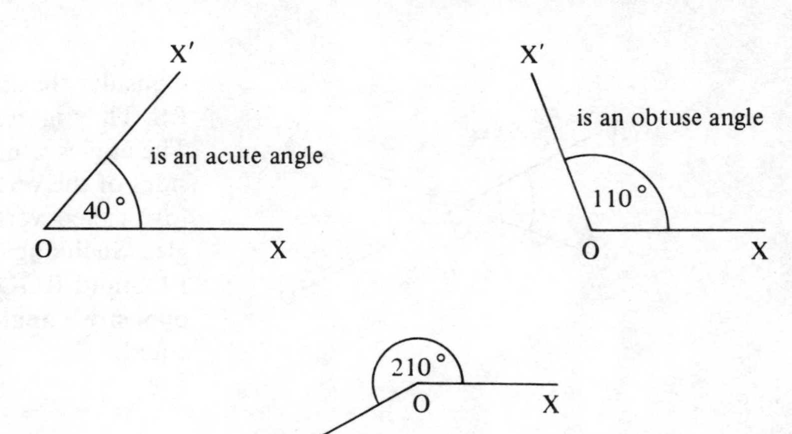

Note The point where the lines OX′ and OX meet is called the **vertex**.

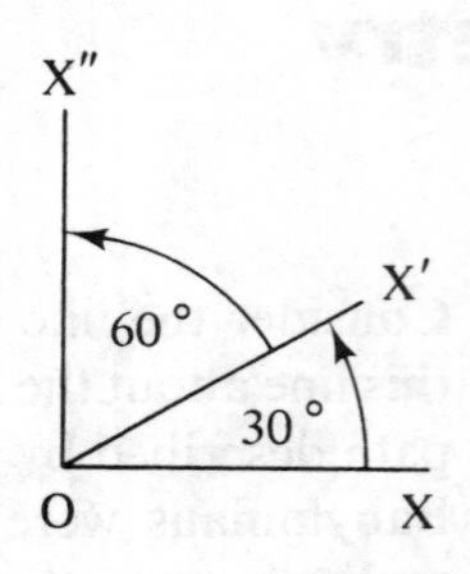

Consider a rotation of +30° represented by $X\hat{O}X'$, followed by a rotation of +60° represented by $X'\hat{O}X''$. The combined rotation is a quarter turn of +90°, represented by $X\hat{O}X''$.

i.e. $X\hat{O}X' + X'\hat{O}X'' = 90°$,

the angles $X\hat{O}X'$ and $X'\hat{O}X''$ are said to be **complementary**.

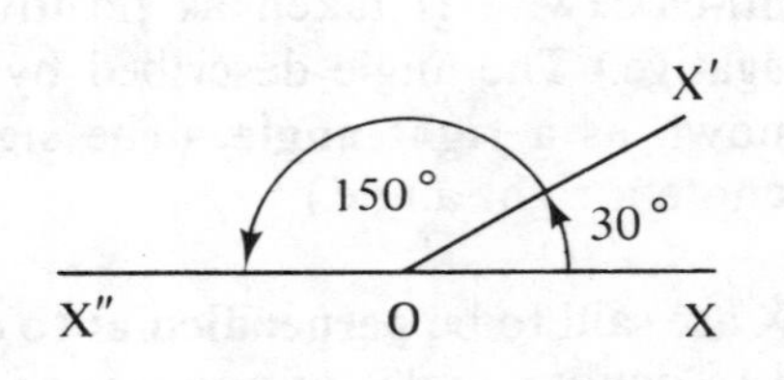

Consider this time a rotation of +30° represented by $X\hat{O}X'$ followed by a rotation of +150° represented by $X'\hat{O}X''$. The combined rotation is a half turn of +180° represented by $X\hat{O}X''$.

i.e. $X\hat{O}X' + X'\hat{O}X'' = 180°$,

the angles $X\hat{O}X'$ and $X'\hat{O}X''$ are in this case said to be **supplementary.**

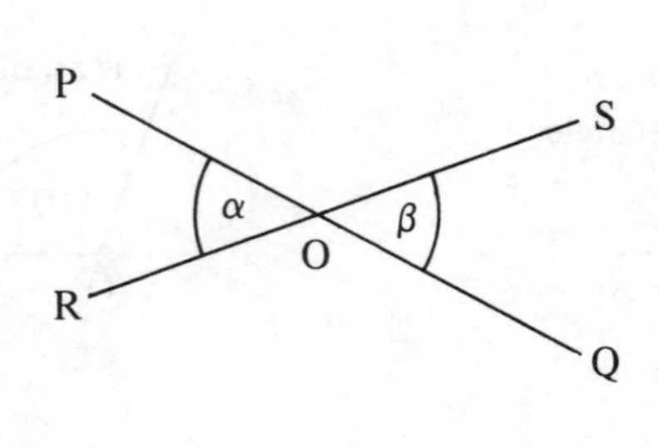

Consider the straight lines PQ and RS. They intersect at O. The angles α and β are on opposite sides of the vertex O and are therefore called vertically opposite angles. Such angles are equal. Angles $P\hat{O}S$ and $R\hat{O}Q$ are also **vertically opposite angles** and therefore equal.

Exercise 77 **1.** Select one of the following titles to describe each angle in turn; right angle, acute angle, obtuse angle, reflex angle.

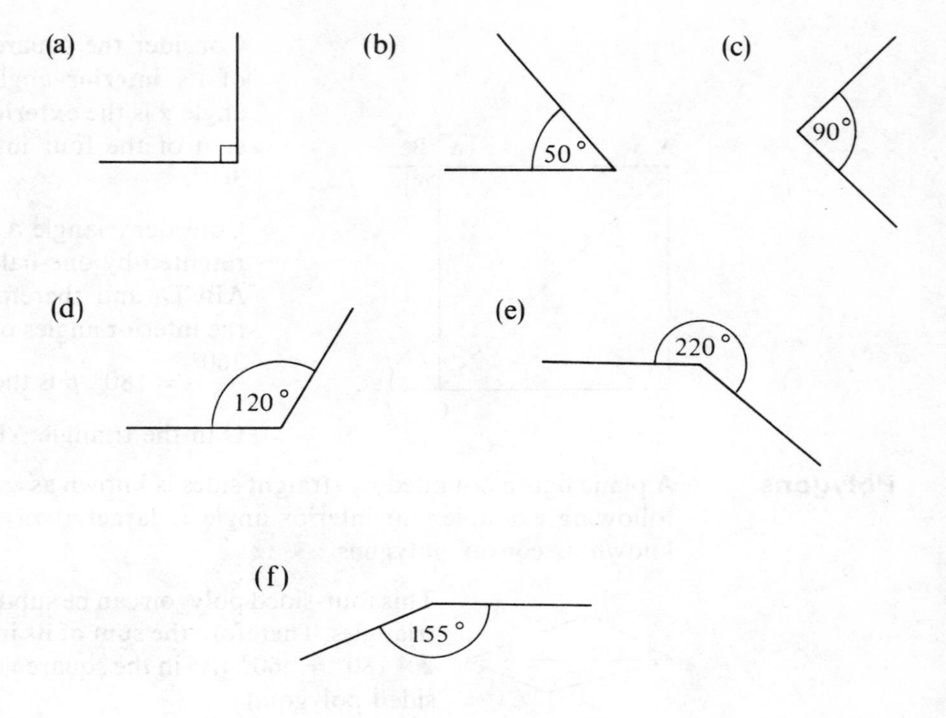

2. Find (i) the complement, and (ii) the supplement of each of the following angles:

(a) 40° (b) 75° (c) 10° (d) 89° (e) $44\frac{1}{2}°$ (f) $09\frac{1}{2}°$

3.

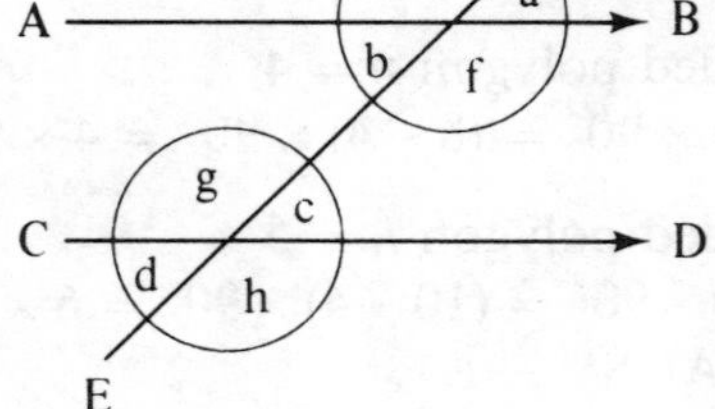

Draw this diagram carefully, using a protractor to measure the angles. The lines AB and CD can be drawn 4 cm apart, the arrows indicating that they are parallel. The line EF is drawn at an angle of 40° to the lines AB and CD. Label the angles as shown, and use a protractor to check your answers to the following.

(a) Name four pairs of vertically opposite angles.
(b) Name eight pairs of supplementary angles.
(c) (i) Does $\hat{a} = \hat{c}$? (ii) Does $\hat{e} = \hat{g}$? (iii) Does $\hat{f} = \hat{h}$? (iv) Does $\hat{b} = \hat{d}$?

Each of these is known as a pair of **corresponding angles**.
(d) (i) Does $\hat{b} = \hat{c}$? (ii) Does $\hat{f} = \hat{g}$?

Each of these is known as a pair of **alternate angles**.

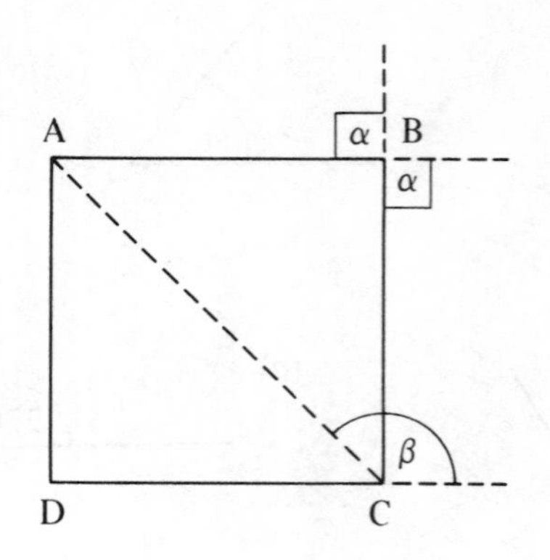

Consider the square ABCD. Each of its **interior** angles is 90°. The angle α is the **exterior** angle B. The sum of the four interior angles is 360°.

Consider triangle ADC; this is represented by one-half of the square ABCD, and therefore the sum of the interior angles of the triangle is $\frac{360°}{2} = 180°$. β is the exterior angle C in the triangle ADC.

Polygons

A plane figure bounded by straight sides is known as a **polygon**. In the following examples no interior angle is larger than 180°, these are known as **convex polygons**.

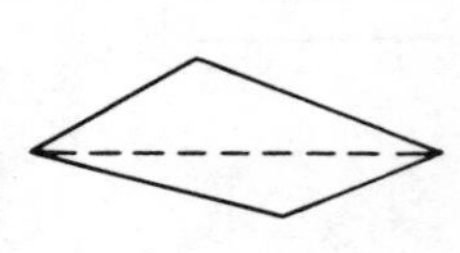

This four-sided polygon can be subdivided into two triangles. Therefore the sum of its interior angles is $2 \times 180° = 360°$. (As in the square which is also a 4-sided polygon.)

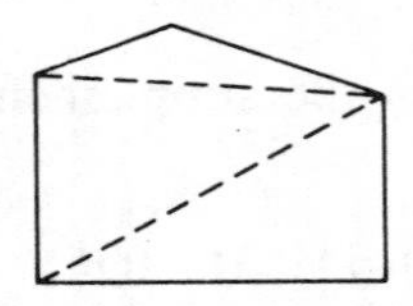

This five-sided polygon can be subdivided into three triangles. Therefore the sum of its interior angles is $3 \times 180° = 540°$.

A formula for calculating the sum of the interior angles of a convex polygon is

$$(2n - 4) \times 90°$$

where n is the number of sides.

e.g. For a 4-sided polygon $n = 4$
$\therefore (2n - 4) \times 90° = (8 - 4) \times 90° = 4 \times 90° = \underline{360°}$ ANS

For a 5-sided polygon $n = 5$
$\therefore (2n - 4) \times 90° = (10 - 4) \times 90° = 6 \times 90°$
$= \underline{540°}$ ANS

A polygon with all sides equal in length is known as a **regular polygon** In such a polygon each interior angle is the same, and can be found by dividing the sum of the angles by n.

e.g. For a 5-sided regular polygon
sum of angles = 540°

$$\therefore \text{size of each angle} = \frac{540°}{n} = \frac{540°}{5} = \underline{108°} \quad \text{ANS}$$

Note The sum of the **exterior** angles of a convex polygon is 360°.

Example

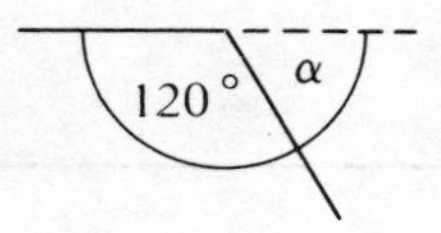

The diagram shows part of a regular polygon whose interior angles are each 120°. Find the number of sides of the polygon.

Let α represent an exterior angle.

$$\therefore \alpha = 180^\circ - 120^\circ = 60^\circ$$

The sum of the exterior angles is 360°,

$$\therefore n = \frac{360}{\alpha} = \frac{360}{60} = \underline{6 \text{ sides}} \quad \text{ANS}$$

Exercise 78

1. Find the sum of the interior angles of the following polygons:
(a) A 6-sided polygon, (b) An 8-sided polygon, (c) A 20-sided polygon, (d) A 10-sided polygon.
2. Find the size of each of the interior angles of the following regular polygons:
(a) A 12-sided polygon, (b) A 15-sided polygon, (c) A 9-sided polygon, (d) A 7-sided polygon.
3. Find the number of sides of the following regular polygons given that each interior angle is:
(a) 144° (b) 170° (c) 135° (d) 162°.
4. Find the size of the lettered angle in each of the following triangles:

(a)

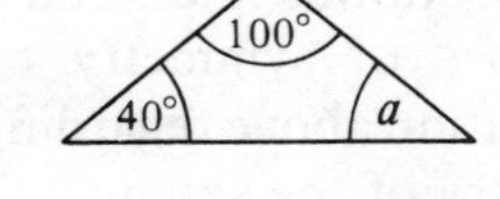

(b)

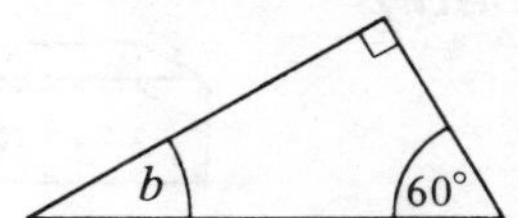

(c)

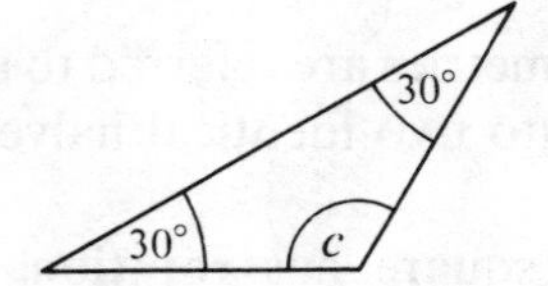

(d)

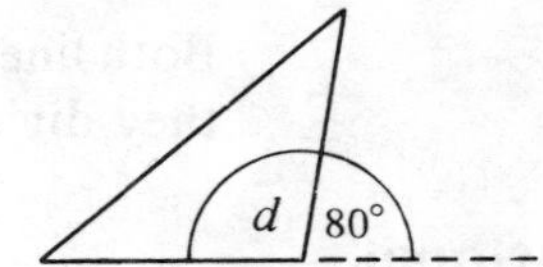

(e)

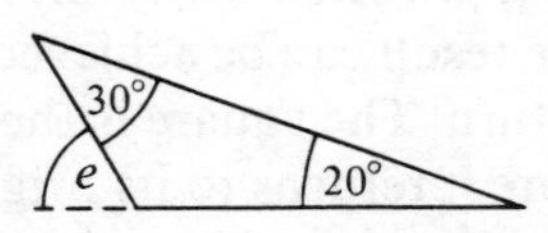

(f)

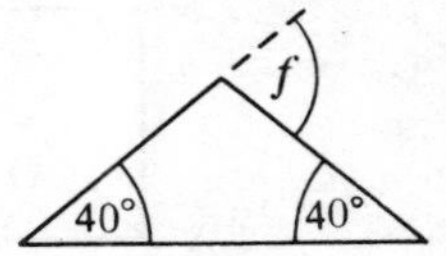

(g)

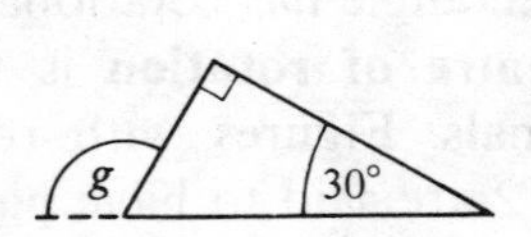

(h)

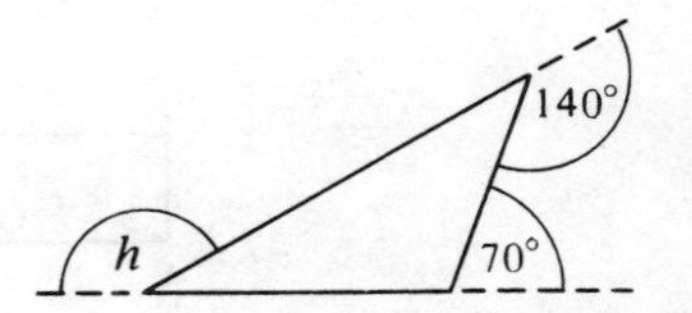

Note

In parts (e), (f), and (g) of question **4** above you should find that the **exterior angle of a triangle is equal to the sum of the two interior opposite angles**. This is always so.

Symmetry

Line symmetry

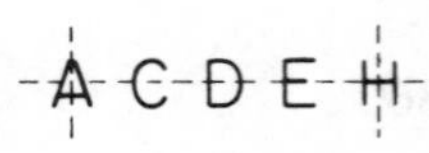

The block letters shown here all have symmetry about the dotted line, called the **axis of symmetry**. The letter H has two axes of symmetry.

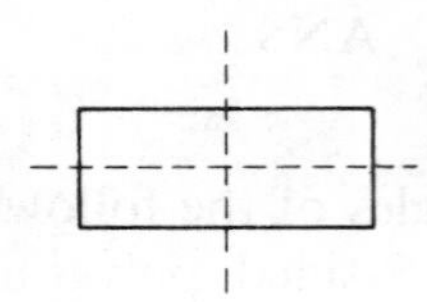

A shape has **line symmetry** about an axis if 'folding' the figure along the axis will fit one half exactly onto the other.

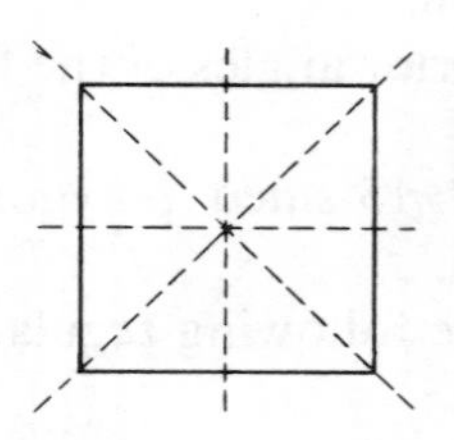

The rectangle shown has two axes of symmetry, the square has four.

Plane symmetry

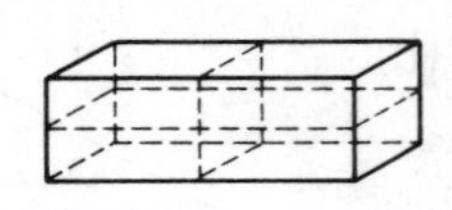

The cuboid shown here can be divided into two identical halves by either of the **planes of symmetry** shown by dotted lines. Can you deduce the number of planes of symmetry of the cube using the information above regarding the number of axes of symmetry of the square?

Both line and plane symmetries are referred to as **bilateral symmetries** they divide the figure into two identical halves.

Rotational symmetry

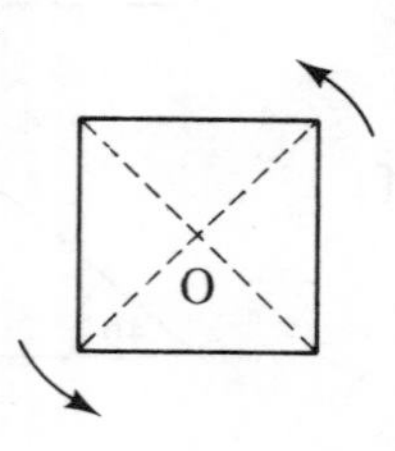

The square has **rotational symmetry** about the point O. The square can be given a $\frac{1}{4}$ turn about O and it will fit exactly into its original shape. The same result can be achieved by a $\frac{1}{2}$ turn, a $\frac{3}{4}$ turn or a full turn. The square is therefore turned four times before it returns to its original position, and is said to have **order** of rotational symmetry 4.

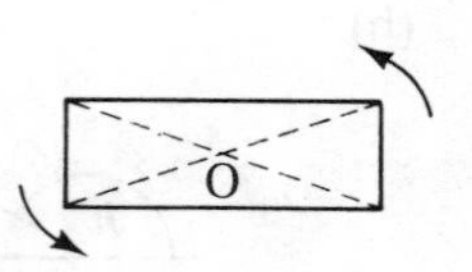

The rectangle has rotational symmetry of order 2. Its **centre of rotation** is the intersection of its diagonals. Figures with rotational symmetry of order 2 are said to have **point symmetry**.

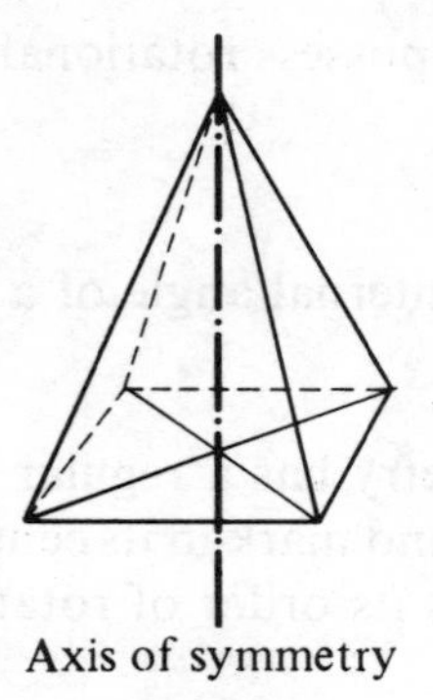
Axis of symmetry

The square based pyramid has an axis of symmetry as shown, and like the square has rotational symmetry of order 4. Can you deduce the order of rotational symmetry of a cuboid, using the information above regarding the rectangle? The cuboid has three axes of symmetry, draw a diagram and mark these in.

Exercise 79

1. 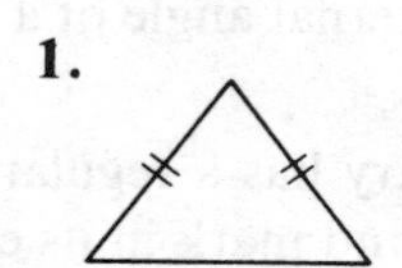

Draw carefully an isosceles triangle and mark in with dotted lines any lines of symmetry it possesses. How many such lines are there?

2. Does the isosceles triangle in question **1** have point symmetry?

3. 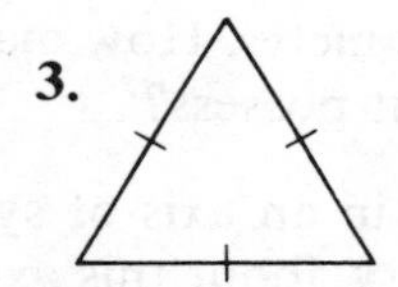

Draw carefully an equilateral triangle and mark in with dotted lines any lines of symmetry it possesses. How many such lines are there?

4. Does the equilateral triangle in question **3** have rotational symmetry? If so, of what order?

5. 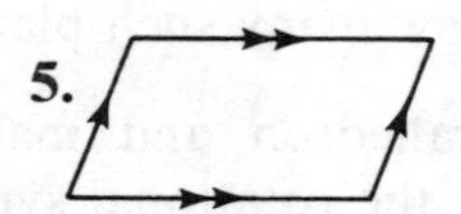

Draw carefully a parallelogram and mark in any axes of symmetry.

6. (a) Does the parallelogram possess rotational symmetry?
(b) Does the parallelogram possess point symmetry?

7. 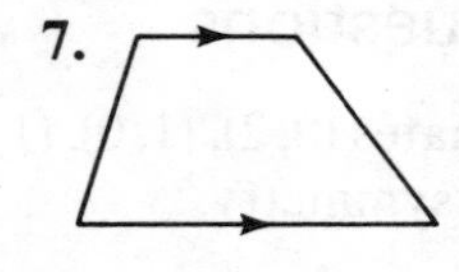

(a) How many axes of symmetry has this trapezium?
(b) Does this trapezium possess rotational symmetry?

8. S What type of symmetry does this letter possess?

9. X Copy this letter of the alphabet and mark in with dotted lines any lines of symmetry.

10. Does the letter in question **9** possess rotational symmetry? If so, of what order?

11. N How many lines of symmetry does this letter possess?

12. Does the letter in question **11** possess rotational symmetry? If so, of what order?

13. 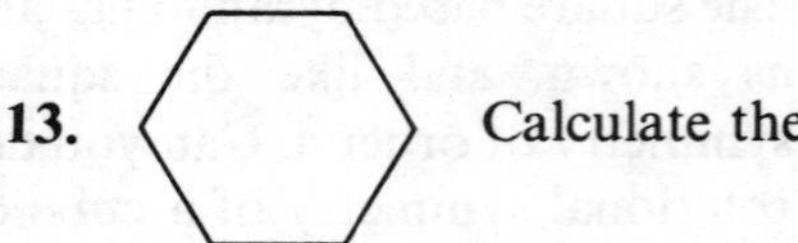 Calculate the internal angle of a regular hexagon.

14. (a) How many axes of symmetry has a regular hexagon?
(b) Sketch a regular hexagon and mark in its centre. Taking this as the centre of rotation, what is its order of rotational symmetry?

15. 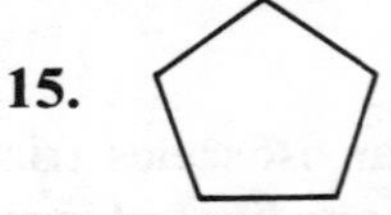 Calculate the internal angle of a regular pentagon.

16. (a) How many axes of symmetry has a regular pentagon?
(b) Sketch a regular pentagon and mark in its centre. Taking this as the centre of rotation, what is its order of rotational symmetry?

17. 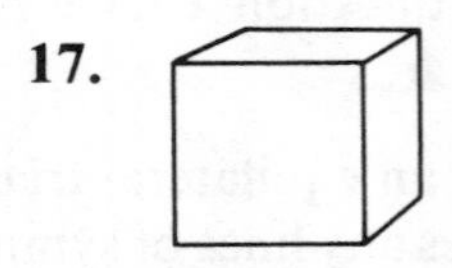 Draw carefully a cube and mark in with dotted lines its planes of symmetry. How many such planes of symmetry does it possess?

18. Draw another cube and mark in an axis of symmetry. Of what order is the rotational symmetry about this axis?

19. Draw carefully a regular tetrahedron, (its four faces are equilateral triangles) and mark in its planes of symmetry. How many such planes are there?

20. Draw another regular tetrahedron and mark in an axis of symmetry. Of what order is the rotational symmetry about this axis?

Examination Standard Questions

Worked examples

1. A kite has vertices with coordinates (0, 2), (1, 0), (1, 3) and (2, 2). Find the equation of the kite's line of symmetry.

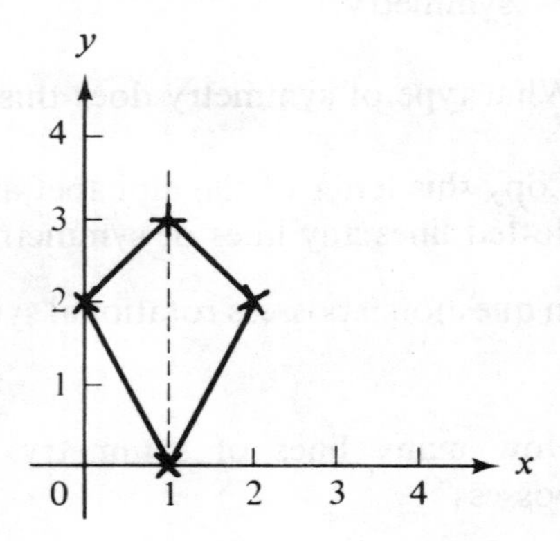

The line of symmetry is shown by the dotted line.
The axis of symmetry is parallel to the y-axis and intersects the x-axis at the point $x = 1$.
The equation of the line of symmetry is therefore

$\underline{x = 1}$ ANS

2.

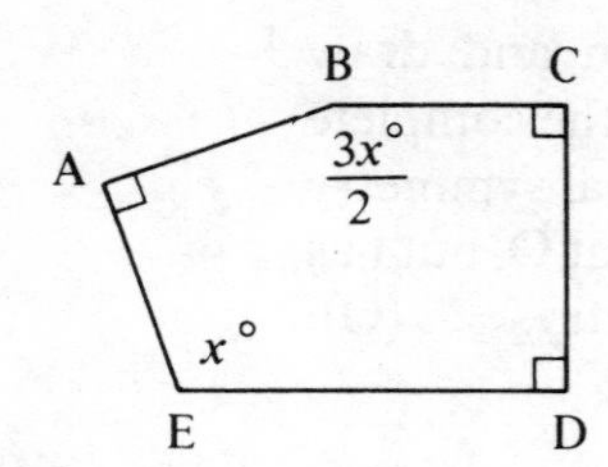

In the pentagon ABCDE the angles FAB, BCD and CDE are 90°.

Angle ABC $= \dfrac{3x°}{2}$ and angle AED $= x°$.
Calculate the value of x.

Sum of the angles $= (2n - 4) \times 90°$, where n is the number of sides.

i.e. $(10 - 4) \times 90° = 6 \times 90°$

We are told that three of the angles are 90°,

$\therefore$ the remaining two angles $= (6 \times 90°) - (3 \times 90°)$
$= 3 \times 90°$

$$\therefore \frac{3x}{2} + x = 270°$$

[Multiply through by 2] $\therefore 5x = 540°$

$\therefore x = 108°$ ANS

Exercise 80

1. A trapezium has vertices with coordinates (0, 0), (1, 2), (4, 2) and (5, 0). Which one of the following equations represents the line of symmetry for the trapezium?
A: $x = 2\frac{1}{2}$ B: $x = 2y$ C: $y = 0$ D: $y = 1$ E: $y = 2\frac{1}{2}$. (L)

2. The straight lines AB, CD and EF are parallel. Given that angle EKG = 47° and that the straight line LHG is at right angles to GK, calculate angle LHB (marked as $x°$). (AEB)

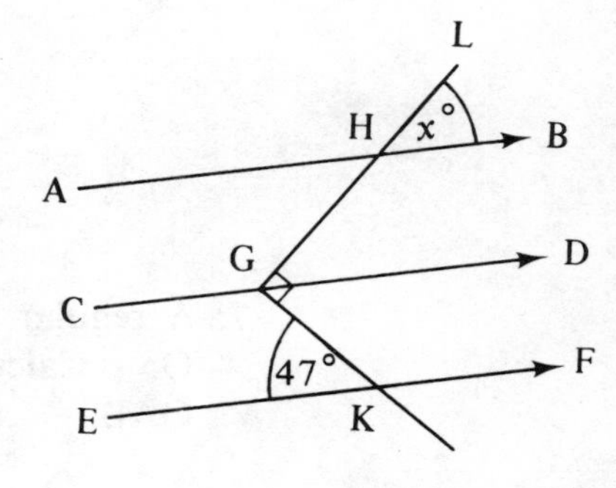

3. In the diagram the lines l and l' are parallel. Calculate the value of x. (O)

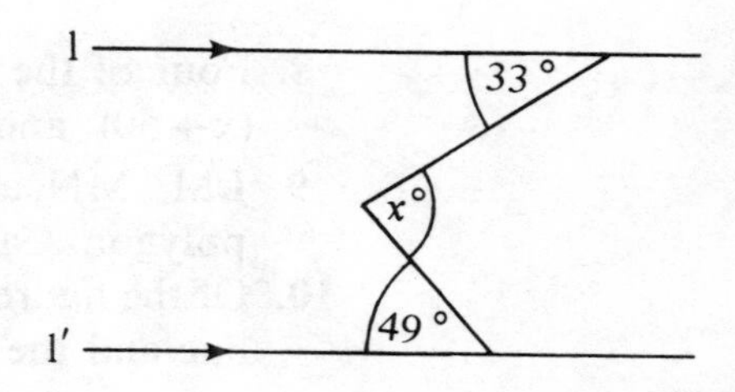

4. Copy the diagram and draw extra lines so that the complete figure has rotational symmetry of order three about O, but has no axes of symmetry. (O)

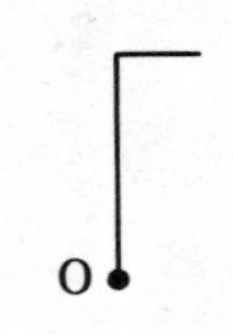

5. The diagram shows one quarter of a figure which is symmetrical about the lines X_1X_2 and Y_1Y_2. Complete the figure. (JMB)

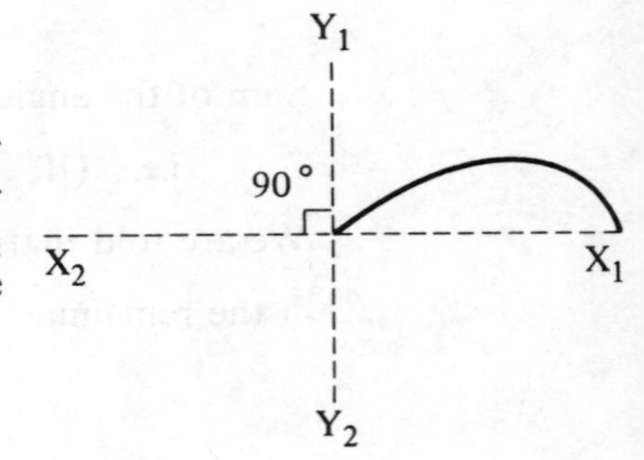

6. The diagram shows a circle with centre C and two parallel chords which are equal in length. Draw the lines of symmetry of the figure. (JMB)

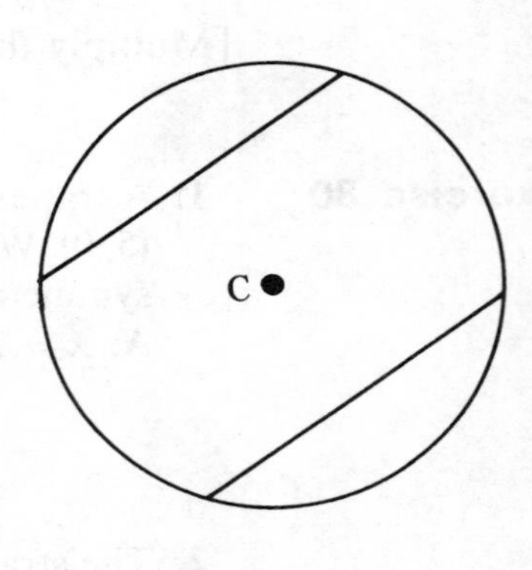

7. A regular octagon has centre O. Calculate the angle OAB. (C)

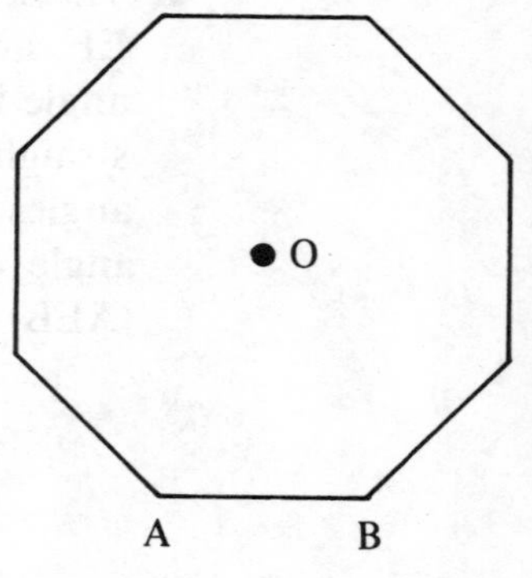

8. Four of the exterior angles of a hexagon are equal to $x°$, one is $(x + 50)°$ and the other is 60°. Calculate the value of x. (JMB)

9. LM, MN and NP are adjacent sides of a regular nine-sided polygon. Calculate $L\hat{M}N$ and $L\hat{N}P$. (C)

10. Of the figures shown on the next page, one has symmetry about a line and the other has symmetry about a point.

Copy the figure which has point symmetry and mark the centre of symmetry with the letter P. Copy the figure which has line symmetry and draw a line of symmetry and label it as 1. (AEB)

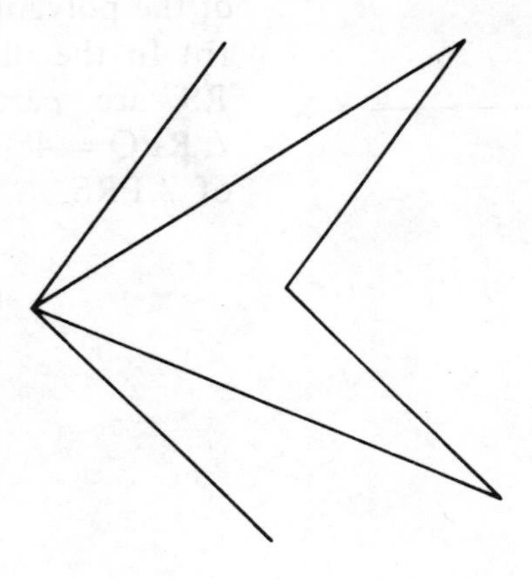

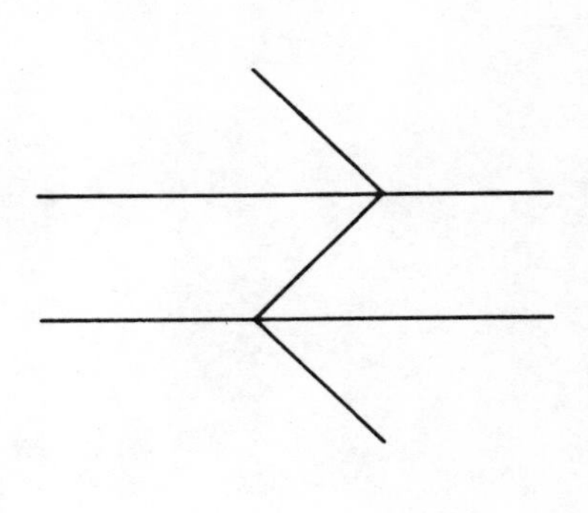

11.

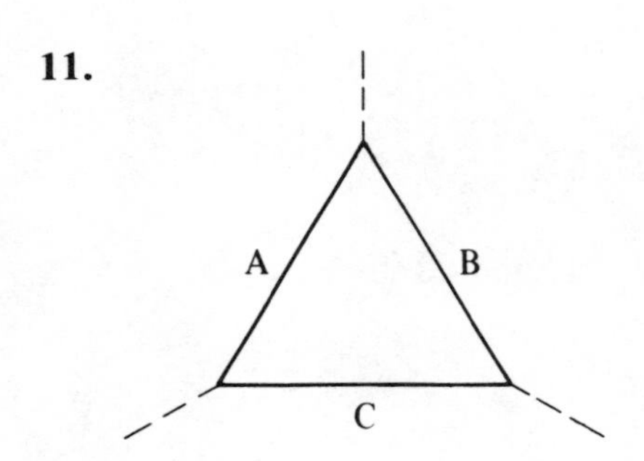

Three identical regular polygons A, B and C fit on the sides of an equilateral triangle as shown. How many sides has each polygon? (AEB)

12.

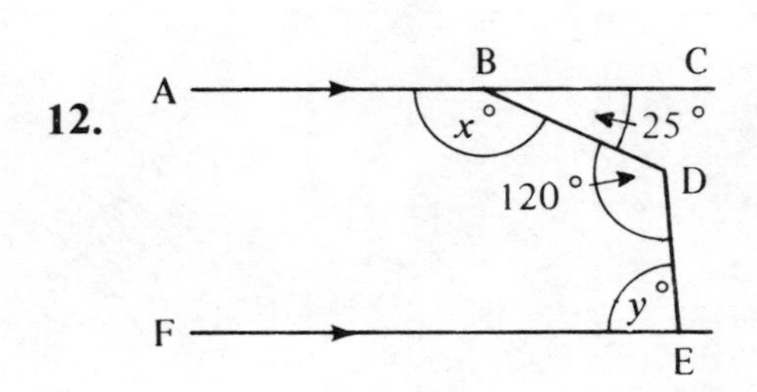

The straight lines ABC and FE are parallel, angle CBD = 25° and angle BDE = 120°. Find the values of (a) $\hat{x}$, (b) $\hat{y}$. (AEB)

13.

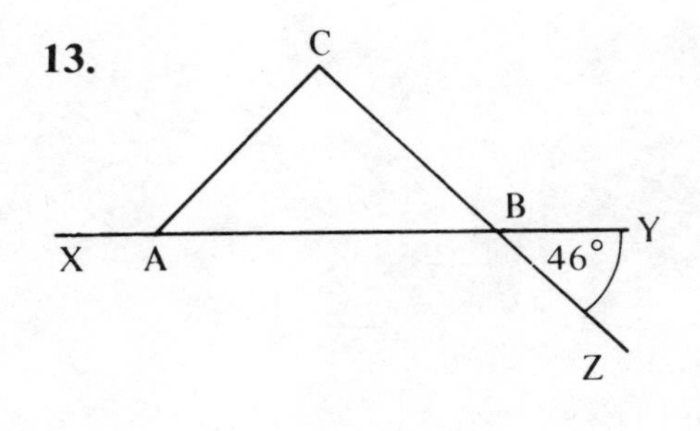

(a) Calculate, in degrees, the interior angle of a regular 12-sided polygon.

(b) In the diagram, A and B are points on the line XY and CB is produced to Z. If $\angle CAX = 137°$ and $\angle YBZ = 46°$, calculate $\angle ACB$. (L)

14.

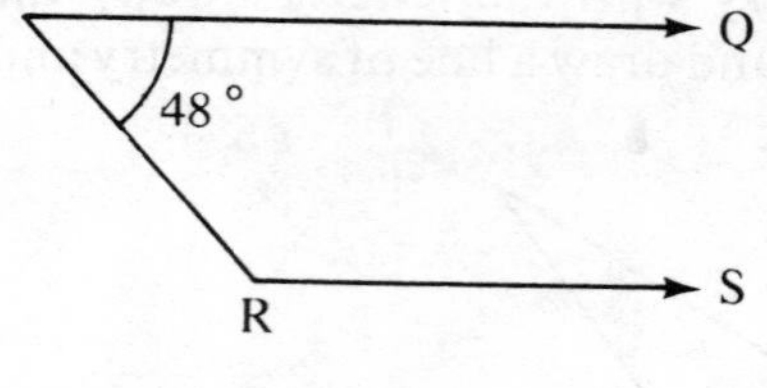

(a) Each interior angle of a regular polygon is 160°. Calculate the number of sides of the polygon.

(b) In the diagram, PQ and RS are parallel lines with $\angle RPQ = 48°$. State the value of $\angle PRS$. (L)

11
Plane Figures

Triangles

In Chapter 10 the names of various types of angle were noted. Such angles can be used to describe triangles.
e.g.

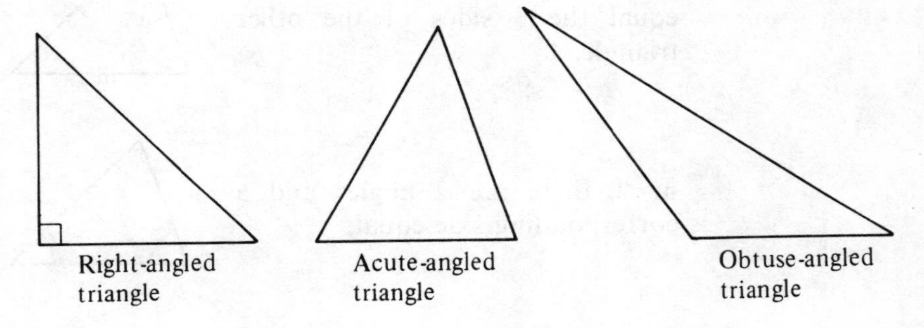

The right-angled triangle has one angle of 90°, the other two being less than 90° and therefore acute.

The acute-angled triangle has all three of its angles less than 90°.

The obtuse-angled triangle has one angle obtuse, the other two are acute.

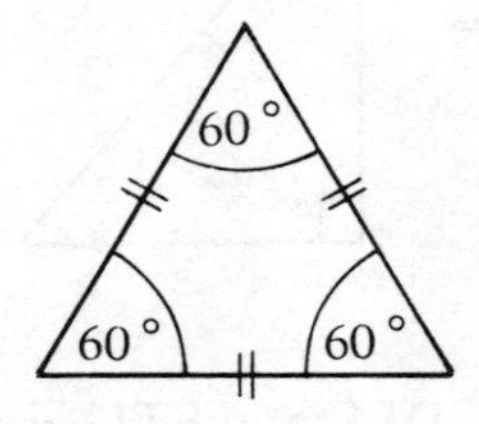

An acute-angled triangle with its three sides equal in length (shown by //) is known as an **equilateral triangle**. Its three angles are also equal in size.

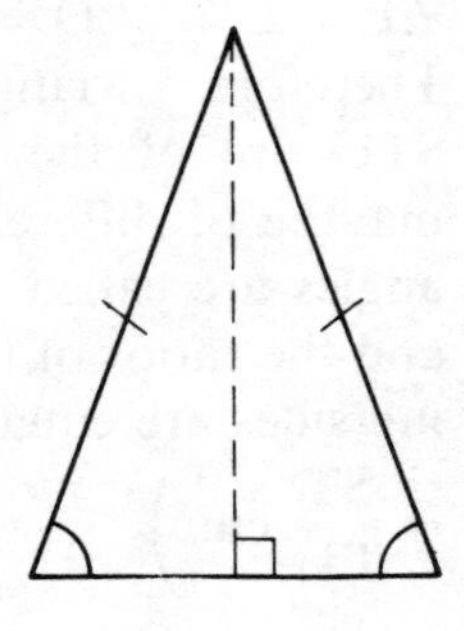

A triangle with two sides equal in length (shown by /) is known as an **isosceles triangle. The angles** opposite the equal sides are also equal (shown by an arc). The triangle is symmetrical about its perpendicular height.

Congruent triangles

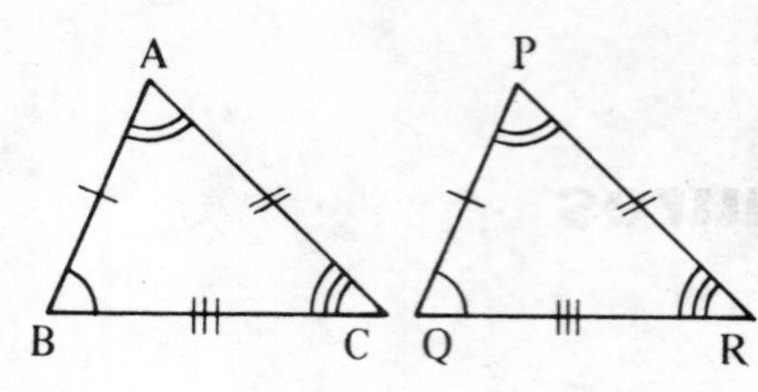

ABC and PQR are two triangles as shown. AB = PQ, AC = PR, BC = QR and ∠B = ∠Q, ∠A = ∠P, ∠C = ∠R. Therefore, triangles ABC and PQR are of the *same shape* and *size.* Such triangles are called **congruent triangles.**

To prove that two triangles are congruent you must know that either:

1. the 3 sides of the one triangle equal the 3 sides of the other triangle,

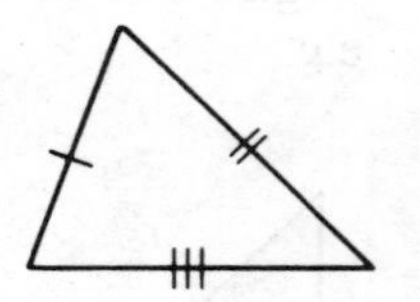

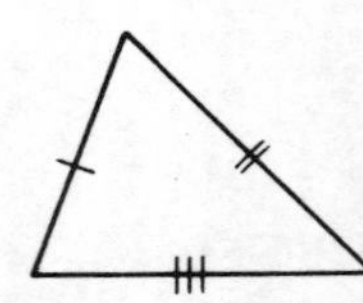

or 2. there are 2 angles and a corresponding side equal,

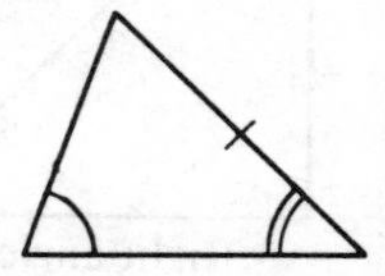

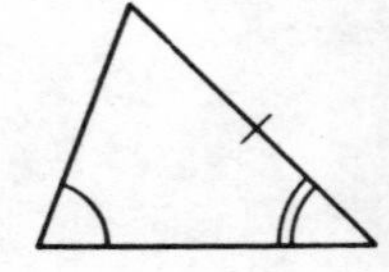

or 3. there are 2 sides and the angle included between them equal,

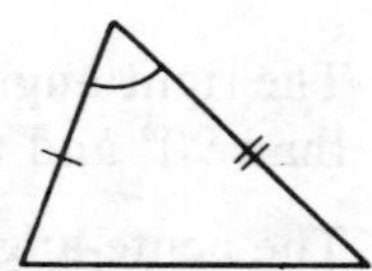

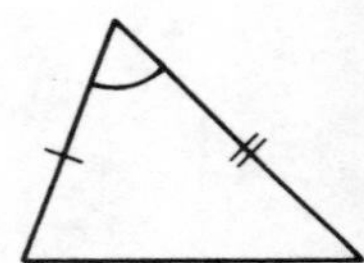

or 4. in right-angled triangles the hypotenuse and another side are equal.

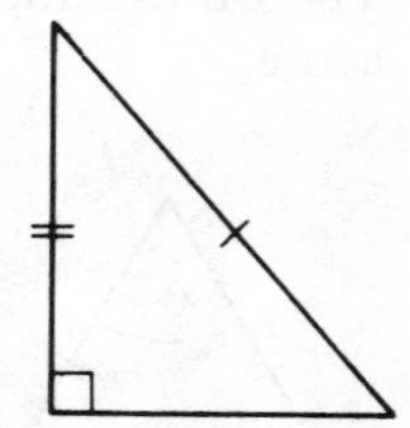

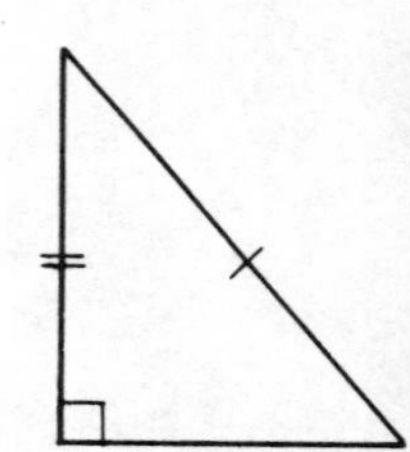

Similar triangles

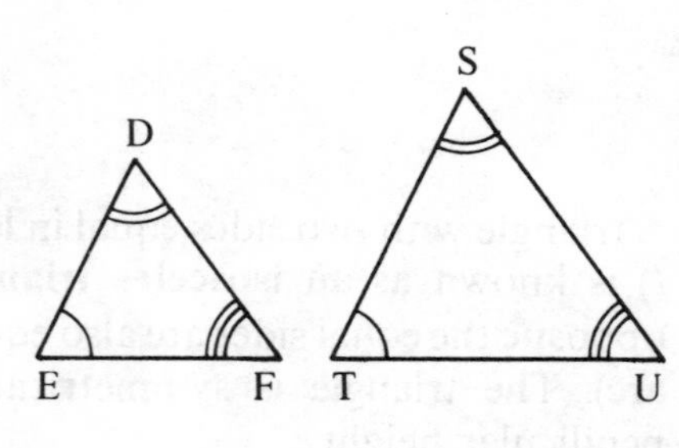

DEF and STU are two triangles as shown.

∠E = ∠T, ∠D = ∠S, ∠F = ∠U. Therefore, triangles DEF and STU are of the same shape but may be of different size. Such triangles are called similar triangles, and the ratios of their corresponding sides are equal.

$$\left(\text{i.e. } \frac{DE}{ST} = \frac{DF}{SU} = \frac{EF}{TU} \quad \text{also} \quad \frac{DE}{EF} = \frac{ST}{TU} \text{ etc.}\right)$$

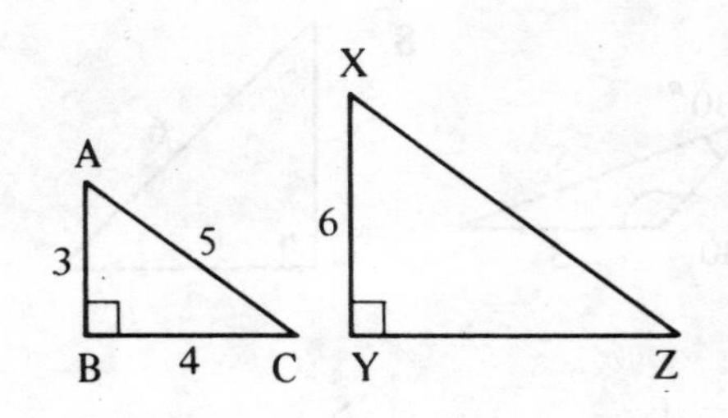

Triangles ABC and XYZ shown here are similar.

$$\frac{AB}{XY} = \frac{3}{6} = \frac{1}{2},$$

therefore the other pairs of corresponding sides are in the same ratio.

i.e. $\frac{BC}{YZ} = \frac{4}{YZ} = \frac{1}{2}$, therefore YZ = 8

$\frac{AC}{XZ} = \frac{5}{XZ} = \frac{1}{2}$, therefore XZ = 10.

To prove that two triangles are similar you must know that either:

1. the 3 angles of the one triangle equal the 3 angles of the other triangle,

or 2. the corresponding sides are proportional,

or 3. there are 2 corresponding sides which are proportional, with the angle between them equal.

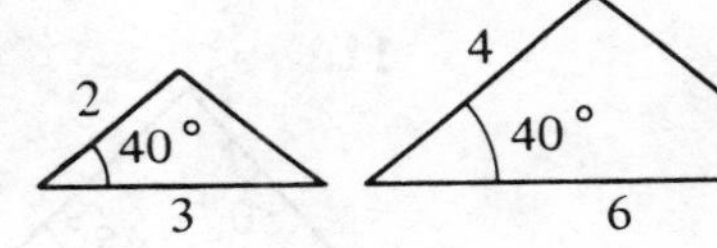

Exercise 81 State in questions **1** to **9** if the pairs of triangles are congruent, giving reasons for your answer.
Note the triangles are not drawn to scale.

1.

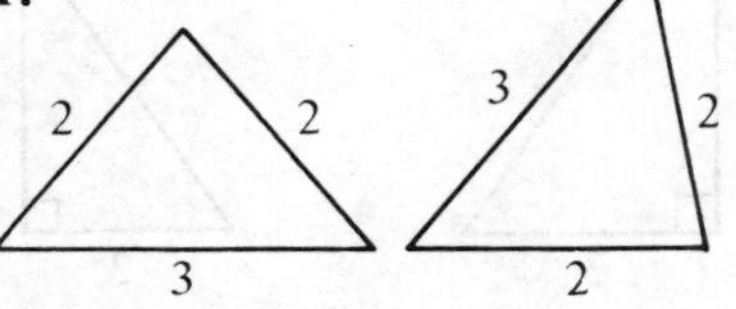

2.

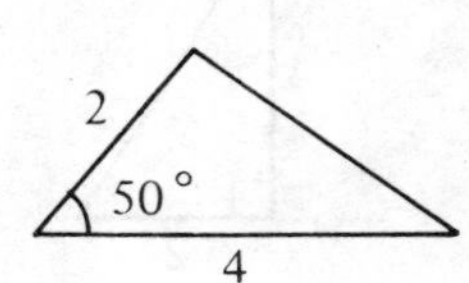

3.

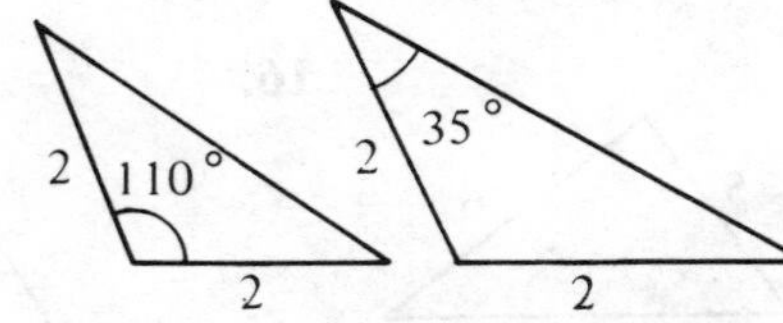

4.

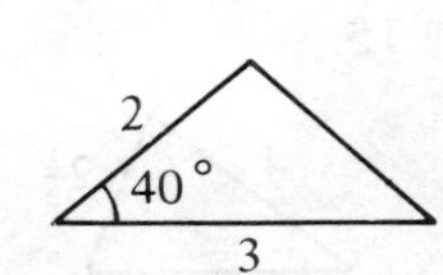

5.

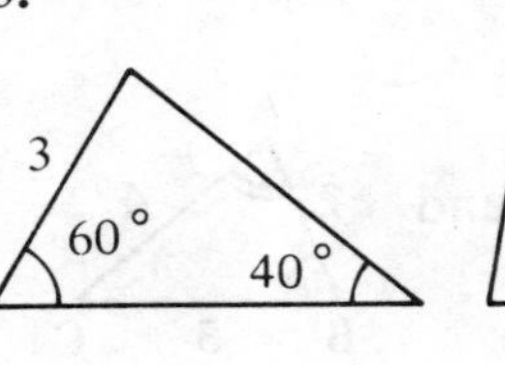

6.

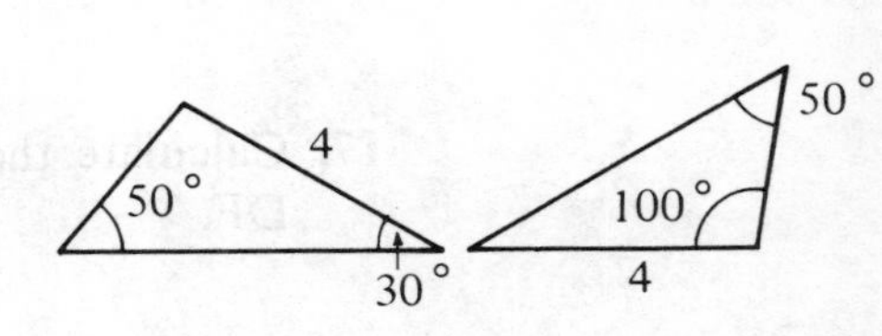

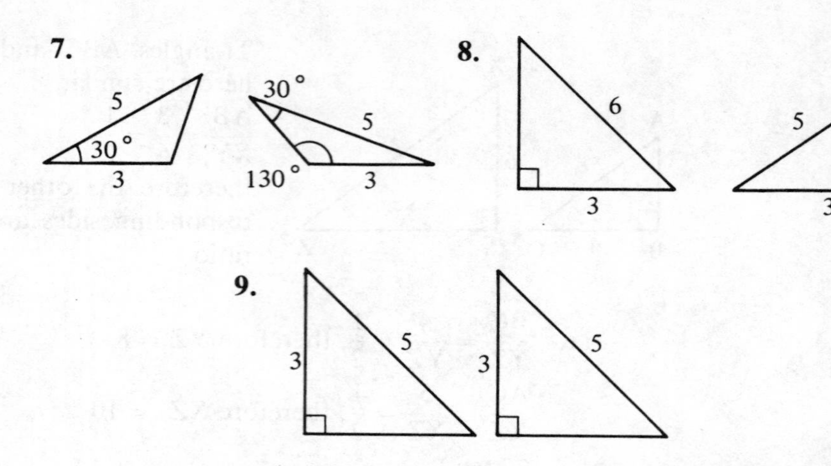

10. In questions **1** to **9** which pairs of triangles are isosceles?

State in questions **11** to **16** if the pairs of triangles are similar, giving reasons for your answer.

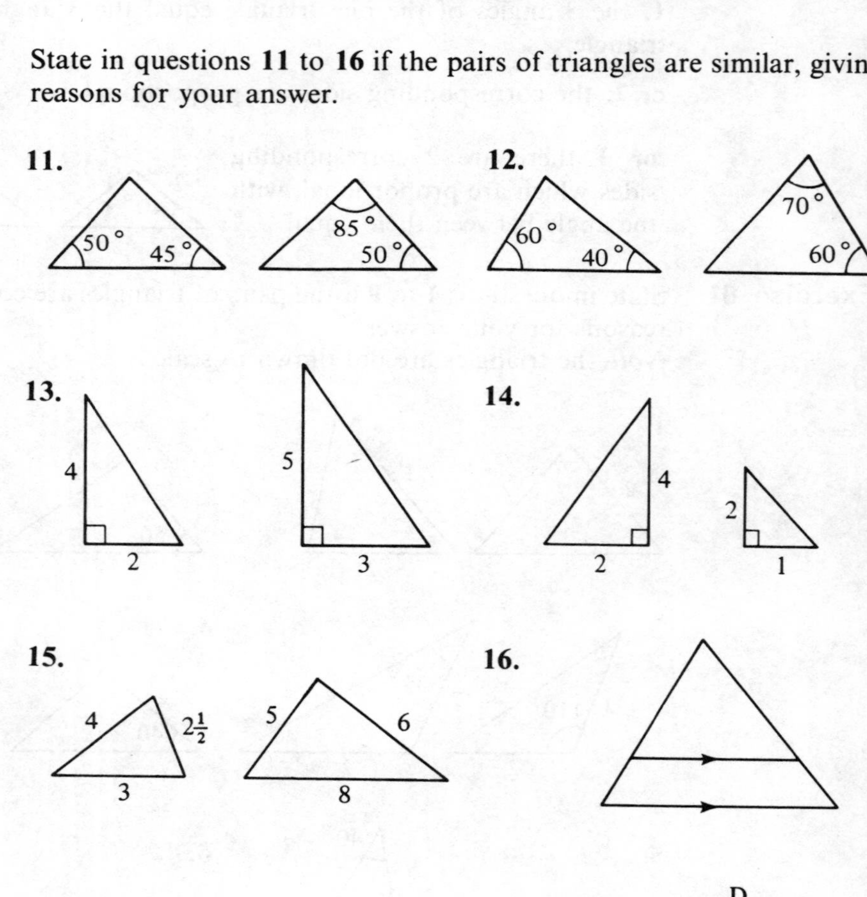

17. Calculate the lengths EF and DF.

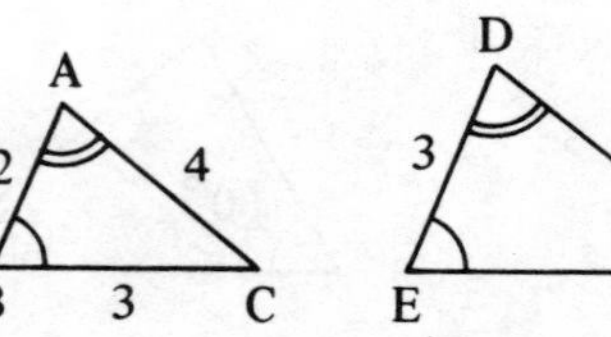

18.

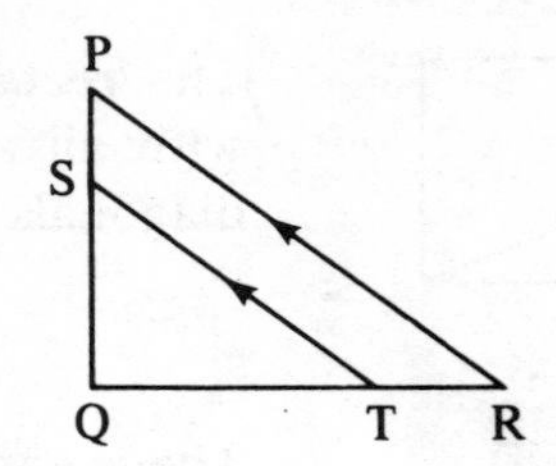

Given that QS = 3, SP = 1,
QT = 4, ST = 5,
calculate the lengths TR and PR.

19.

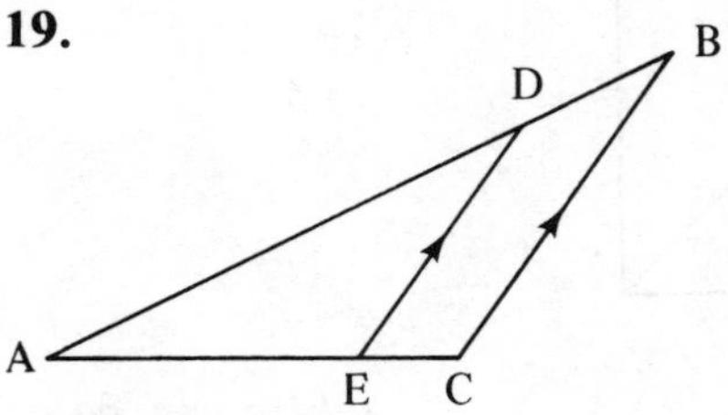

Given that AB = 10,
DB = 2,
BC = 6,
AC = 4,
calculate the lengths EC and DE.

20.

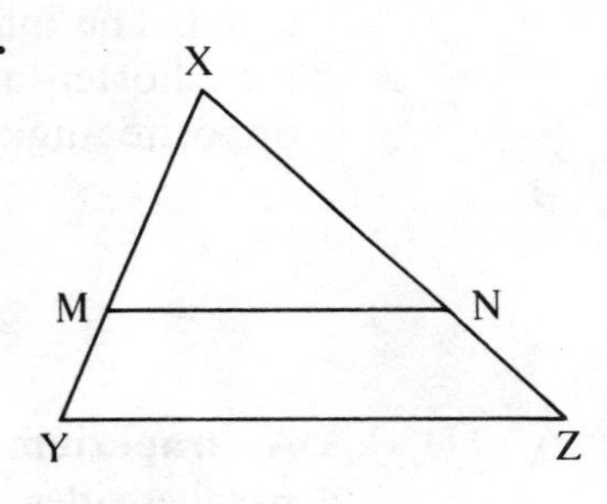

Given that XY = 5, XM = 3,
XZ = 8, NZ = $3\frac{1}{5}$,
prove that MN is parallel to YZ.

Quadrilaterals

We shall consider some types of convex quadrilateral.

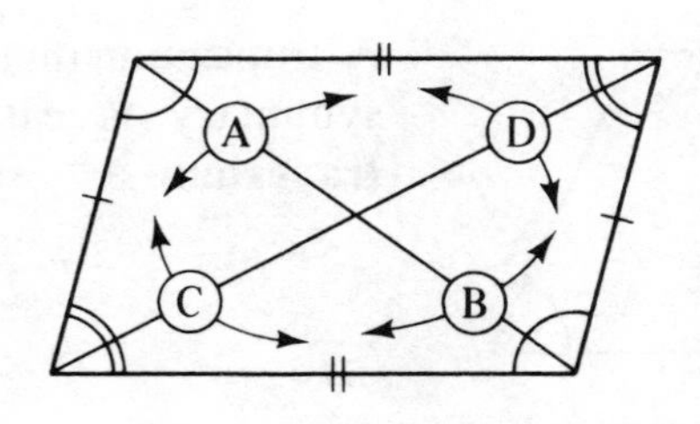

The **parallelogram** can be constructed from 2 congruent triangles, A and B or C and D. Its opposite sides are parallel and equal in length. Its opposite angles are equal. The diagonals bisect the parallelogram and each other.

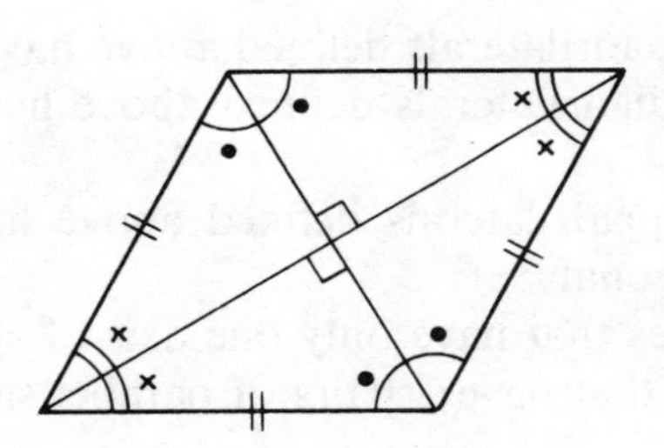

A rhombus is a parallelogram with all four sides equal in length and can be constructed from 2 congruent, isosceles, triangles. Its diagonals bisect each other at right angles, and bisect the angles of the rhombus.

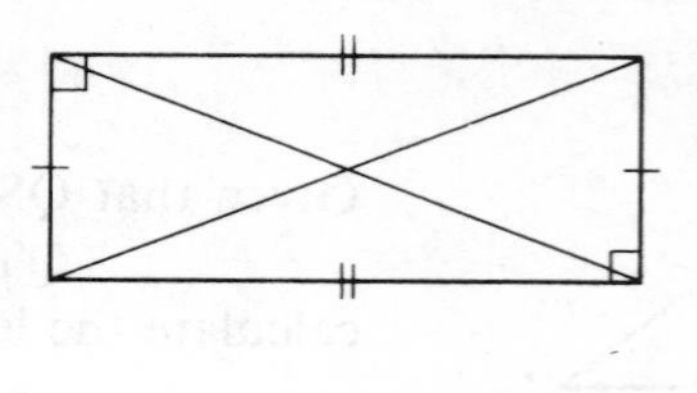

The **rectangle** is a parallelogram with all angles a right angle. Its diagonals are equal in length.

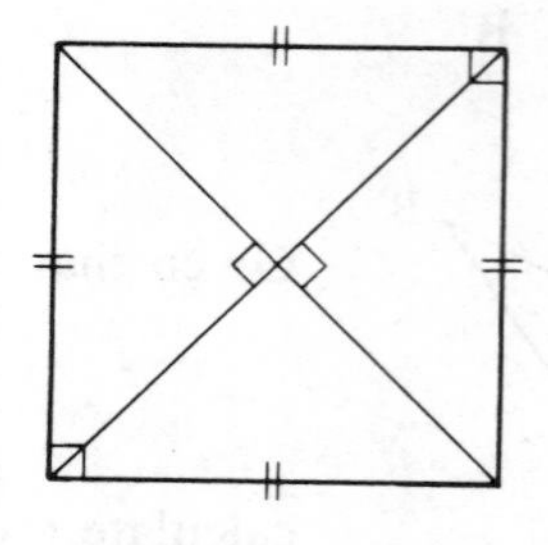

The **square** is a rectangle with the same properties as the rhombus.

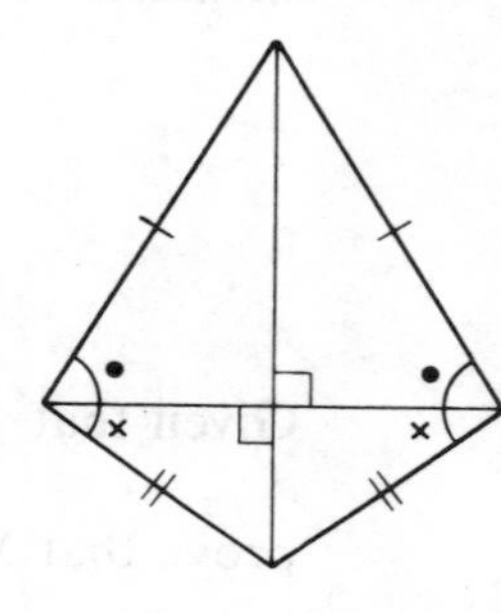

A **kite** can be constructed from 2 isosceles triangles whose bases are equal. The longer diagonal bisects the shorter at right angles. The opposite angles marked are equal.

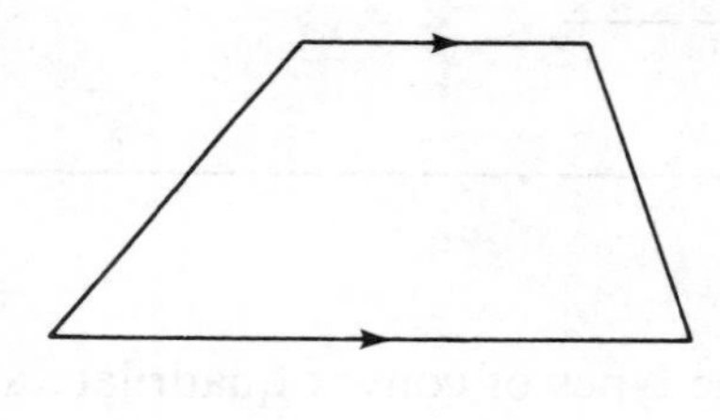

A **trapezium** has one pair of parallel sides.

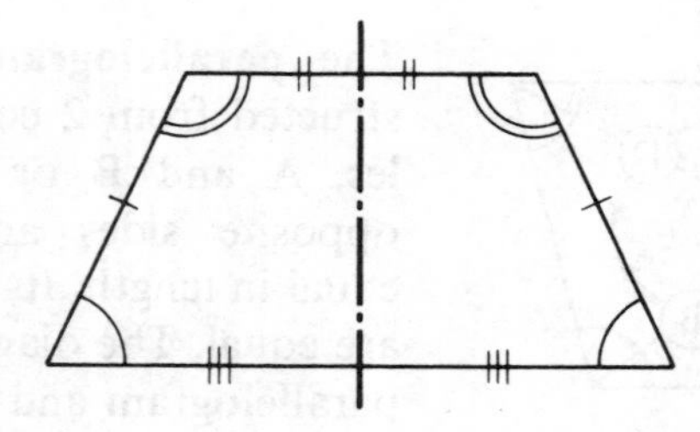

A trapezium that has one axis of symmetry is called an **isosceles trapezium**.

Exercise 82

1. Which of the quadrilaterals defined above have point symmetry?
2. Which of the quadrilaterals defined above have rotational symmetry of order 4?
3. Which of the quadrilaterals defined above have line symmetry about their diagonals?
4. Name the figures that have only one axis of symmetry.
5. List the figures that have 2 pairs of parallel sides.

6. List the figures that have no sides parallel.
7. List the figures whose diagonals bisect each other.
8. List the figures whose diagonals intersect at right angles.
9. Name the figures that can be constructed from 4 congruent triangles.
10. Name the figures that can be constructed from 4 right angled triangles.

Pythagoras's Theorem

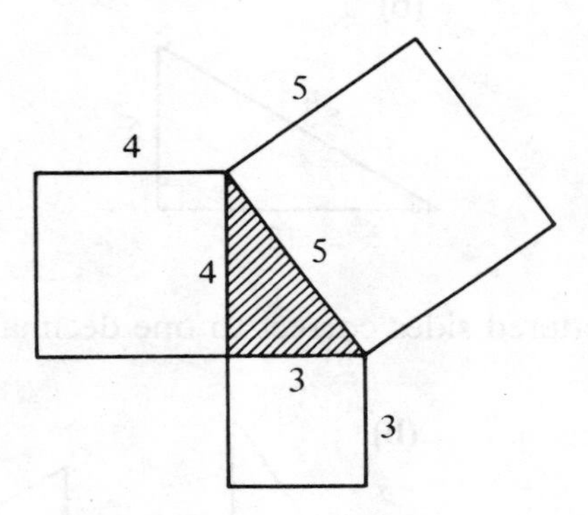

Consider the right angled triangle (shaded) with sides of length 3cm, 4cm, and 5cm. A square is constructed on each side. Calculate for yourself the area of each square. You should find that the area of the two smaller squares added together equals the area of the largest square.

i.e. $4^2 + 3^2 = 5^2$

i.e. $16 + 9 = 25$

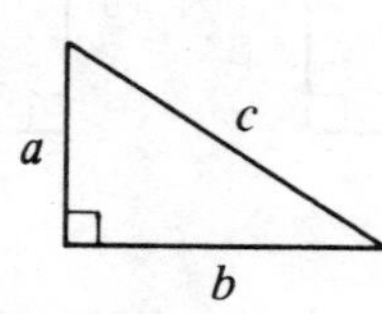

In any right angled triangle where a and b represent the lengths of the shorter sides and c the hypotenuse,

$a^2 + b^2 = c^2$

An early proof of this relationship was formulated by a Greek mathematician named Pythagoras, and the theorem has been given his name.

Example

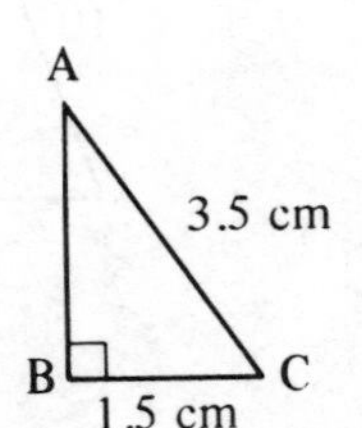

Given a right angled triangle ABC with AC = 3.5cm and BC = 1.5cm as shown, calculate the length AB correct to one decimal place.

Using Pythagoras theorem

$$AB^2 + BC^2 = AC^2$$

[Subtracting BC^2 from both sides]

$$\therefore AB^2 = AC^2 - BC^2$$

$$\text{i.e. } AB^2 = (3.5)^2 - (1.5)^2$$

[Factorising the 'difference of two squares']	$\therefore$ AB^2	$= (3.5 - 1.5)(3.5 + 1.5)$
		$= 2 \times 5$
		$= 10$
	$\therefore$ AB	$= \sqrt{10}$
[From tables]		$= \underline{3.2\,\text{cm}}$ (correct to 1 d.p.) ANS

Exercise 83

1. Calculate the lettered sides:

(a)

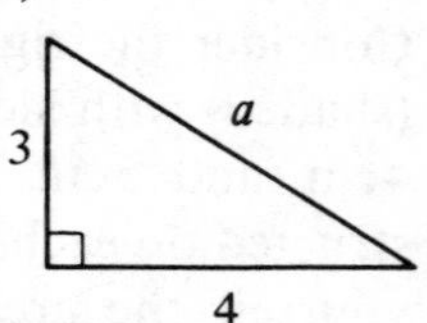

(b)

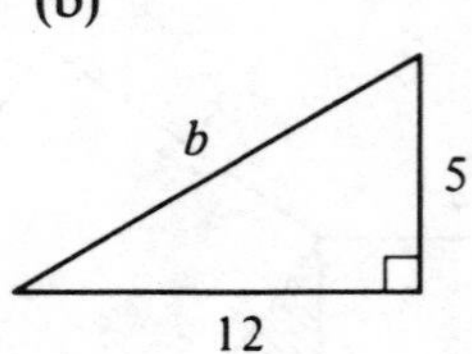

(c)

c 10 6

2. Calculate the lettered sides correct to one decimal place:

(a)

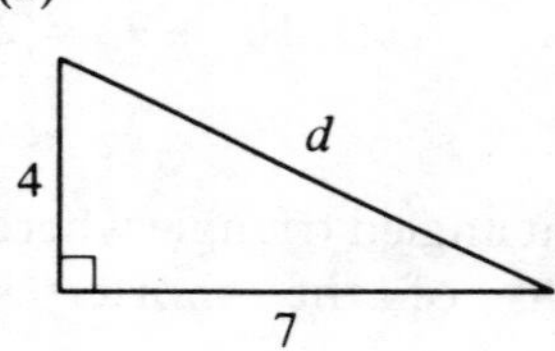

(b)

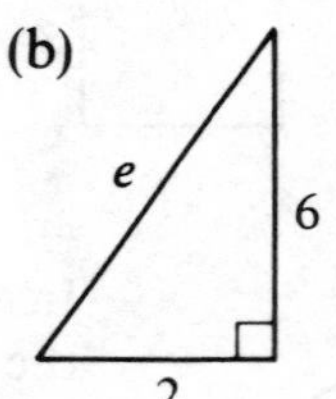

(c)

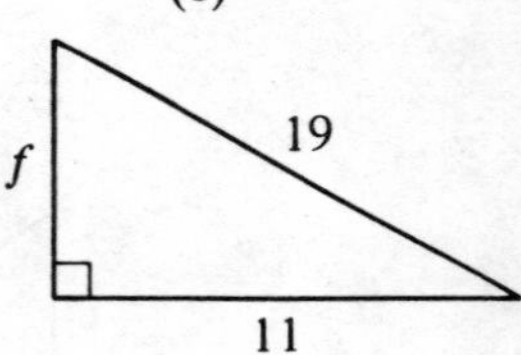

3. Find the perpendicular heights h of these isosceles triangles:

(a)

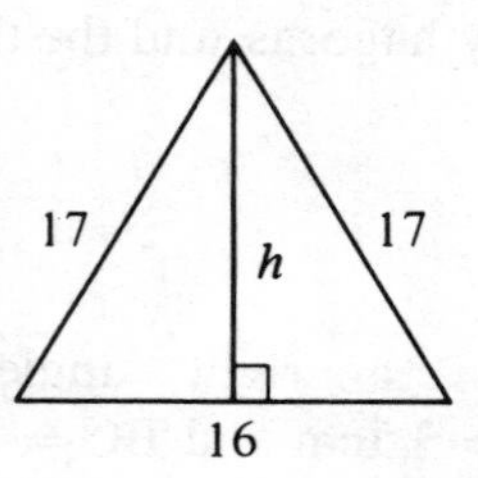

(b)

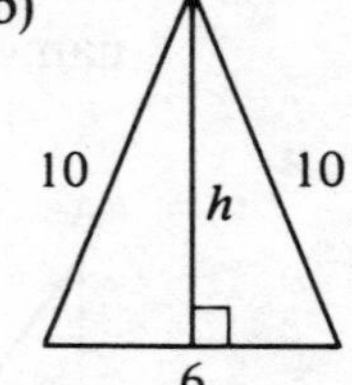

4.

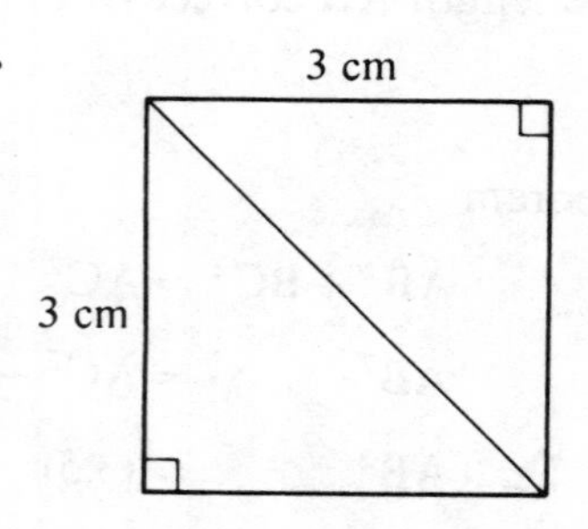

Find the length of the diagonal of this square correct to two decimal places.

5.

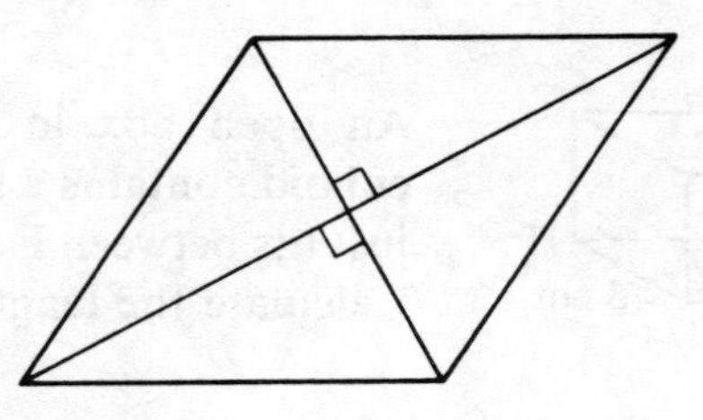

Find the length of the side of this rhombus whose diagonals are of lengths 24cm and 10cm.

6.

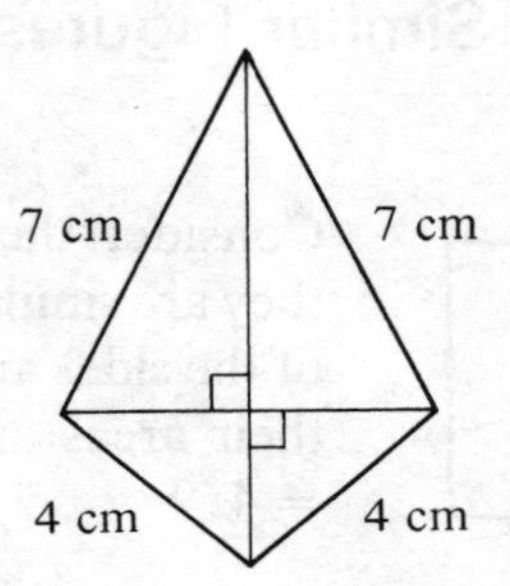

Find the length of the longer diagonal of this kite correct to two decimal places, given that its shorter diagonal is 5cm.

7. A ladder of length 12.5m has one end against a wall and its base on level ground 2.5m from the wall. Calculate the distance the ladder reaches up the wall correct to the nearest metre.

8.

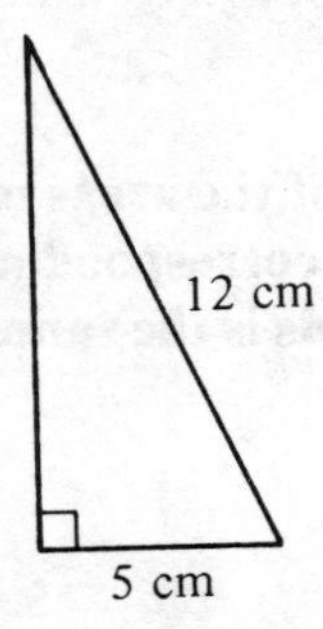

Calculate the area of the triangle correct to two decimal places.

9.

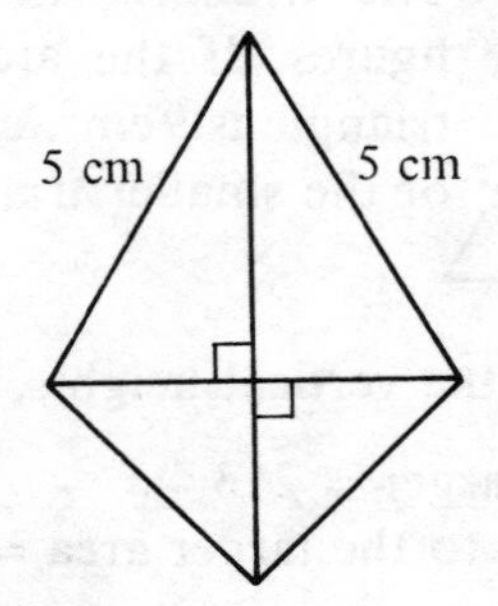

Calculate the area of this kite given that the lengths of its longer and shorter diagonals are 7cm and 6cm respectively.

10.

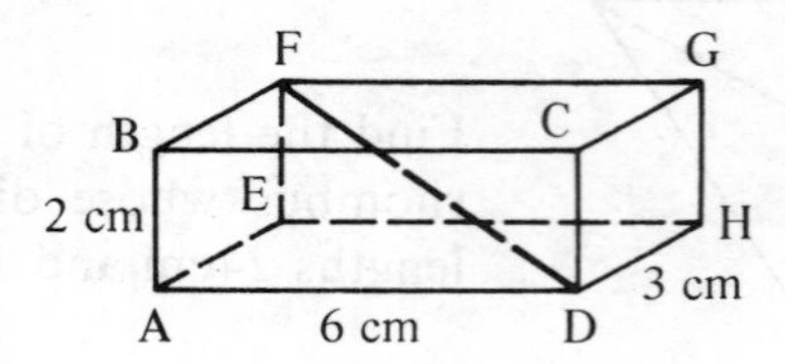

An open box in the shape of a cuboid contains a straight rod that just fits between F and D as shown. Calculate the length of this rod.

Area and Volume of Similar Figures and Solids

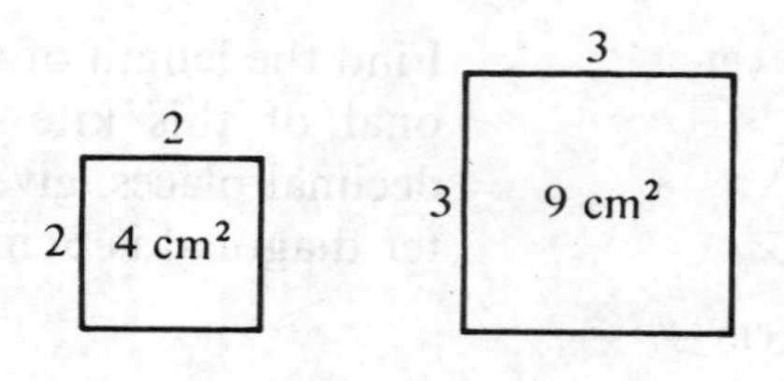

Consider the two squares shown, they are similar figures. The lengths of the sides are in the ratio 2 : 3, but their areas are in the ratio $2^2 : 3^2 = 4 : 9$

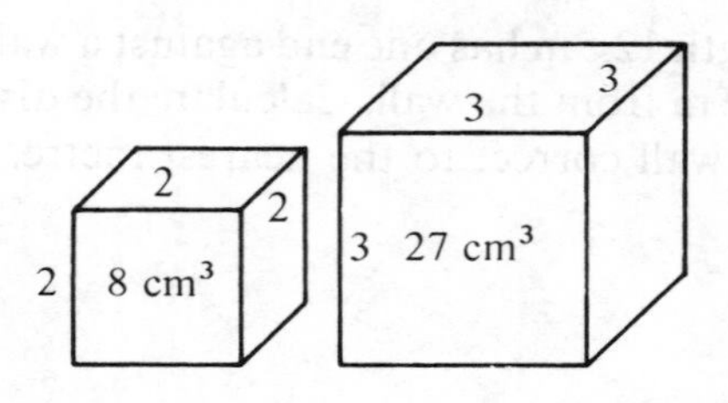

Consider the two cubes shown, they are similar solids. The lengths of the edges are in the ratio 2 : 3, but their volumes are in the ratio $2^3 : 3^3 = 8 : 27$.

It can therefore be said that the **ratio of the areas of similar figures is the same as the ratio of the squares of corresponding lengths**, and that the **ratio of the volumes of similar solids is the same as the ratio of the cubes of corresponding lengths**.

Examples **1.**

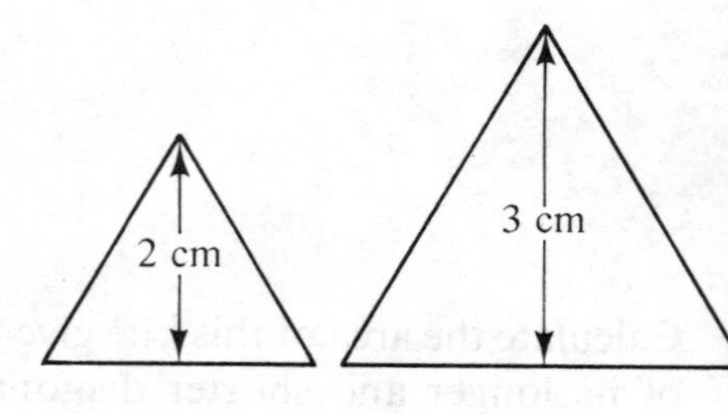

The triangles shown are similar figures. If the area of the larger triangle is 9cm^2, calculate the area of the smaller triangle.

[The corresponding lengths are the vertical heights.]

The ratio of the smaller to the larger = 2 : 3

$\therefore$ The ratio of the smaller area to the larger area $= 2^2 : 3^2 = 4 : 9$

i.e. the smaller triangle's area is $\frac{4}{9}$ that of the larger triangle.

$$= \frac{4}{9} \times \frac{9}{1} = \underline{4\text{cm}^2} \quad \text{ANS}$$

2.

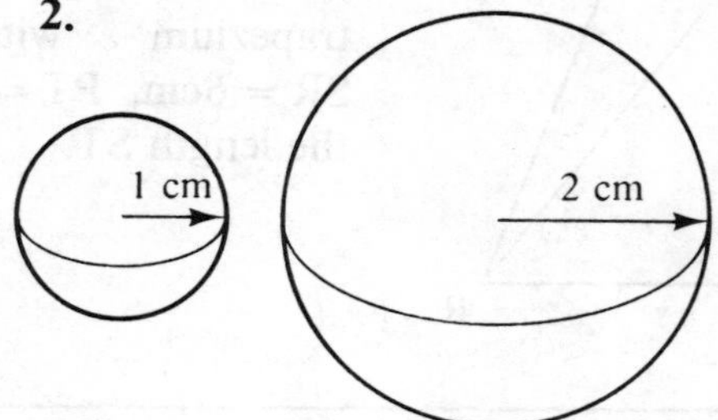

The spheres shown are similar solids. If the volume of the smaller sphere is 10cm^3, calculate the volume of the larger sphere.

[The corresponding lengths given are the radii.]

The ratio of the smaller to the larger $= 1:2$

$\therefore$ The ratio of the smaller volume to the larger volume $= 1^3:2^3 = 1:8$

i.e. the larger sphere's volume is 8 times that of the smaller sphere.

$$= 8 \times 10 = \underline{80\text{cm}^3} \quad \text{ANS}$$

Exercise 84

1. A triangle of base length 3cm has an area of 6cm^2. What is the area of a similar triangle of base length 6cm?
2. A trapezium of area 8cm^2 has a perpendicular height 4cm. A similar trapezium has an area of 18cm^2. Calculate the perpendicular height of the larger trapezium.
3. A cylinder of radius 4cm has a volume of 64cm^3. Find the volume of a similar cylinder of radius 3cm.
4. A cube of edge 3cm has a volume of 27cm^3. What length is the edge of a cube whose volume is 125cm^3?
5. An accurate model is made of a van using a scale of 1 : 30. If the area of the rear doors on the model is 33cm^2, what is the area of the actual doors on the van?
6. If the full-size van in question 5 has a fuel tank of capacity 135 litres, what would be the capacity of the model's fuel tank in cubic centimetres? (1 litre $= 1000\text{cm}^3$.)
7. 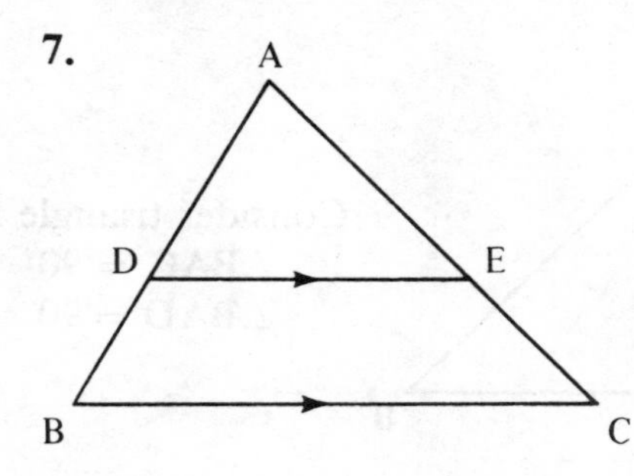

Given that AB = 7cm, and that the area of triangle ADE is $\frac{4}{9}$ of the area of triangle ABC, calculate the length of AD.

8.

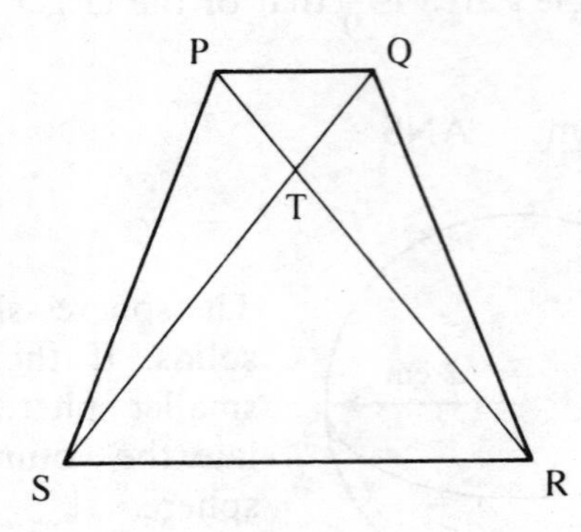

Given that PQRS is an isosceles trapezium with PQ = 2cm, SR = 6cm, PT = $1\frac{1}{2}$cm, calculate the length ST.

Examination Standard Questions

Worked examples

1.

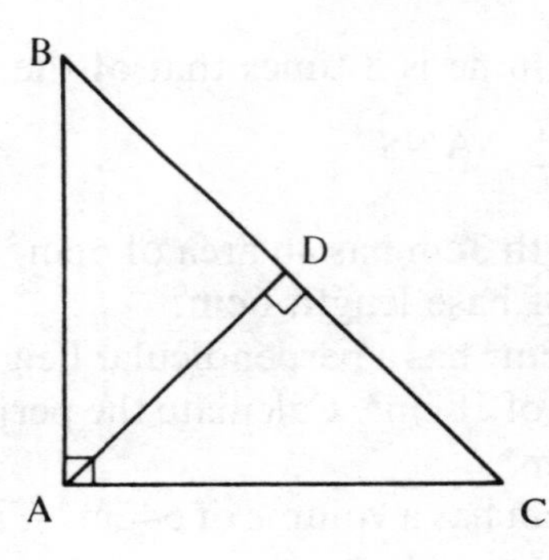

ABC is a triangle right angled at A.
D is a point on BC, such that $\angle ADB = \angle BAC$.
Prove that $\angle ACB = \angle BAD$.

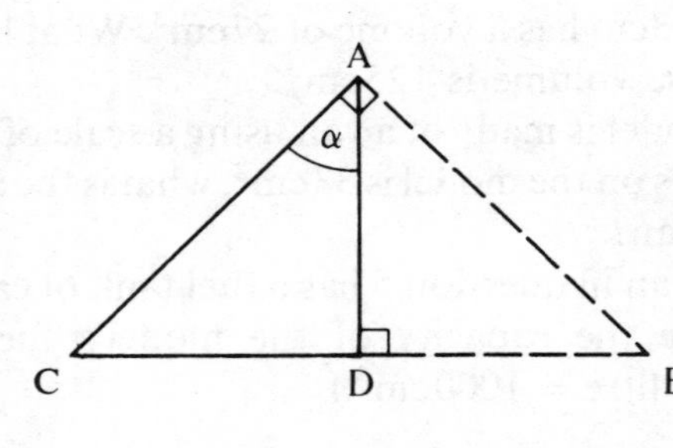

Consider triangle ACD: Let $\angle DAC = \alpha$, then
$\angle ACD = 90° - \alpha = \angle ACB$

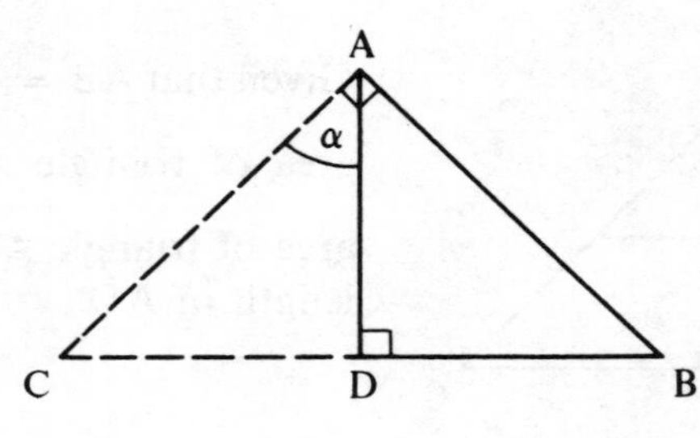

Consider triangle ABD:
$\angle BAC = 90°$
$\therefore\ \angle BAD = 90° - \alpha = \angle ACB$

2.

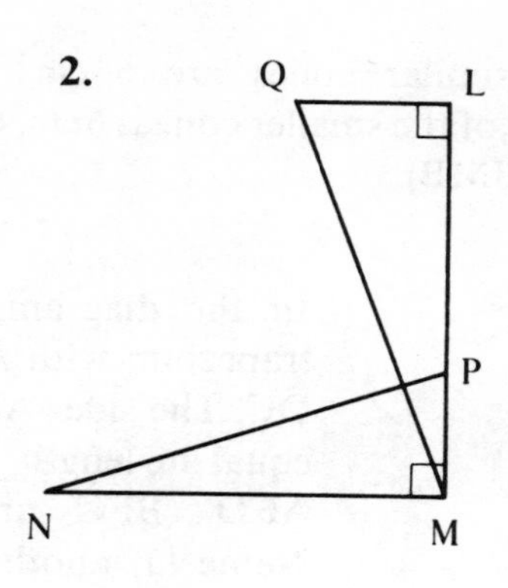

In the diagram QL and NM are perpendicular to LM, and LM = MN. Given also that LP = 2PM and QL = $\frac{1}{2}$LP, prove that triangles LMQ and PNM are congruent.

We know that ∠NMP and ∠QLM are right angled, that QL = PM and that LM = MN.
i.e. there are two sides and the angle included between them equal.
∴ Triangles LMQ and PNM are congruent.

3. The volumes of two similar cylinders are 27cm³ and 125cm³ respectively. If the length of the longer cylinder is 10cm, find the length of the shorter cylinder.

The ratio of the volumes = 27 : 125 = the ratio of the cubes of the respective lengths.

$$\therefore \text{ The ratio of the lengths } = \sqrt[3]{27} : \sqrt[3]{125} = 3 : 5$$

i.e. the smaller cylinder has length equal to $\frac{3}{5}$ that of the longer cylinder.

i.e. $\frac{3}{5} \times \frac{10}{1} = \underline{6\text{cm}}$ ANS

Exercise 85

1. In the triangle ABC, ∠BAC is an obtuse angle, D is a point on BC, between B and C, such that ∠ADB = ∠BAC.
Prove that ∠ACB = ∠BAD. (O)

2.

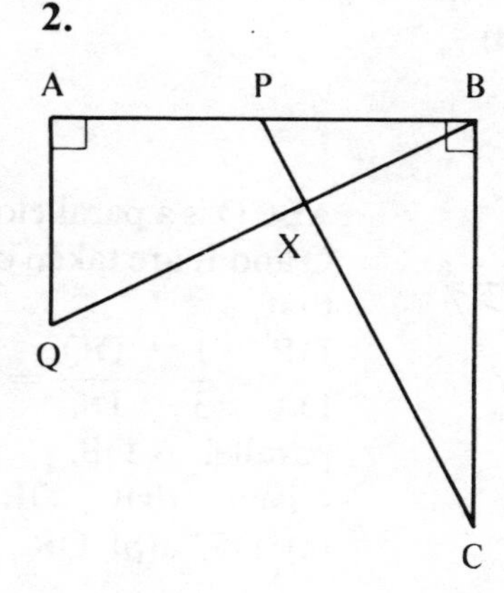

In the diagram QA and CB are both perpendicular to AB, and AB = BC. The point P is the mid-point of AB and QA = $\frac{1}{2}$AB. The straight lines PC and QB intersect at X.
Prove that:
(a) the triangles ABQ and BCP are congruent,
(b) $C\hat{P}B = Q\hat{B}C$,
(c) $P\hat{X}B = 90°$. (C)

3. The volumes of two similar cones are 64cm^3 and 216cm^3 respectively. If the height of the smaller cone is 6cm, find the height of the larger cone. (JMB).

4. In the diagram, ABCD is a trapezium with AB parallel to DC. The sides AD and BC are equal in length. The triangles AFD, BFC are congruent. Name (a) another pair of triangles which are congruent to one another, (b) a pair of triangles which are similar but not congruent. (JMB)

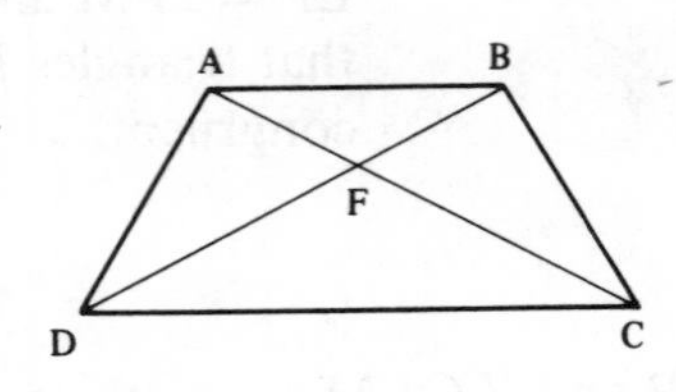

5. A and B are the mid-points of the sides PQ and PR of the triangle PQR. Express the area of the quadrilateral ABRQ as a fraction of the area of the triangle PQR. (JMB)

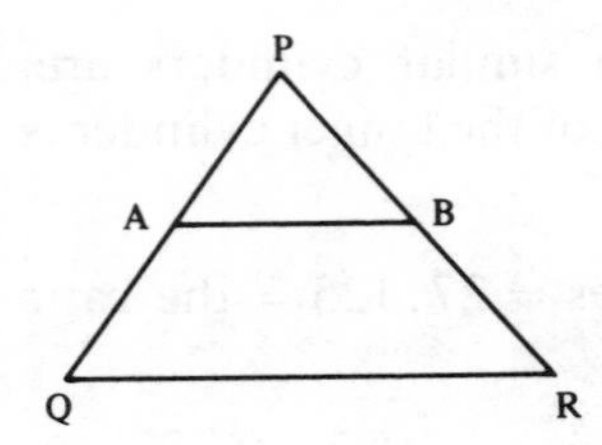

6. ABC is a triangle right-angled at B. The point E is the foot of the perpendicular from B to AC and the bisector of angle ABC meets AC at D.

   (a) Prove that triangles BEC, ABC are similar.

   (b) Prove that $\dfrac{AD}{DC} = \dfrac{AE}{EB}$.

   Given that AB = 28cm, BC = 21cm, calculate the length of DE. (JMB)

7. In an isosceles triangle ABC, the size of angle A is 95°. Find the size of angle B. (JMB)

8. ABCD is a parallelogram. Points P, Q and R are taken on the sides such that $\dfrac{DP}{DA} = \dfrac{1}{3}$, $\dfrac{DQ}{DC} = \dfrac{1}{3}$ and QR is parallel to DB.
Given that DB = 30cm, find (a) DS, (b) QR. (C)

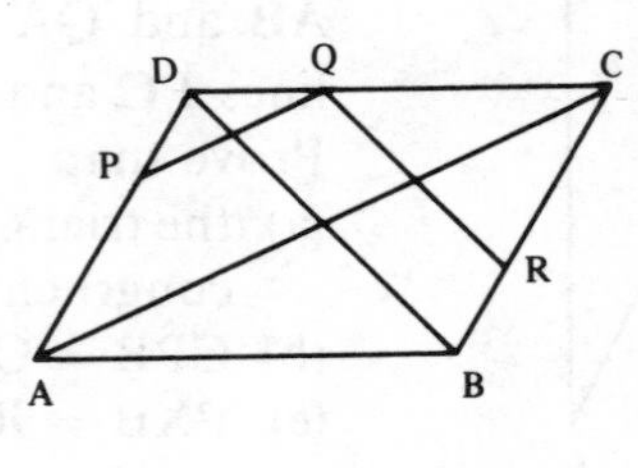

9.

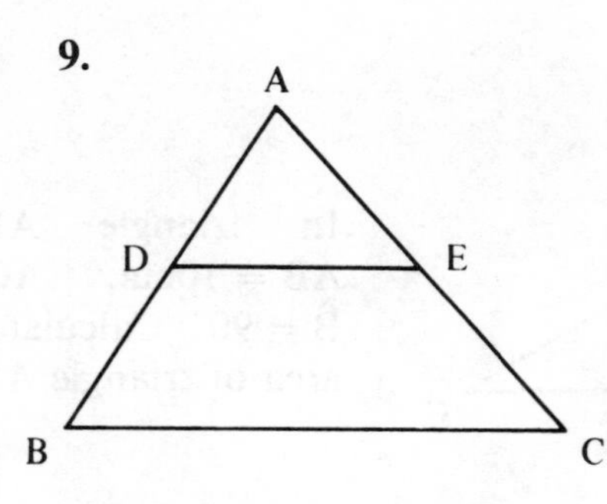

In the diagram the points D and E are such that

$$\frac{AD}{DB} = \frac{AE}{EC} = \frac{1}{2}$$

Given that the area of triangle ABC is $36 cm^2$, find the area of BDEC. (C)

10.

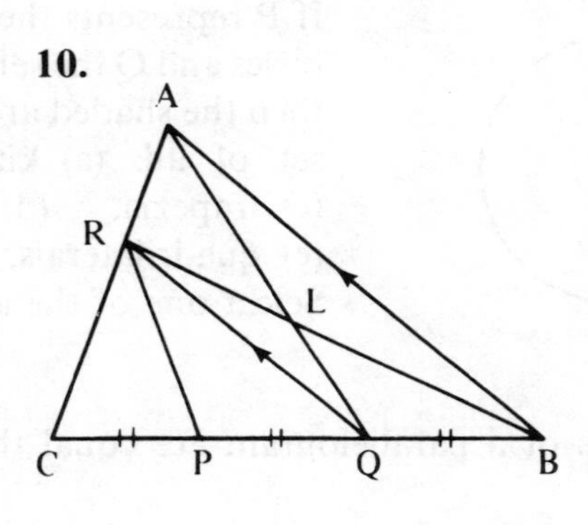

In the diagram, BA and QR are parallel and CP = PQ = QB, thus making AR = $\frac{1}{3}$AC. The lines AQ and BR intersect at L.

(a) Prove that the area of triangle ARL = the area of triangle BLQ.

(b) Prove that the area of triangle BRP = the area of quadrilateral ARPQ. (C)

11. A model of a lorry is made on a scale of 1 to 10.

(a) The windscreen of the model has an area of $100 cm^2$. Calculate the area, in square centimetres, of the windscreen of the lorry.

(b) The fuel tank of the lorry, when full, holds 100 litres. Calculate the capacity, in cubic centimetres, of the fuel tank on the model. (AEB)

12. The volume of a cylindrical can is $50 cm^3$. A second cylindrical can has the same height as the first but its diameter is three times the diameter of the first. Calculate the volume of the second can. (JMB).

13.

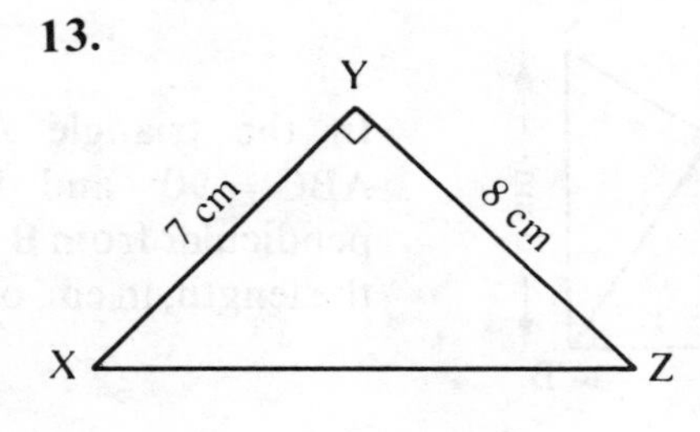

Between which pair of consecutive integers does the length of XZ, measured in cm, lie? (L)

14.

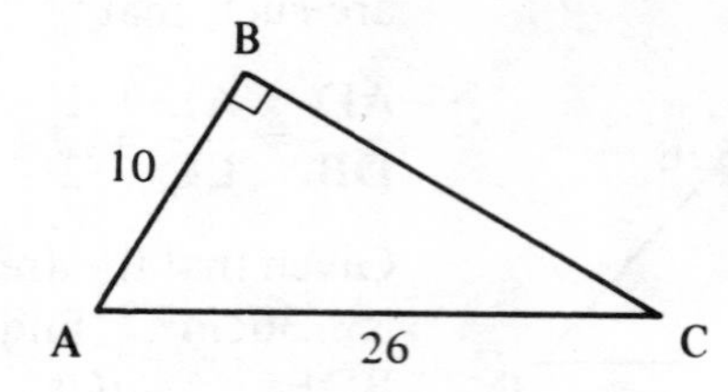

In triangle ABC, the side AB = 10cm, AC = 26cm and $\hat{B} = 90°$. Calculate (a) BC, (b) the area of triangle ABC. (C)

15.

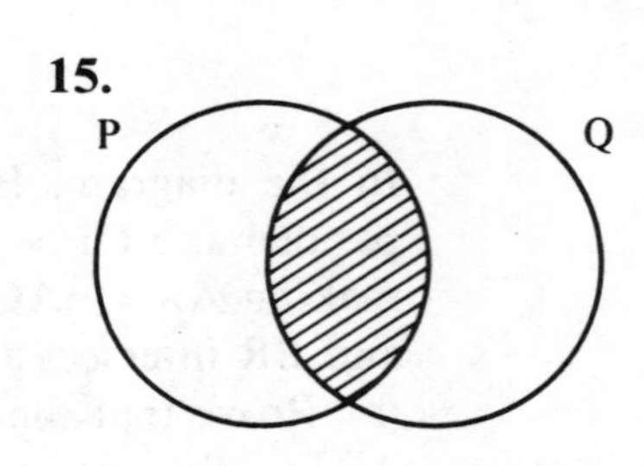

If P represents the set of all rhombuses and Q the set of all rectangles, then the shaded area represents the set of all: (a) kites, (b) squares, (c) trapezia, (d) parallelograms, (e) quadrilaterals.
Select one of the above. (L)

16. If the diagonals of a parallelogram are equal then the figure is necessarily:
(a) a square, (b) a rhombus, (c) a rectangle, (d) none of these.
Select one of the above. (O)

17. Draw a labelled Venn diagram to illustrate:
$\mathscr{E}$ = {all triangles}, O = {obtuse-angled triangles}, R = {right-angled triangles}, Q = {acute-angled triangles}, I = {isosceles triangles}, S = {equilateral triangles}. (O)

18. The universal set is the set of plane quadrilaterals.
A = {quadrilaterals with half turn symmetry}.
B = {quadrilaterals with one or more axes of symmetry}.
Draw a member of (a) $A \cap B'$ (b) $A' \cap B$ (c) $A' \cap B'$ (O)

19. (a) Draw a quadrilateral which has just one axis of symmetry; in addition this axis must not pass through a vertex of the quadrilateral. (O)
(b) Draw a quadrilateral which has rotational symmetry of order two but not of order four. (AEB)

20.

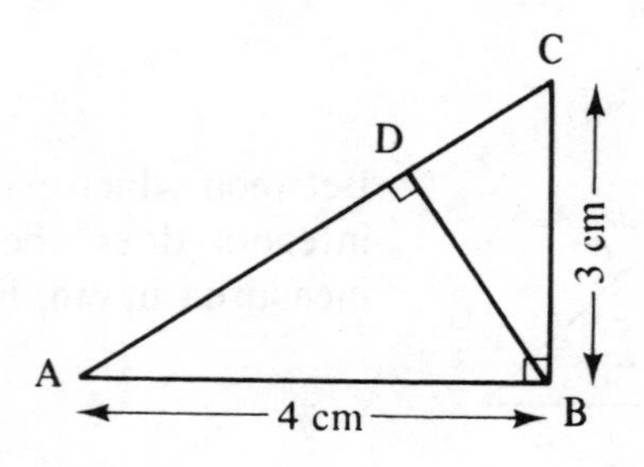

In the triangle ABC, the angle ABC = 90° and BD is the perpendicular from B to AC. Calculate the length, in cm, of BD. (AEB)

12
The Circle

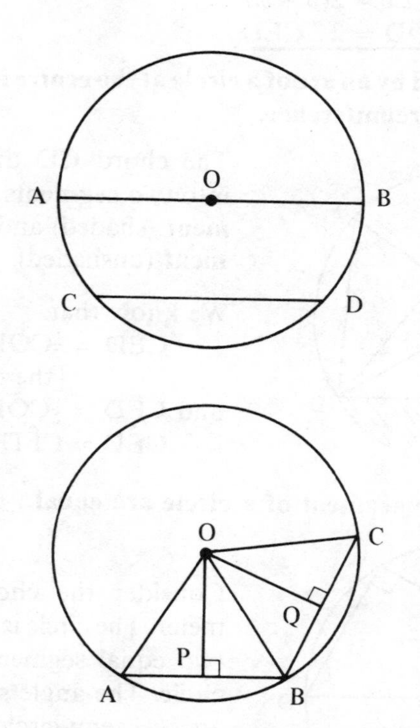

In the diagram O is the centre of the circle. OB is a **radius** of the circle, OA and OB are both **radii** (plural). A **chord** of a circle is any straight line joining two points on the circumference. CD is a chord, as is the **diameter** AB. Any chord which is not a diameter divides the circle into two arcs, the longer **major arc** and the shorter **minor arc** as shown earlier in Chapter 3.

Radii OA, OB and OC are drawn as shown in the diagram so that $A\hat{O}B = B\hat{O}C$ giving two equal sectors of the circle. The triangles AOB and BOC are congruent (two sides and the included angle) and chords AB and BC are equal in length, as are arcs AB and BC. The angle AOB is said to be **subtended** by the arc AB, (and the chord AB) and angle BOC is subtended by the arc BC (and the chord BC). P is the foot of the perpendicular drawn from O to BC. Triangles AOP, POB, BOQ and QOC are now congruent, therefore OP = OQ.

Note

1. **The line drawn from the centre of a circle perpendicular to a chord bisects that chord.**
2. **Equal chords are the same distance from the centre of the circle.**
3. **Equal arcs of a circle subtend equal angles at the centre.**

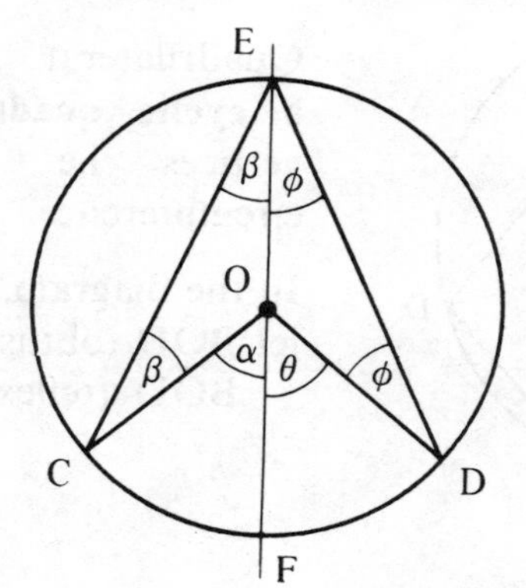

Consider the arc CFD subtending angle COD at the centre of the circle and angle CED at the circumference. A line drawn from E through the centre of the circle cuts the circle at F.

Triangle ECO is isosceles [EO = CO radii]

$\therefore \angle ECO = \angle CEO [\angle\beta]$

$\angle\alpha = \angle\beta + \angle\beta$ [External angle = sum of 2 interior opposite angles of triangle.]

$\therefore \underline{\angle\alpha = 2\angle\beta}$

Similarly $\underline{\angle\theta = 2\angle\phi}$

$\therefore \angle\alpha + \angle\theta = 2\angle\beta + 2\angle\phi$

$\therefore \angle\alpha + \angle\theta = 2(\beta + \phi)$

$\therefore \underline{\angle COD = 2\angle CED}$

Note **The angle subtended by an arc of a circle at the centre is twice the angle subtended at the circumference.**

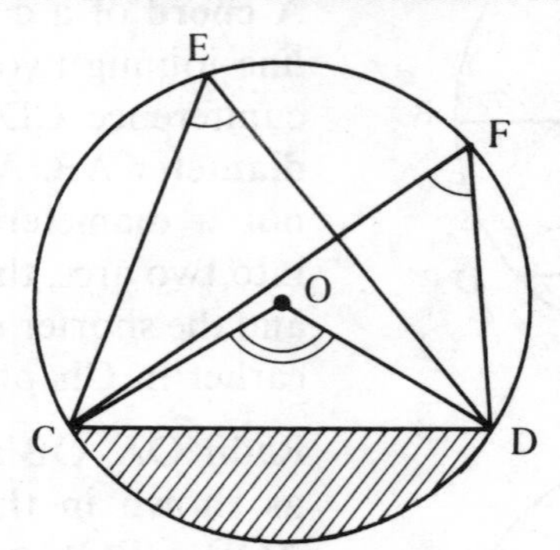

The chord CD divides the circle into two **segments**. The **minor segment** (shaded) and the **major segment** (unshaded).

We know that

$\hat{CED} = \frac{1}{2}\hat{COD}$

[the convex $\angle$]

and $\hat{CFD} = \frac{1}{2}\hat{COD}$

$\therefore \hat{CED} = \hat{CFD}$

Note **Angles in the same segment of a circle are equal.**

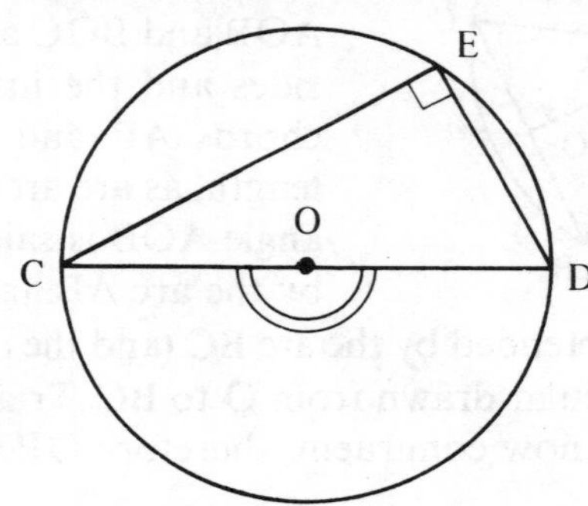

Consider the chord CD as diameter. The circle is now divided into two equal segments, each a semi-circle. The angle subtended by the arc (a semi-circle) at the centre ($\hat{COD}$) = 180°.

$\therefore \hat{CED} = \frac{1}{2}$ of $180° = \underline{90°}$

Note **The angle in a semi-circle is always 90°.**

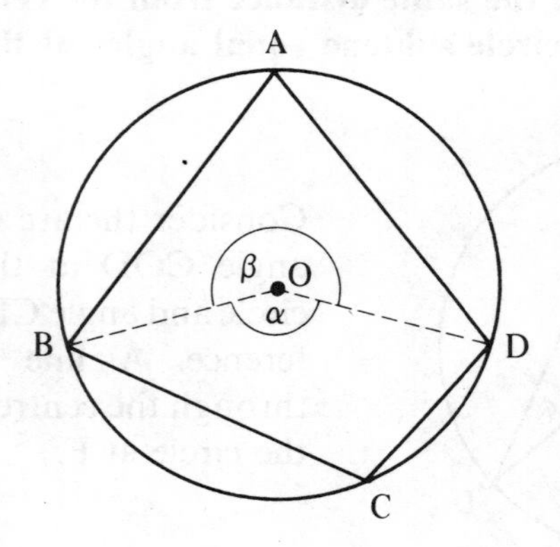

Quadrilateral ABCD is called a **cyclic quadrilateral,** its four vertices lie on the circle's circumference.

In the diagram,
let $\hat{BOD}$ (obtuse) = α
$\hat{BOD}$ (reflex) = β

[Angle subtended at the centre . . .]

We know that $\angle\alpha = 2\angle BAD$

and that $\angle\beta = 2\angle BCD$

But $\angle\alpha + \angle\beta = 360°$

$\therefore 2\angle BAD + 2\angle BCD = 360°$

[Divide both sides by 2]

$\therefore \underline{\angle BAD + \angle BCD = 180°}$

Note **The opposite angles of a cyclic quadrilateral are supplementary (i.e. total 180°).**

Exercise 86 Find the lettered angles in each case where O is the centre of the circle:

1.

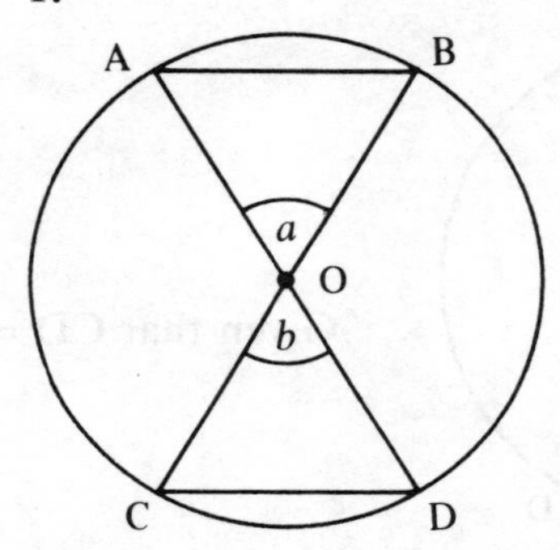

Given AB = OB = CD.

2.

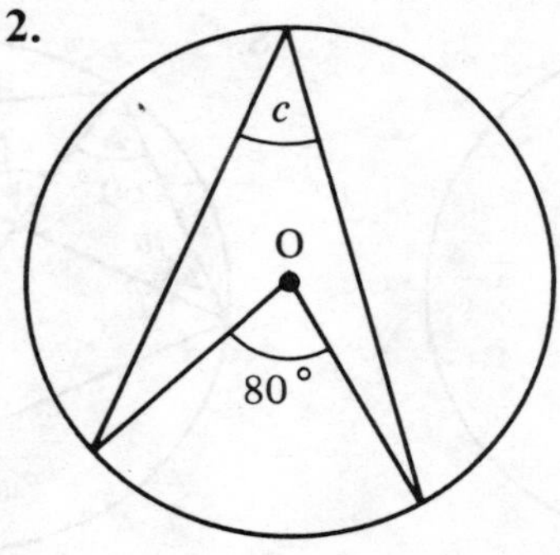

3.

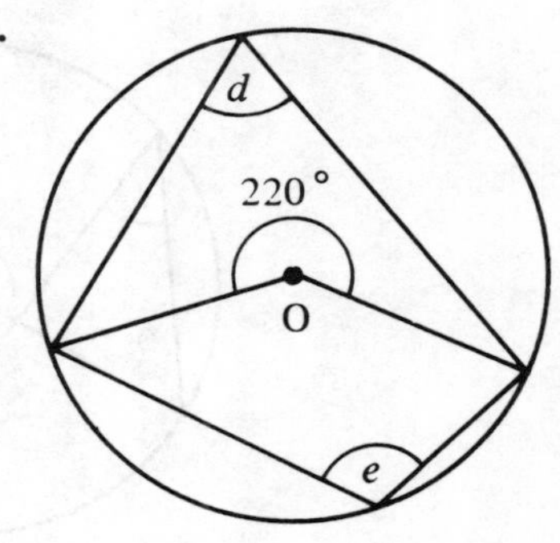

4.

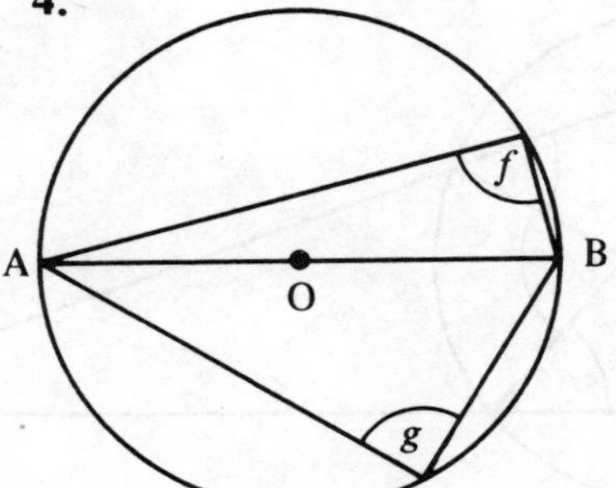

Given AOB is a diameter.

5.

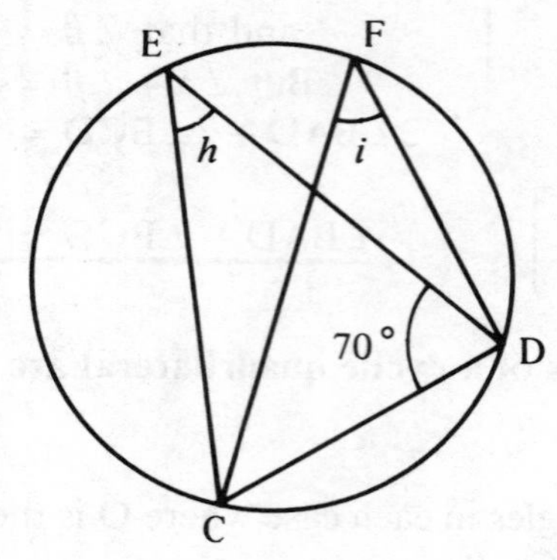

Given that chord CE = chord ED.

6.

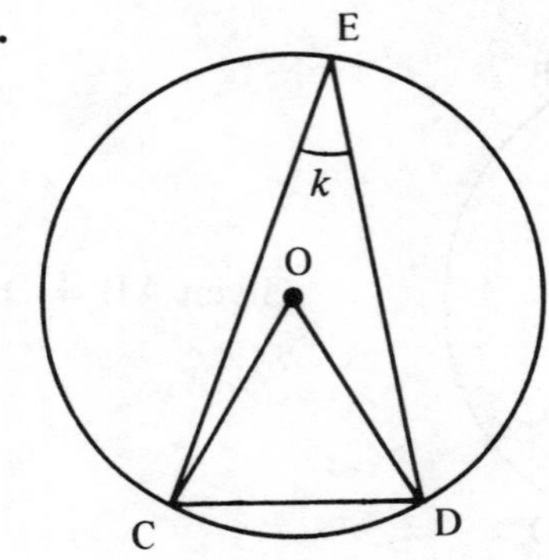

Given that CD = OC = OD.

7.

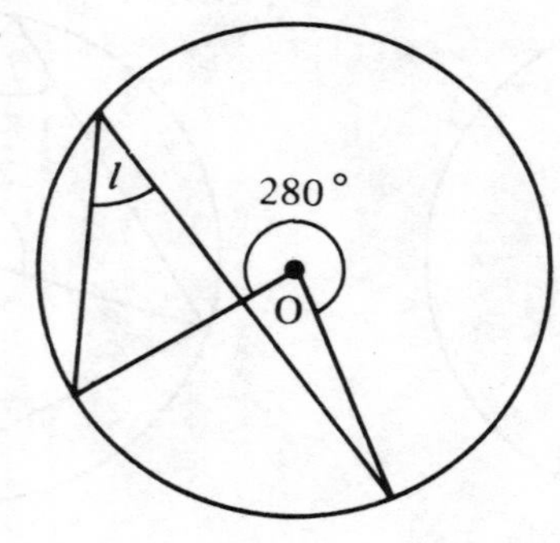

8.

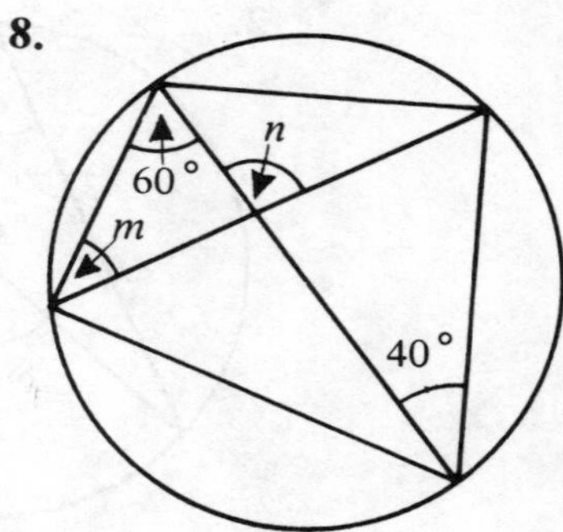

9.

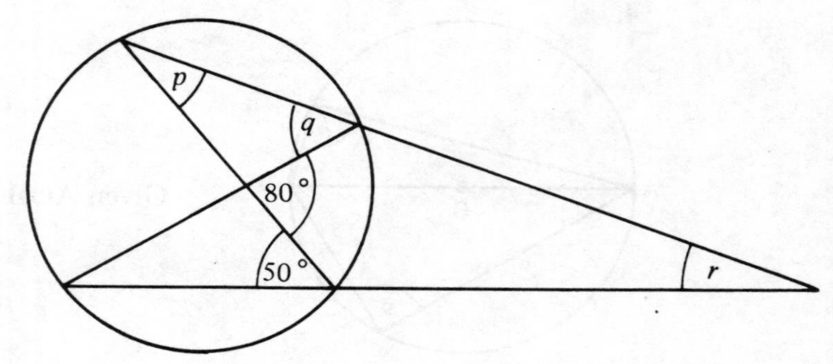

10.

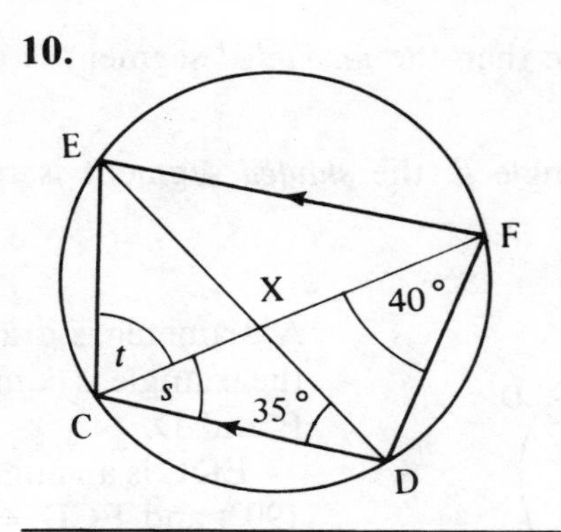

Given that chords EF and CD are parallel, and that $C\hat{D}E = 35°$, $C\hat{F}D = 40°$.

(a) Calculate angles *s* and *t*.

(b) Prove that triangles CXD and EXF are similar and isosceles.

(c) What can you say about triangles CXE and DXF?

Tangents

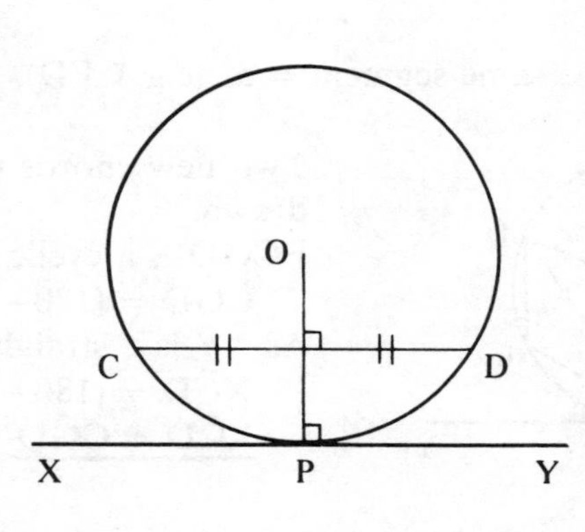

In the diagram XY is called a **tangent** to the circle, where P is its only point of contact.

We already know that the line joining the centre of a circle to the centre of a chord is at right angles to that chord. If the line from the centre is extended to the circumference of the circle it becomes a radius (OP). XY is a tangent drawn parallel to the chord CD. The radius OP is therefore at right angles to the tangent XY.

Note **A radius is perpendicular to a tangent at its point of contact.**

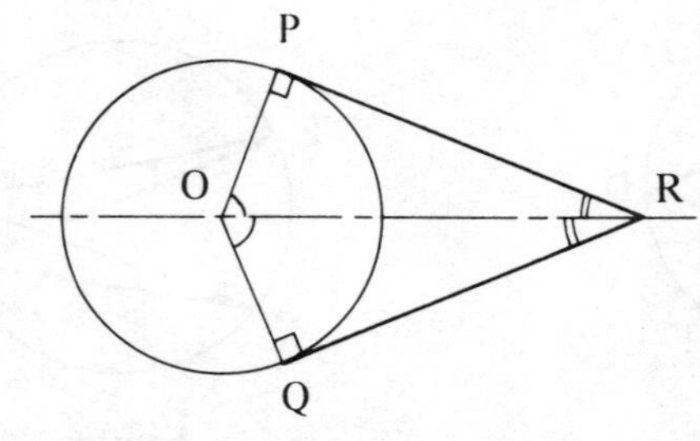

In the diagram OP is a radius and PR a tangent to the circle. A point Q is located on the circumference of the circle such that figure POQR has line symmetry about the broken line through O (i.e. triangles OPR and OQR are congruent).

Note **Two tangents drawn from the same point outside a circle are equal in length.**

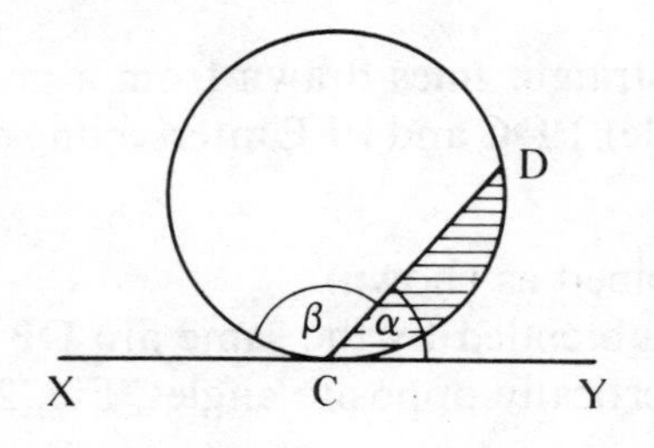

In the diagram XY is a tangent to the circle at C, and CD is a chord.

Considering angle α: then the *unshaded* segment is referred to as the **alternate segment**.

Now considering angle β: the *shaded* segment is referred to as the **alternate segment**.

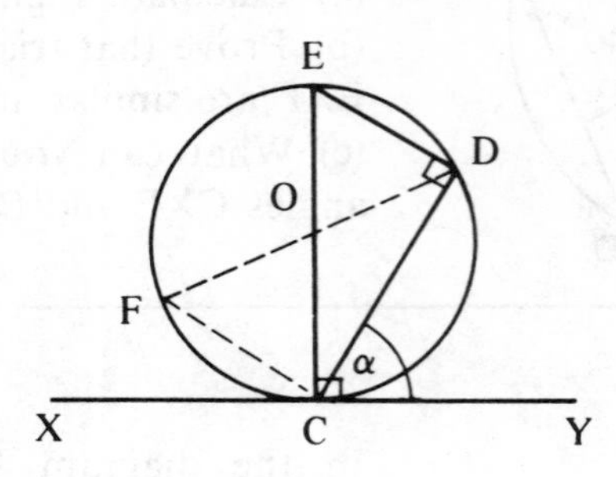

A diameter is drawn from C to E, the triangle is completed by joining E and D.

$\therefore$ $E\hat{D}C$ is an angle in a semi-circle (90°) and $E\hat{C}D = (90 - \alpha)°$

$\therefore$ $C\hat{E}D = 180° - [(90 - \alpha) + 90]$
$= \underline{\alpha°}$

$\therefore$ Any angle in the same segment $= \alpha°$ (e.g. $C\hat{F}D$).

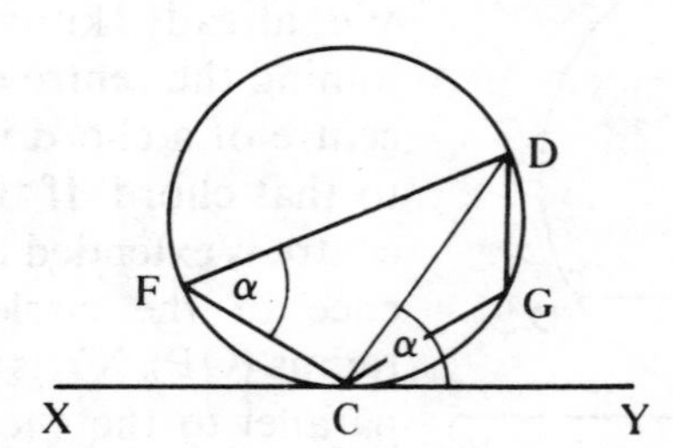

Two new chords CG and DG are drawn.

FCGD is a cyclic quadrilateral.

$\therefore$ $C\hat{G}D = (180 - \alpha)°$

As XY is a straight line,
$X\hat{C}D = (180 - \alpha)°$

$\therefore$ $\underline{X\hat{C}D = C\hat{G}D}$

Note **The angle between a chord and tangent at the point of contact is equal to any angle in the alternate segment.**

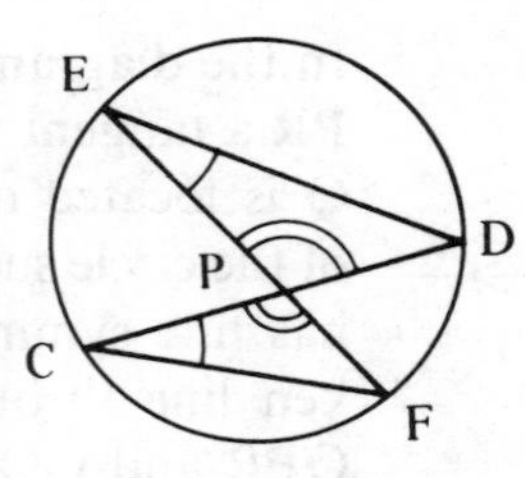

Fig. 1

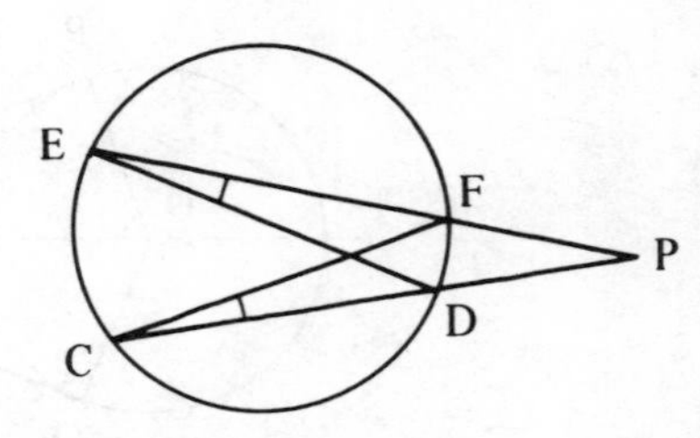

Fig. 2

In the diagrams Fig. 1 shows chords CD and EF intersecting inside circle at P.

Fig. 2 shows **secants** (straight lines drawn from a point outside, and cutting through the circle), PDC and PFE intersecting outside the circle at P.

CF and ED are also joined as shown.

$F\hat{C}D = F\hat{E}D$ [angles subtended by the same arc DF]

$E\hat{P}D = C\hat{P}F$ [Fig. 1 vertically opposite angles, Fig. 2 same angle]

∴ Triangles EPD and CPF are similar
∴ Corresponding sides are proportional.

i.e. $\frac{PF}{PD} = \frac{PC}{PE}$

$\left[\begin{array}{l}\text{Cross-}\\ \text{multiplying}\end{array}\right]$ **PF × PE = PD × PC**

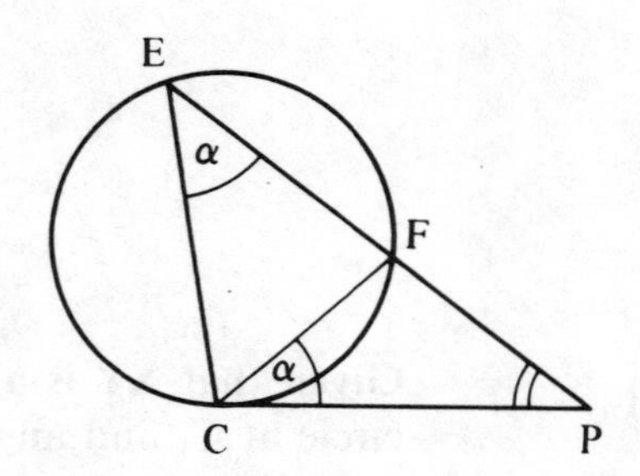

In the diagram PC is a tangent to the circle at C, and PFE is a secant. $F\hat{C}P = F\hat{E}C = \alpha°$ [alternate segment]

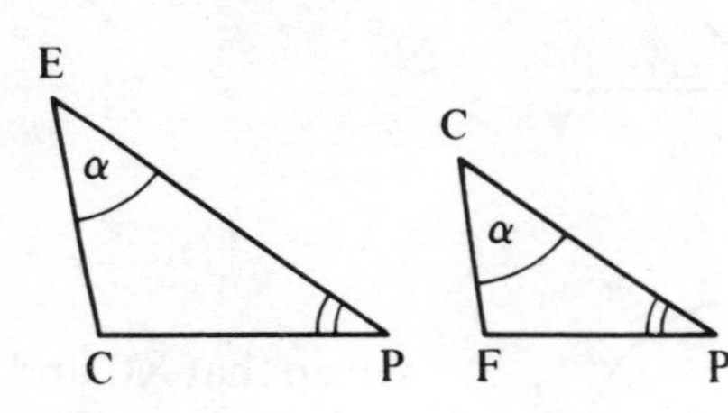

∴ Triangles ECP and CFP are similar [$\hat{P}$ is common]

$\therefore \frac{PC}{PE} = \frac{PF}{PC}$

i.e. $\mathbf{PC^2 = PF \times PE}$

Exercise 87

1.

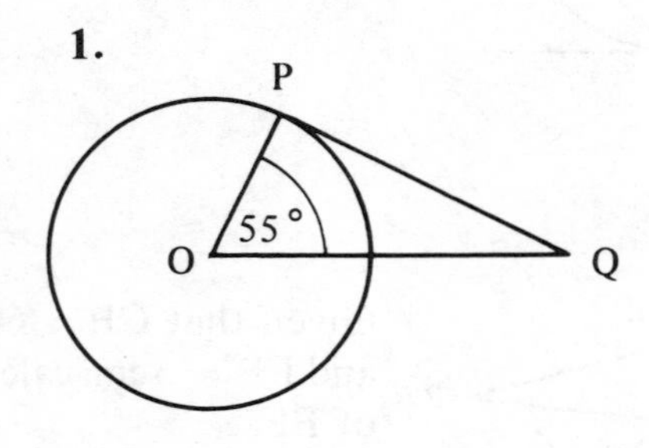

Given OP is a radius, PQ is a tangent to the circle at P, calculate $O\hat{Q}P$.

2.

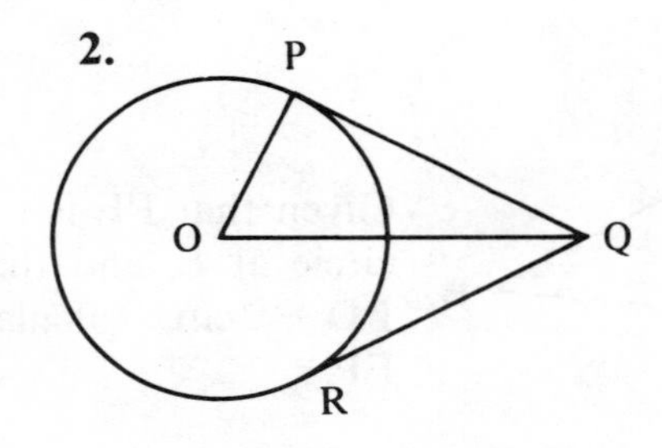

Given OP = 2cm is a radius, PQ is a tangent to the circle at P, and RQ is another tangent to the circle at R. If OQ = 5cm calculate the length of RQ.

3.

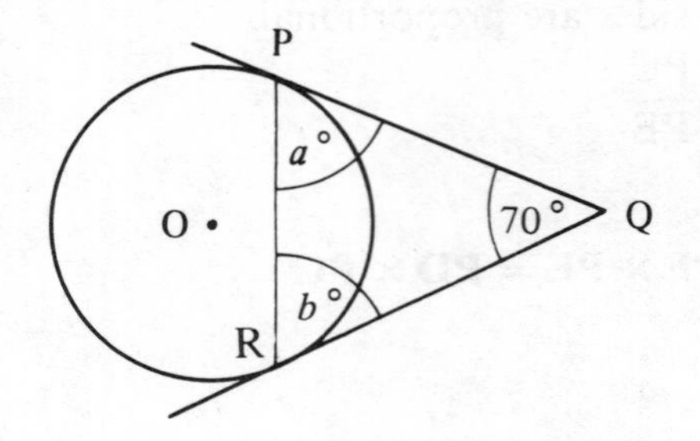

Given that PQ and QR are tangents to the circle at the points of intersection with the chord PR. Calculate angles a and b.

4.

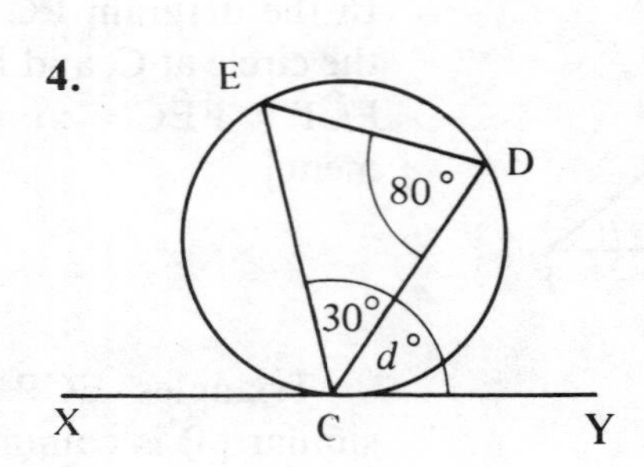

Given that XY is a tangent to the circle at C, find angle d.

5.

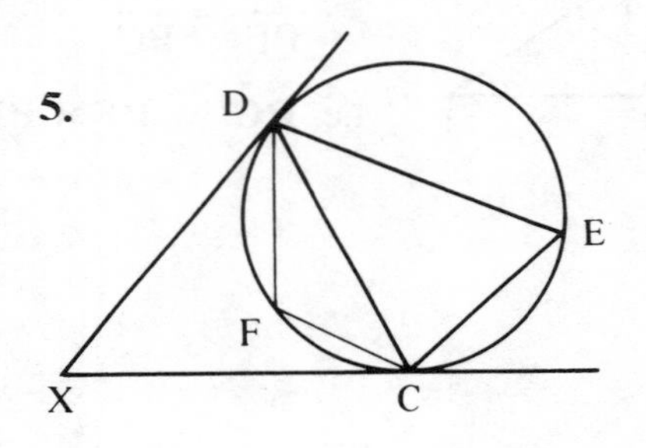

Given that XC and XD are tangents to the circle at C and D respectively, and that $X\hat{C}D = 75°$; calculate $C\hat{F}D$.

6.

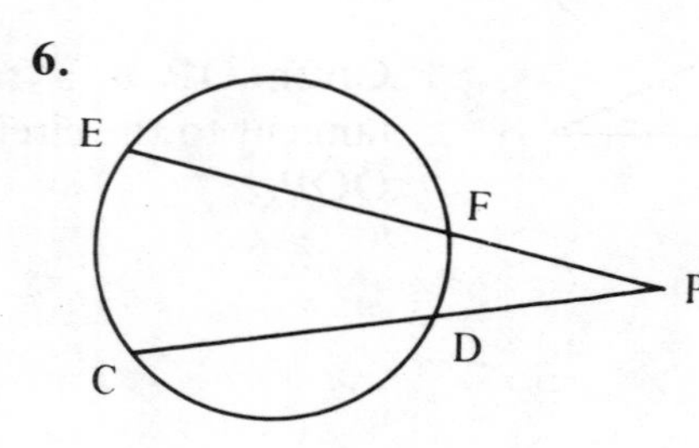

Given that CP = 6cm, DP = 2cm and FP = 3cm, calculate the length of EF.

7.

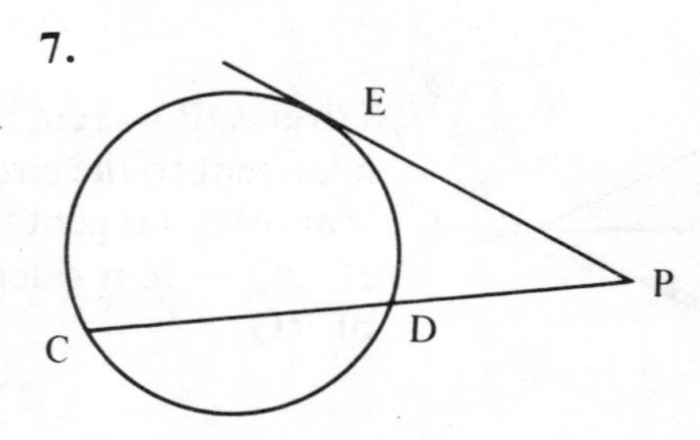

Given that PE is a tangent to the circle at E, and that PC = 18cm, PD = 2cm, calculate the length EP.

8.

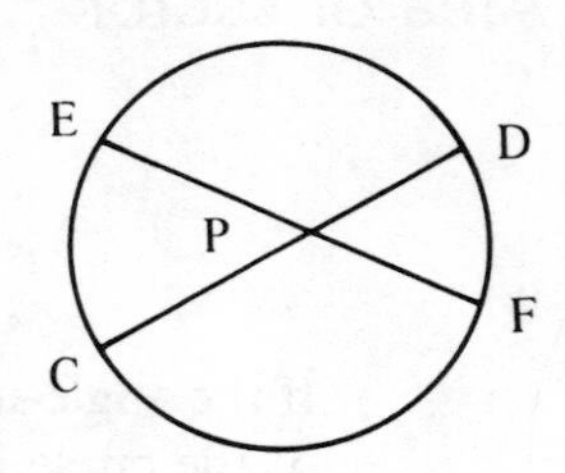

Given that EP = 3cm, PF = 2cm, PD = $1\frac{1}{2}$cm, calculate the length CP.

9.

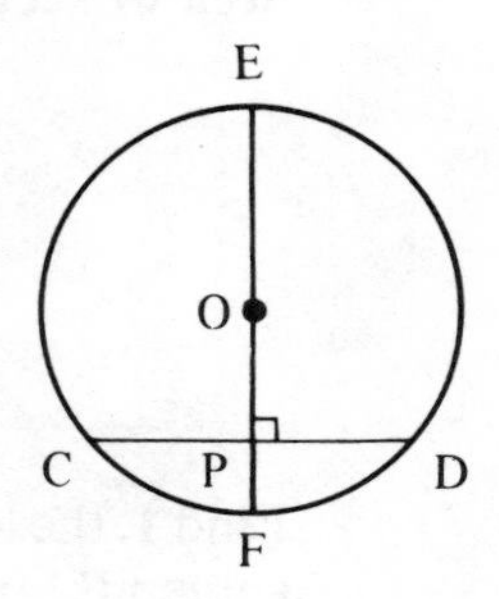

Given that EF is a diameter, and CD is a chord perpendicular to EF; if PF = 1cm and the circle is of radius $2\frac{1}{2}$cm, calculate the length CP.

10.

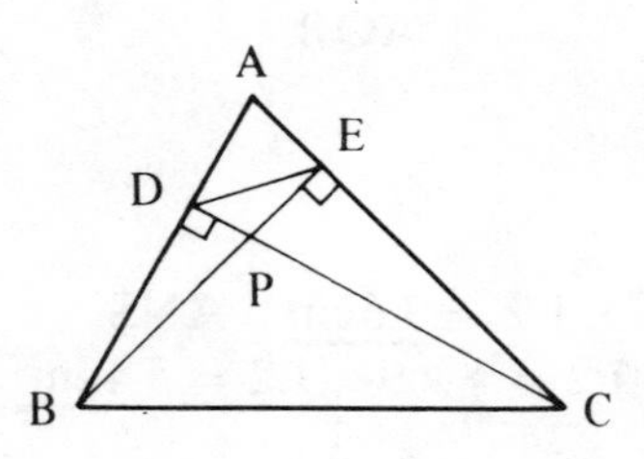

ABC is a triangle with altitudes CD and BE intersecting at P. If D and E are joined prove that BDEC is a cyclic quadrilateral.
If BP = 3cm, PE = 2cm,
DP = 1cm, calculate the altitude of the triangle BDC.

Circular Measure

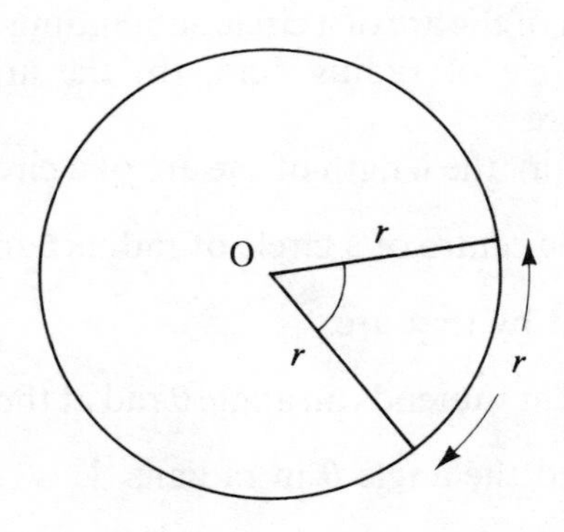

A radian is the angle subtended at the centre of a circle by an arc equal in length to the radius. In the diagram the angle is 1 radian (written 1 rad, or 1^c).
We know that the circumference of a circle $= 2\pi r$,

$\therefore 2\pi$ radians $= 360°$

$\therefore$ 1 radian (1 rad) $= \dfrac{360°}{2\pi}$

Taking π as 3.142, 1 rad $\simeq 57.3°$.

Length of arc and Area of sector

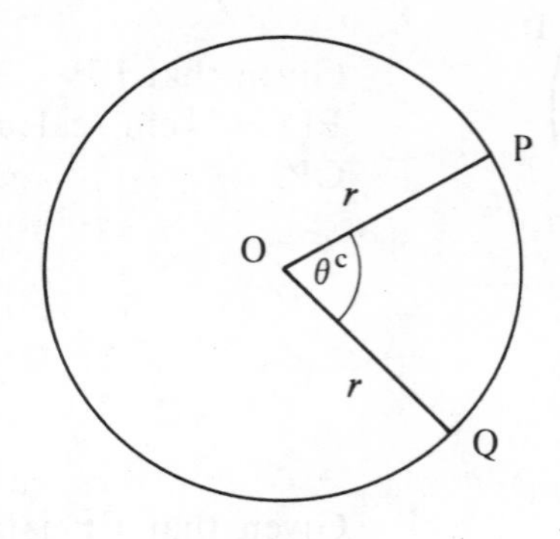

If the angle subtended at the centre of the circle is θ^c, the **length of arc PQ = rθ,**
area of sector POQ $= \frac{1}{2}r^2\,\theta$

Example

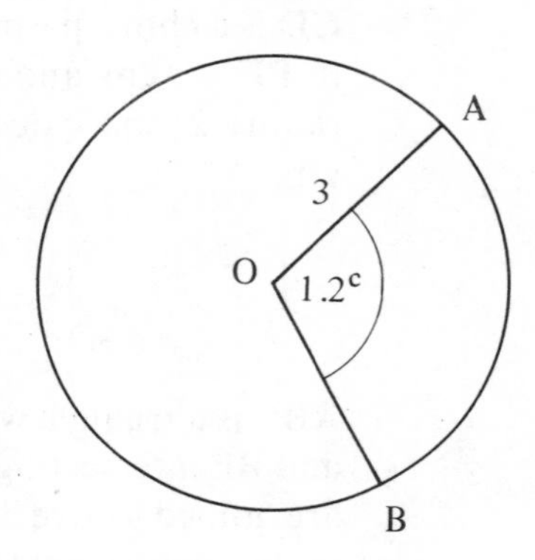

Find **1.** the length of the arc AB that subtends an angle of 1.2^c at the centre of a circle of radius 3cm, **2.** the area of the minor sector AOB.

1. Length of arc AB $= 3 \times 1.2 = \underline{3.6\text{cm}}$ ANS
2. Area AOB $= \frac{1}{2} \times 3^2 \times 1.2 = \frac{1}{2} \times 9 \times 1.2 = \underline{5.4\text{cm}^2}$ ANS

Exercise 88

1. Using a table of 'degrees to radians' convert: (a) 20° to radians, (b) 65.5° to radians, (c) 1.1 rad to degrees, (d) 0.7 rad to degrees.
2. Find (a) the length of an arc subtending an angle of 0.7 rad at the centre of a circle of radius 10cm, (b) the area of the sector bounded by that arc.
3. Find (a) the length of the arc of a circle subtending an angle of 20° at the centre of a circle of radius 4cm, (b) the area of the sector bounded by that arc.
4. Find in terms of π (a) the length of the arc of a circle subtending an angle of $\frac{\pi}{4}$ rad at the centre of a circle of radius 6cm, (b) the area of the sector bounded by that arc.
5. An arc of length $\frac{\pi}{3}$cm subtends an angle θ rad at the centre of a circle of radius 2cm. Find the angle θ in radians.

Examination Standard Questions

Worked example

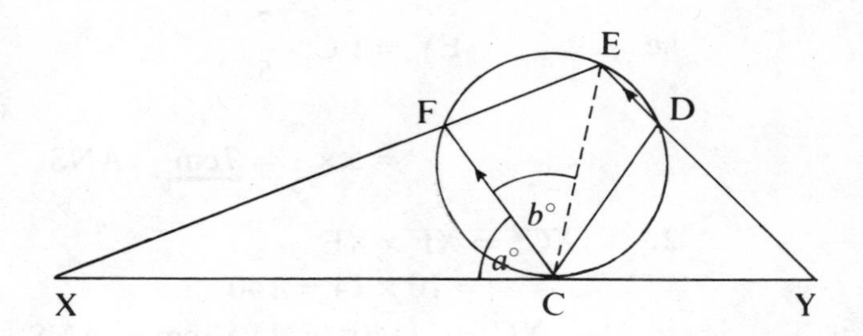

CDEF is a cyclic quadrilateral, with CF parallel to DE. The tangent to the circle at C meets EF produced at X and ED produced at Y.

If $X\hat{C}F = a°$ and $F\hat{C}E = b°$, calculate in terms of a and b angles $C\hat{E}F$, $E\hat{F}C$ and $C\hat{D}E$.
Given that XF = 10cm, XE = 14cm, and CF = 5cm, calculate **1.** the length of EY, **2.** the length XC.

Considering angle XCF = $a°$:

$C\hat{E}F$ is an angle in the alternate segment

$\therefore C\hat{E}F = \underline{a°}$ ANS

Considering triangle CFE:

$$E\hat{F}C + C\hat{E}F + E\hat{C}F = 180°$$
$$\text{i.e. } E\hat{F}C + a° + b° = 180°$$
$$\therefore E\hat{F}C = \underline{(180 - a - b)°} \quad \text{ANS}$$

Considering the cyclic quadrilateral:

$$E\hat{F}C + C\hat{D}E = 180° \quad \text{(opposite angles)}$$
$$\therefore (180 - a - b)° + C\hat{D}E = 180°$$
$$\text{i.e.} \quad C\hat{D}E = 180° - (180 - a - b)°$$
$$= \underline{(a + b)°} \quad \text{ANS}$$

1. Considering triangles XFC and XEY

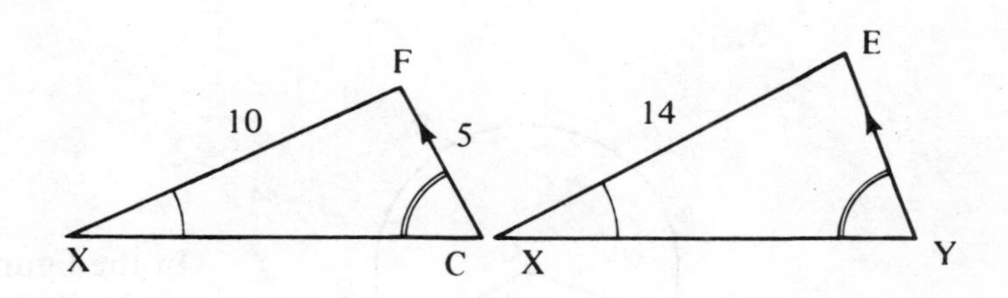

$\hat{X}$ is common, $\hat{C} = \hat{Y}$ (corresponding angles)

$\therefore$ Triangles XFC and XEY are similar.

$$\text{Ratio XF:XE} = 10:14 = 5:7$$

$$\therefore \text{Ratio FC:EY} = 5:7$$

i.e. $$EY = FC \times \frac{7}{5}$$

$$= 5 \times \frac{7}{5} = \underline{7\text{cm}} \quad \text{ANS}$$

2. $$XC^2 = XF \times XE$$

$$= 10 \times 14 = 140$$

$$\therefore XC = \sqrt{140} = \underline{11.83\text{cm}} \quad \text{ANS}$$

Exercise 89

1.

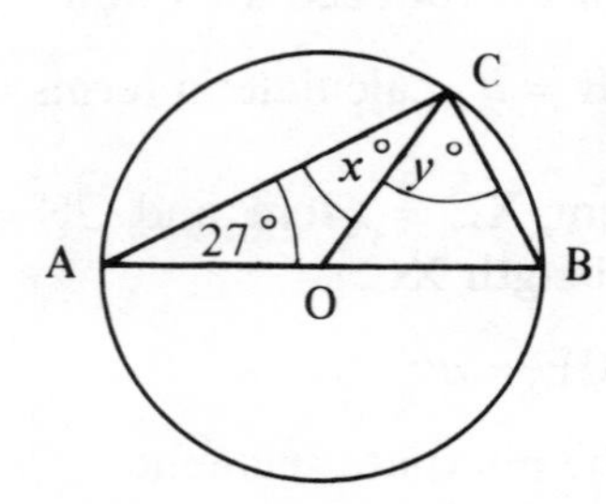

In the diagram, which is not drawn to scale, O is the centre of the circle. The size of the angle CAB is 27°. Calculate the values of x and y. (JMB)

2.

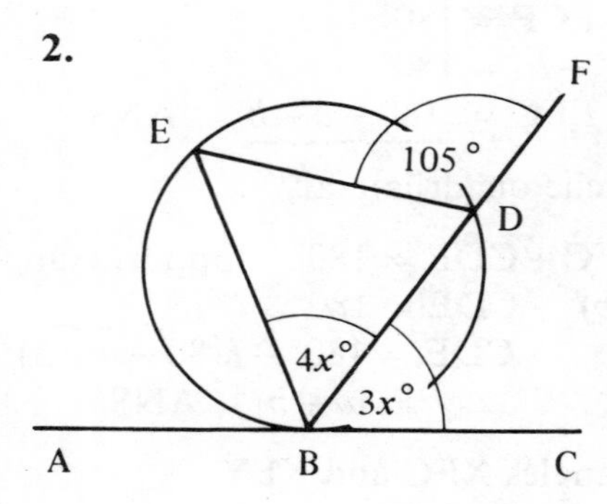

In the figure, ABC is a tangent to the circle. The line BD is produced to F. Angle EDF = 105°, angle EBD = $4x$°, and angle DBC = $3x$°.
(i) Write down in terms of x the value of the angle ABE.
(ii) Calculate the value of x. (O)

3.

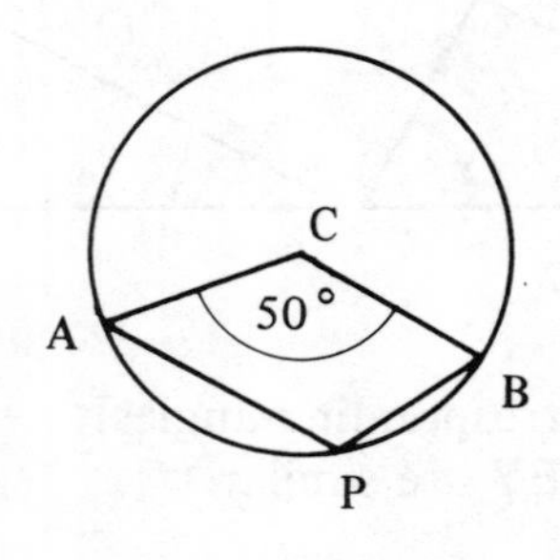

In the figure, C is the centre of the circle. Calculate the size of angle APB. (AEB)

4.

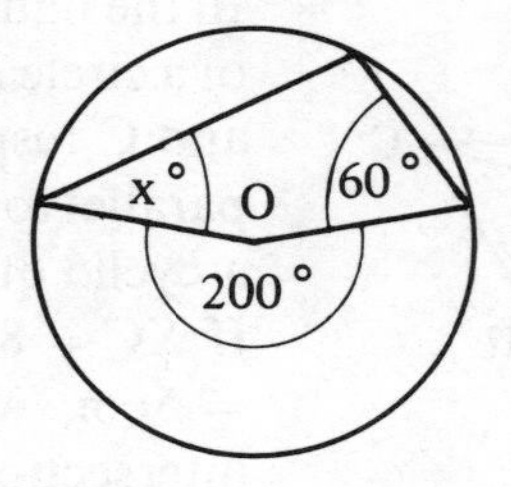

In the figure, O is the centre of the circle. Find x. (AEB)

5.

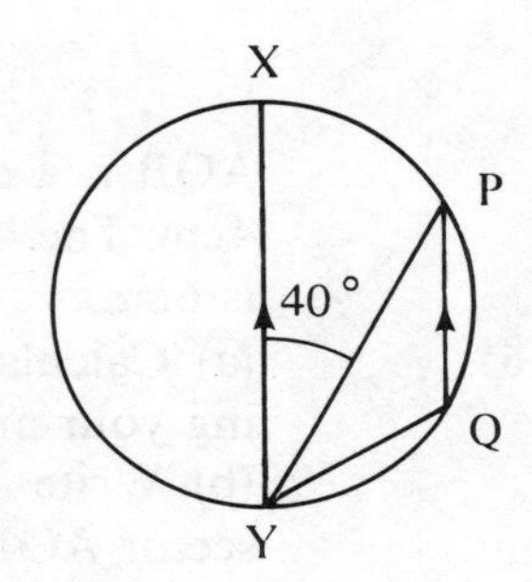

PQ is parallel to the diameter XY. Given that $\angle XYP = 40°$, find $\angle PYQ$. (L)

6.

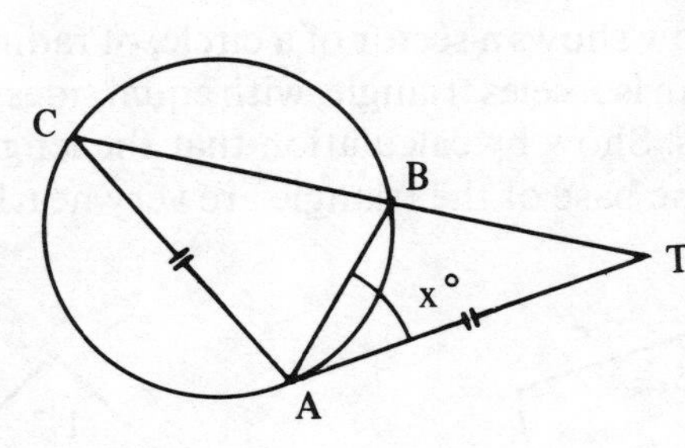

The figure shows a triangle ABC inscribed in a circle, and the tangent to the circle at A meets CB produced at T. Given that AT = AC and angle BAT = $x°$, express in terms of x the angles ACB, BTA, CAB and the angle subtended at the centre of the circle by the chord BC. State why the triangle ABT is isosceles.

Given that AT = AC = 12cm, TB = 8cm and that the bisector of the angle ATB meets BA at Y, calculate (a) the length BC, (b) the length BY, (c) the ratio of the areas of the triangles TCA and TBY giving your answer in the form $n{:}1$ (AEB)

7.

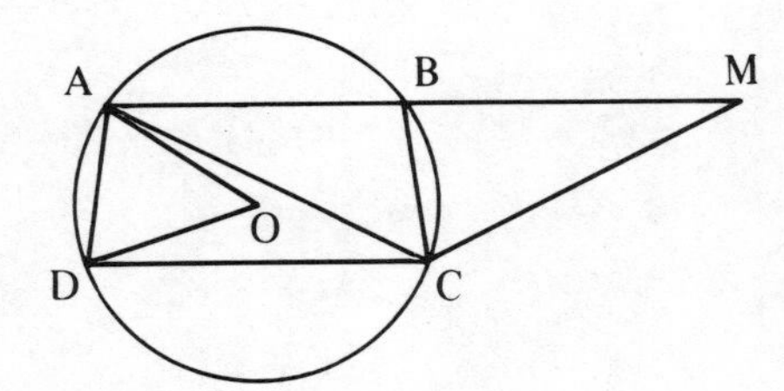

In the diagram ABM is a straight line parallel to DC. The centre of the circle is O and $A\hat{O}D = 2A\hat{M}C$. Prove that (a) $M\hat{A}C = A\hat{M}C$, (b) the triangles BMC and DCA are similar. (C)

8.

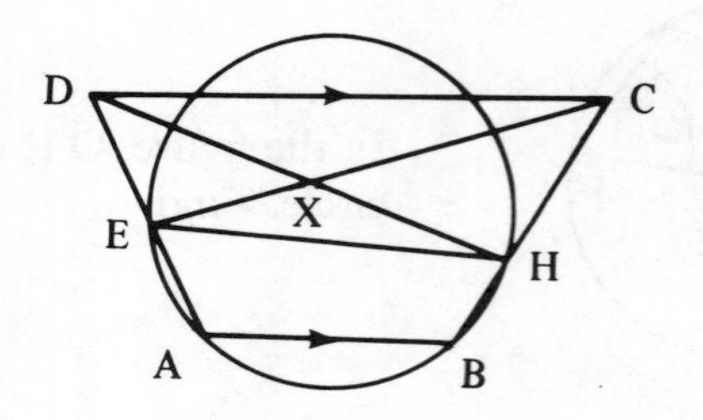

In the figure the chords AE and BH of a circle AEHB are produced to D and C respectively such that DC is parallel to AB. Prove that DCHE is a cyclic quadrilateral.
If XC = 8cm, XE = 3cm and XD = 5cm, where X is the point of intersection of DH and CE, calculate the length of DH. (L)

9.

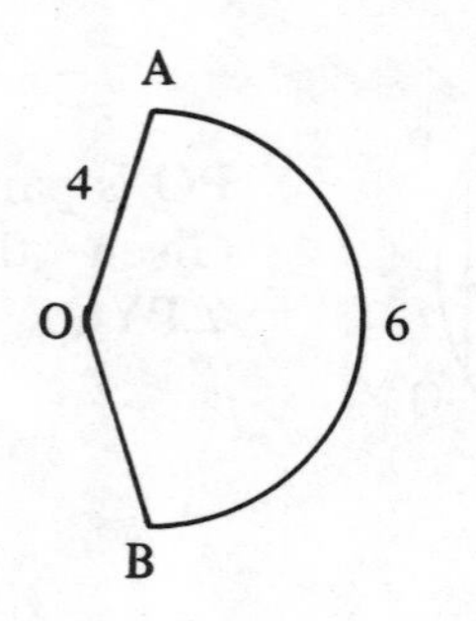

AOB is a circular sector of radius 4cm. The length of the circular arc is 6cm.
(a) Calculate the angle AOB, giving your answer in radians.
(b) Write down the area of the sector AOB. (O)

10. The figure below shows a sector of a circle, of radius r, whose angle is 1.5 radians, and an isosceles triangle, with equal sides r, whose vertical angle is 1.7 radians. Show by calculation that the lengths of the arc of the sector and of the base of the triangle are very nearly equal. (O)

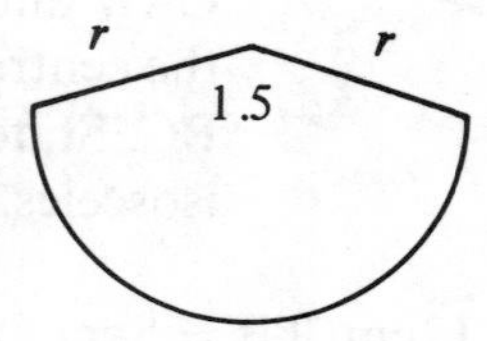

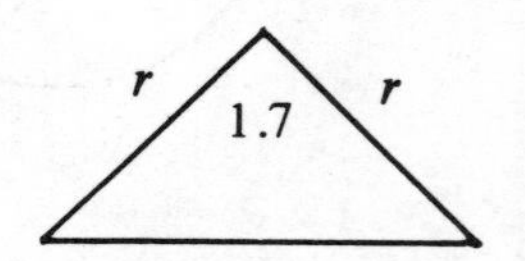

13

Trigonometric Ratios and their Application

Three Basic Ratios

The trigonometric ratios apply *only* to right-angled triangles.
In the diagram angle B is a right-angle, the three ratios to consider are tangent, sine and cosine. These are abbreviated **tan, sin** and **cos.**

The ratios for angle C are:

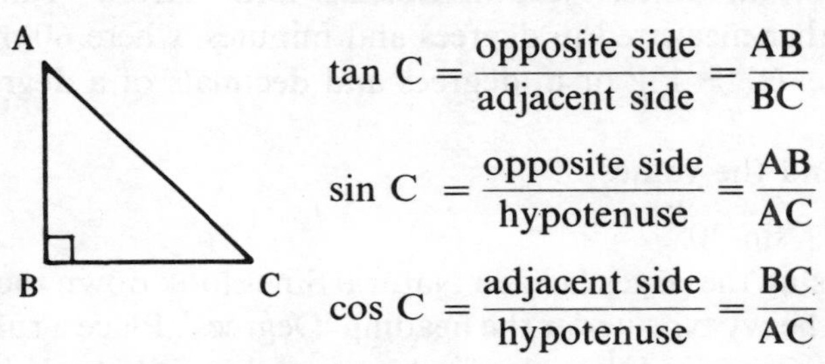

$$\tan C = \frac{\text{opposite side}}{\text{adjacent side}} = \frac{AB}{BC}$$

$$\sin C = \frac{\text{opposite side}}{\text{hypotenuse}} = \frac{AB}{AC}$$

$$\cos C = \frac{\text{adjacent side}}{\text{hypotenuse}} = \frac{BC}{AC}$$

Note The hypotenuse is the side opposite the right-angle (i.e. AC).
The side AB is opposite angle C, but adjacent (next) to angle A.
The side BC is opposite angle A, but adjacent (next) to angle C.
Therefore, the ratios for angle A are:

$$\tan A = \frac{\text{opposite side}}{\text{adjacent side}} = \frac{BC}{AB}$$

$$\sin A = \frac{\text{opposite side}}{\text{hypotenuse}} = \frac{BC}{AC}$$

$$\cos A = \frac{\text{adjacent side}}{\text{hypotenuse}} = \frac{AB}{AC}$$

From the next diagram we can calculate the values of the following ratios: (i) tan A, (ii) tan C, (iii) sin A, (iv) cos C.

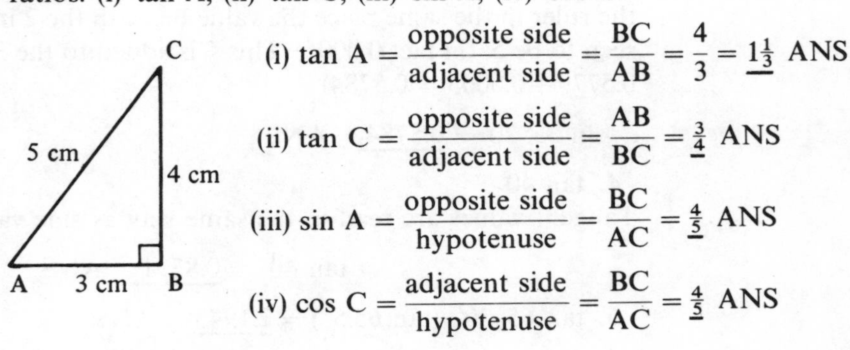

(i) $\tan A = \frac{\text{opposite side}}{\text{adjacent side}} = \frac{BC}{AB} = \frac{4}{3} = \underline{1\frac{1}{3}}$ ANS

(ii) $\tan C = \frac{\text{opposite side}}{\text{adjacent side}} = \frac{AB}{BC} = \underline{\frac{3}{4}}$ ANS

(iii) $\sin A = \frac{\text{opposite side}}{\text{hypotenuse}} = \frac{BC}{AC} = \underline{\frac{4}{5}}$ ANS

(iv) $\cos C = \frac{\text{adjacent side}}{\text{hypotenuse}} = \frac{BC}{AC} = \underline{\frac{4}{5}}$ ANS

Exercise 90 From the diagrams write down the values of:

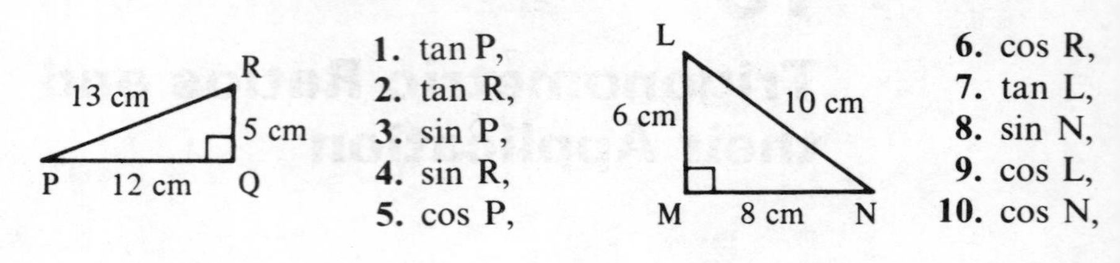

1. tan P,
2. tan R,
3. sin P,
4. sin R,
5. cos P,
6. cos R,
7. tan L,
8. sin N,
9. cos L,
10. cos N,

Using Tables

In a book of four-figure tables there will be found pages headed – Natural Sines, Natural Cosines and Natural Tangents. Angles are either measured in degrees and minutes, where 60 minutes = 1 degree, (i.e. $60' = 1°$); or in degrees and decimals of a degree.

Examples

Find the values of

1. sin 30°

Using the pages headed Natural Sines look down the left-hand column in heavy type under the heading 'Degrees'. Place a ruler beneath the 30° line, and read the value in the next column that is below the O′ heading (i.e. zero minutes).

$$\therefore \sin 30° = \underline{0.5} \quad \text{ANS}$$

2. sin 35° 18′, (sin 35.3°)

Place a ruler beneath the 35° line and read the value in the column that is below the 18′ (0.3) heading.

$$\therefore \sin 35° 18' \text{ (or } 35.3°) = \underline{0.5779} \quad \text{ANS}$$

3. sin 35° 20′

Keep the ruler beneath the 35° line and make a note of the value in the minutes column that is the nearest *below* 20′ (i.e. 35° 18′ = 0.5779). The value of the additional two minutes required is found in the 'mean differences' section on the right-hand side of the same page. Keeping the ruler in the same place the value beneath the 2 in the differences is seen to be 5, (in fact 0.0005). This 5 is added to the 35° 18′ value (thus: $0.5779 + 0.0005 = 0.5784$)

$$\therefore \sin 35° 20' = \underline{0.5784} \quad \text{ANS}$$

4. tan 40°,

Tangent values are read in the same way as sine values.

$$\therefore \tan 40° = \underline{0.8391} \quad \text{ANS}$$

5. tan 65° 30′, (tan 65.5°) = $\underline{2.1943}$ ANS

Note When using tangent tables check the whole number at the beginning of the row. The next example illustrates a variation of this procedure.

6. $\tan 71°48'$, $(\tan 71.8°) = \underline{3.0415}$ ANS

7. $\tan 70°03' = 2.7475 + 0.0078$ [mean difference]
$= \underline{2.7553}$ ANS

8. $\cos 60° = \underline{0.5}$ ANS

9. $\cos 70°24'$, $(\cos 70.4°) = \underline{0.3355}$ ANS

10. $\cos 70°27'$, Note the instruction at the top of the Natural Cosines page.

$\cos 70°27' = 0.3355 - 0.0008$
$= \underline{0.3347}$ ANS

Exercise 91 Use tables to find the values of:

1. sin 45°,	**2.** sin 45.3°,	**3.** sin 45° 20′,	**4.** sin 77.2°,
5. tan 30°,	**6.** tan 30.7°,	**7.** tan 66° 40′,	**8.** tan 01.1°,
9. cos 35°,	**10.** cos 35.3°,	**11.** cos 35° 20′,	**12.** cos 70°,
13. cos 70° 33′,	**14.** cos 70° 35′,	**15.** cos 90°,	**16.** sin 45° 19′,
17. tan 11.3°,	**18.** tan 63° 45′,	**19.** cos 30° 15′,	**20.** cos 11° 11′.

Acute Angles

Examples Find the acute angles equal to

1. $\sin^{-1} 0.5$,

[This means find the angle whose sine is 0.5.]
[Using Natural Sine tables] $\sin^{-1} 0.5 = \underline{30°}$ ANS

2. $\tan^{-1} 2.1543$

[Using Natural Tangent tables] $= \underline{65°06'}$ (or $\underline{65.1°}$) ANS

3. $\tan^{-1} 0.7706$

[Look for the value just *below* that required, then add the extra minutes from the difference columns.]

$= 0.7701 + 0.0005$
$= 37°36' + 1'$ difference
$= \underline{37°37'}$ ANS

4. $\cos^{-1} 0.4110$

[Remember, as the value of the cosine gets smaller, the angle gets larger. Look for the value just below that required, then *subtract*.]

$= 0.4099 + 0.0011$
$= 65°48' - 4'$ difference
$= \underline{65°44'}$ ANS

Exercise 92 Find the acute angles equal to:

1. $\sin^{-1} 0.8192$, **2.** $\sin^{-1} 0.9968$, **3.** $\tan^{-1} 2.8239$,
4. $\tan^{-1} 0.8556$, **5.** $\cos^{-1} 0.766$, **6.** $\cos^{-1} 0.9728$,
7. $\cos^{-1} 0.2683$, **8.** $\cos^{-1} 0.6745$, **9.** $\cos^{-1} 0.9947$,
10. $\sin^{-1} 0.0029$.

Calculation of angles

Examples Find the angles marked α in each case:

1.

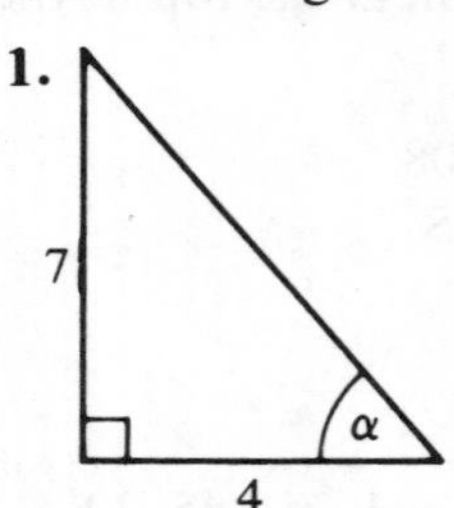

The sides given are those opposite (7) and adjacent to (4) the angle α.

Using $\tan\alpha = \dfrac{\text{opposite side}}{\text{adjacent side}} = \dfrac{7}{4}$

$= 4\overline{)7.000}$ giving 1.75

[Using Natural Tangent tables correct to the nearest 1′] $\therefore \tan^{-1} 1.75 = \underline{60°15'}$ ANS

2.

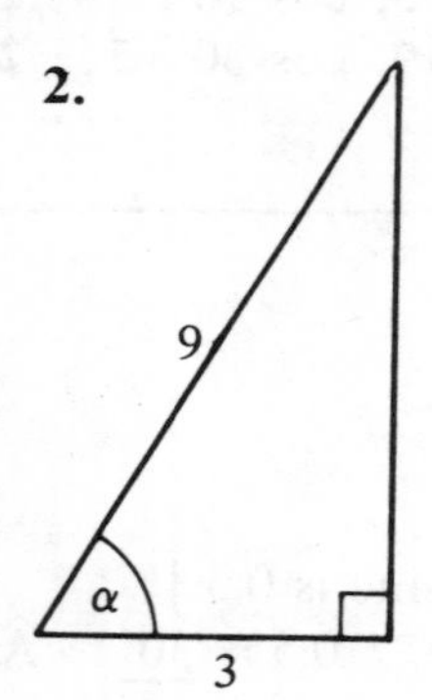

The sides given are those adjacent to (3) the angle and the hypotenuse (9)

Using $\cos\alpha = \dfrac{\text{adjacent side}}{\text{hypotenuse}} = \dfrac{3}{9}$

$= 9\overline{)3.0000}$ giving 0.3333

$\therefore \cos^{-1} 0.3333 = \underline{70°32'}$ ANS

Exercise 93 Find the angle marked α in each case:

1.

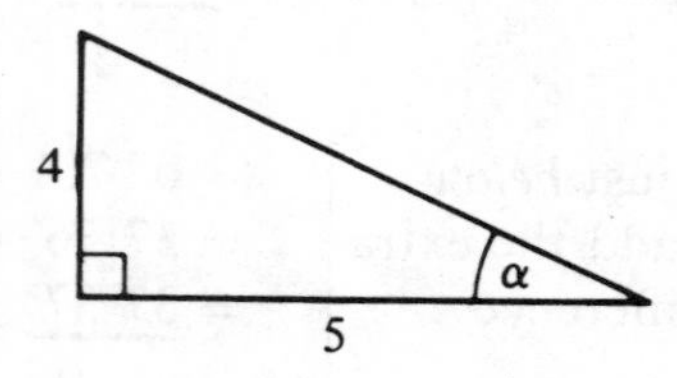

2.

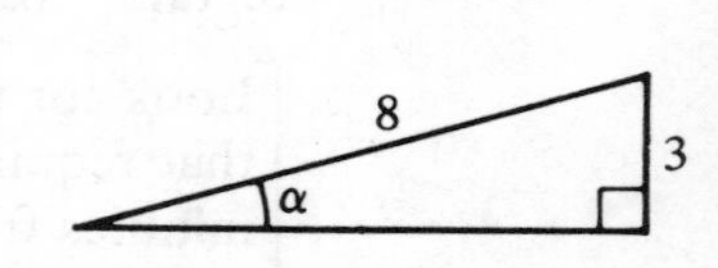

3.

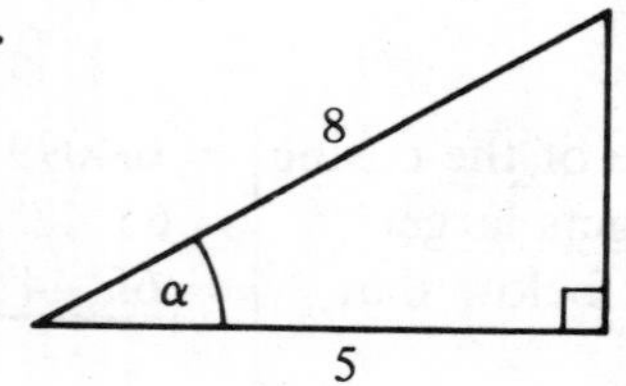

4.

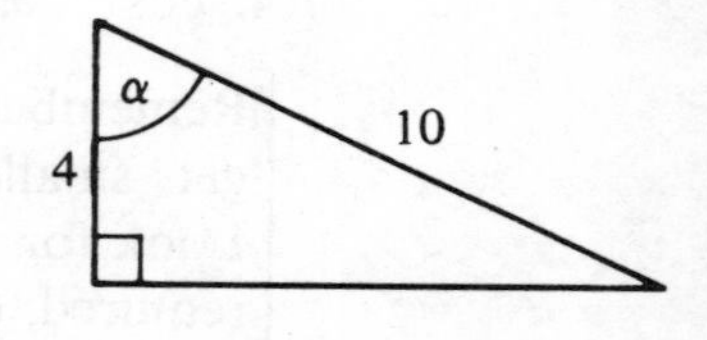

5.

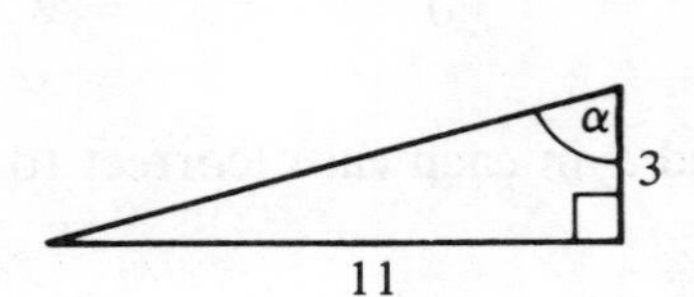

6.

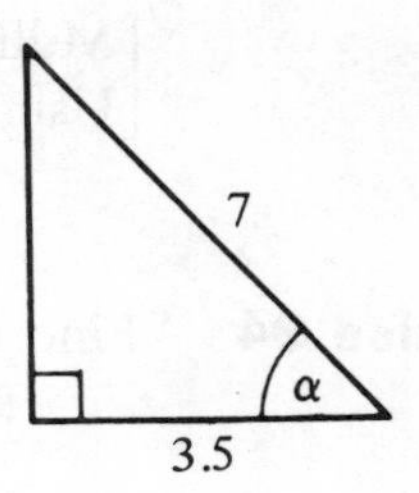

7.

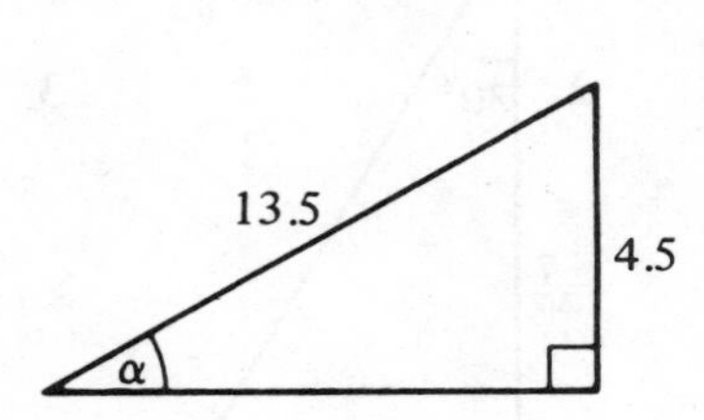

8.

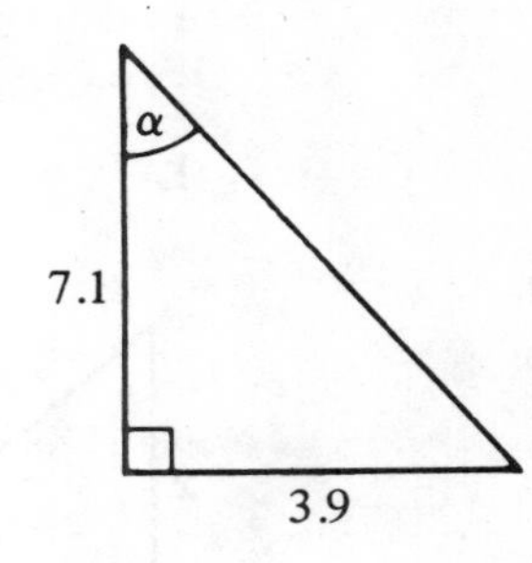

9.

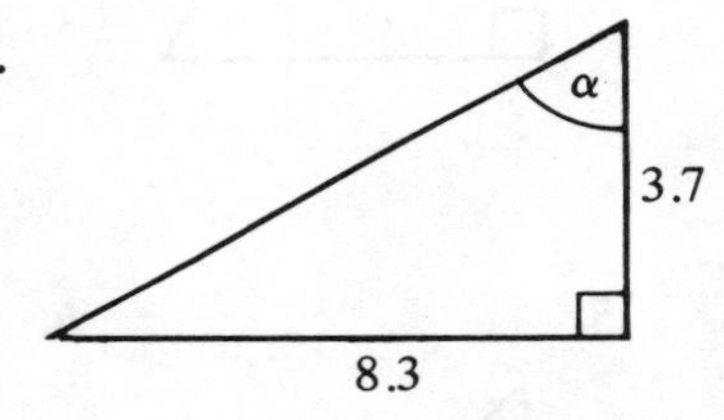

10.

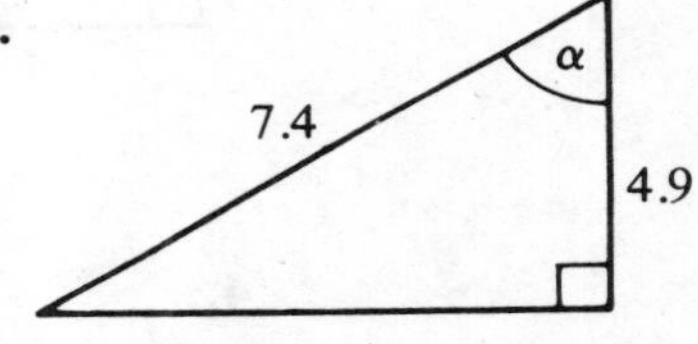

Calculations of lengths

Examples

Find the side marked x in each case:

1.

x, $45°$, 6

The sides involved are those opposite (x) and adjacent to (6) the angle of 45°.

Using $\tan 45° = \dfrac{\text{opposite side}}{\text{adjacent side}} = \dfrac{x}{6}$

[Multiply both sides by 6] $\therefore\ 6 \times \tan 45° = x$

[Using Natural Tangent tables] $\therefore\ 6 \times 1.0 = x$

$\therefore\ x = \underline{6}$ ANS

2.

12, x, $50°$

The sides involved are those opposite (x) to the angle of 50°, and the hypotenuse (12).

Using $\sin 50° = \dfrac{\text{opposite side}}{\text{hypotenuse}} = \dfrac{x}{12}$

[Multiply both sides by 12] $\therefore\ 12 \times \sin 50^\circ = x$
[Using Natural Sine tables] $\therefore\ 12 \times 0.766 = x$

$\therefore\ x = \underline{9.192}$ ANS

Exercise 94 Find the side marked x in each case (correct to 2 d.p.):

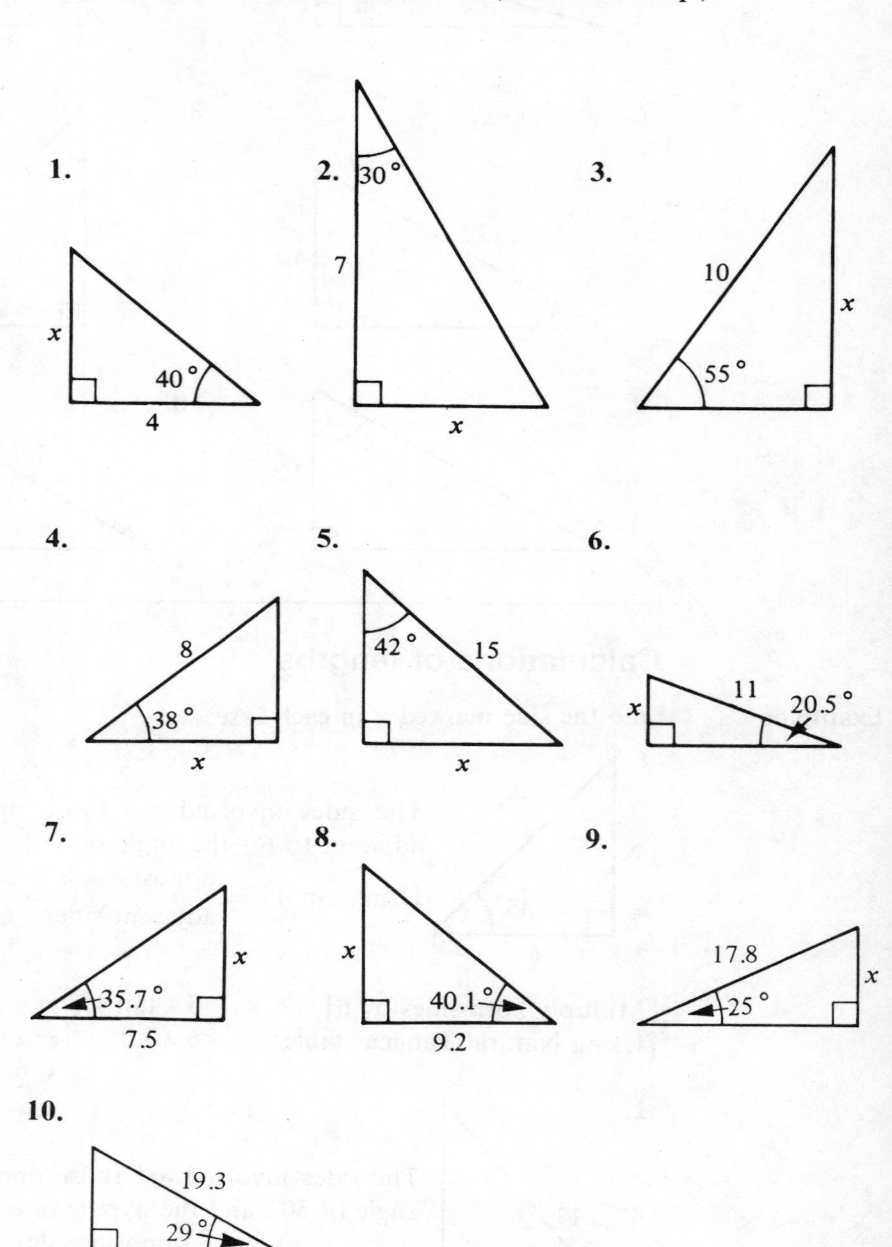

Complementary angle

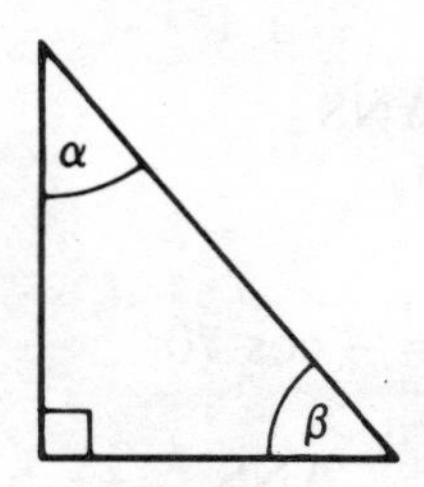

In this diagram the angles $\alpha^\circ + \beta^\circ = 90^\circ$, and angle α is said to be complementary to angle β.
It can be seen from tables that $\sin 30^\circ = \cos 60^\circ$.
(i.e. $\sin x = \cos(90^\circ - x)$ and $\cos x = \sin(90^\circ - x)$.
It is sometimes helpful to use the complementary angle when dealing with the tangent of an angle to avoid performing an awkward division.

Example Find the side marked x:

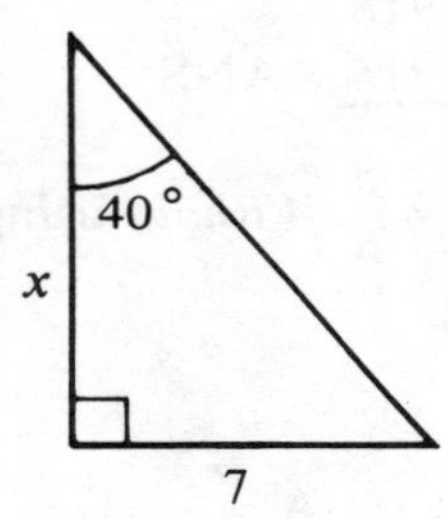

The sides involved are those adjacent to (x) and opposite to (7) the angle of 40°.

$$\text{Using } \tan 40^\circ = \frac{\text{opposite side}}{\text{adjacent side}} = \frac{7}{x}$$

To avoid an awkward division use the complementary angle;

i.e.

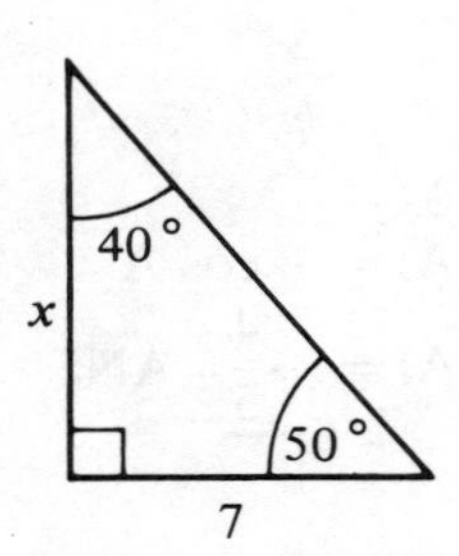

$$\text{Using } \tan 50^\circ = \frac{\text{opposite side}}{\text{adjacent side}} = \frac{x}{7}$$

$$\therefore 7 \times \tan 50^\circ = x$$

$$\therefore x = \underline{8.3426} \quad \text{ANS}$$

Obtuse Angles

Tables of the trigonometric ratios deal only with acute angles, i.e. angles between 0° and 90°. The following rules enable obtuse angles (i.e. those between 90° and 180°) to be found.

$$\sin(180^\circ - x) = \sin x$$
$$\cos(180^\circ - x) = -\cos x$$
$$\tan(180^\circ - x) = -\tan x$$

Examples

1. Find the value of $\sin 120°$:
Let $x = 120°$ in the previous rule,
$\sin(180° - x) = \sin(180° - 120°) = \sin 60°$
from the tables $\sin 60° = 0.866$
$\therefore \sin 120° = \underline{0.866}$ ANS

2. Find the value of $\cos 110°$:
Let $x = 110°$ in the previous rule,
$\cos(180° - x) = \cos(180° - 110°) = -\cos 70°$
from the tables $-\cos 70° = -0.342$
$\therefore \cos 110° = \underline{-0.342}$ ANS

3. Find the value of $\tan 130°$:
Let $x = 130°$ in the previous rule,
$\tan(180° - x) = \tan(180° - 130°) = -\tan 50°$
from the tables $-\tan 50° = -1.1918$
$\therefore \tan 130° = \underline{-1.1918}$ ANS

4. The angle A is acute and $\sin A = \frac{3}{5}$. *Without* using tables, calculate the value of $\cos(180° - A)$.

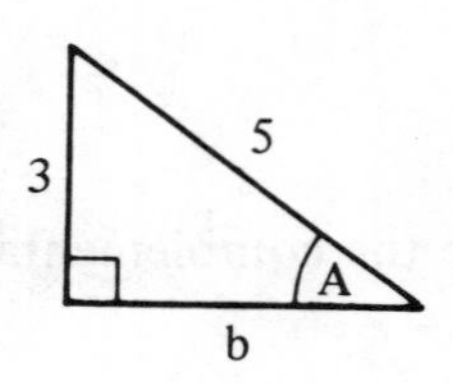

In this diagram $\sin A = \frac{3}{5}$.

By Pythagoras Theorem; the base b can be found
Thus: $b^2 = 5^2 - 3^2 = 25 - 9 = 16$
$\therefore b = \sqrt{16} = \underline{4}$

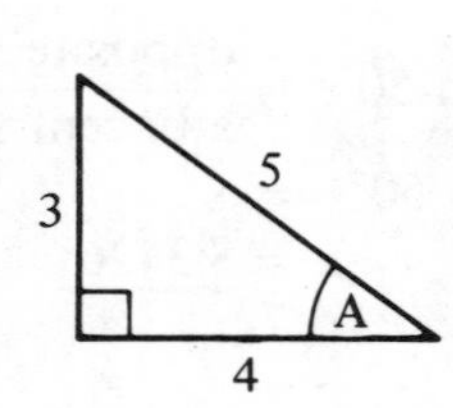

$\cos A = \frac{4}{5}$

As $\cos(180° - A) = -\cos A$

$\cos(180° - A) = \underline{-\frac{4}{5}}$ ANS

5. If $\cos x = -\frac{5}{13}$ and $0° < x < 180°$, find $\sin x$.
The question states that x is between 0° and 180°, and as $\cos x$ is negative, x is an obtuse angle.

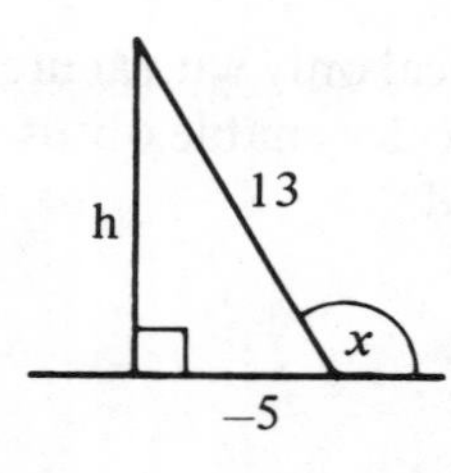

In this diagram x is the obtuse angle.
By Pythagoras Theorem:

$$h^2 = 13^2 - (-5)^2$$
$$= 13^2 - (+25)$$
$$= 169 - 25$$
$$\therefore h = \sqrt{144} = \underline{12}$$

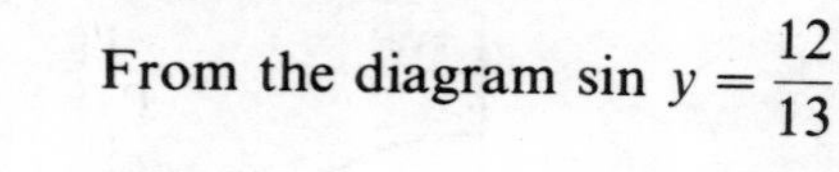

Angle $y = 180° - x°$

From the diagram $\sin y = \dfrac{12}{13}$

i.e. $\sin(180 - x)° = \dfrac{12}{13}$

The rules for obtuse angles state that $\sin(180 - x)° = \sin x$

$\therefore \sin x = \dfrac{12}{13}$ ANS

Exercise 95 Find the values of the following:

Use your tables for questions **1** to **10**.

1. sin 100° **2.** cos 100° **3.** tan 100°
4. sin 165° **5.** sin 179.5° **6.** cos 103.7°
7. cos 157.1° **8.** tan 91° **9.** tan 147.6°
10. tan 111°

Solve numbers **11** to **15** without using tables:

11. The angle x is acute and $\sin x = \frac{4}{5}$. Calculate the value of $\cos(180 - x)°$.

12. The angle y is acute and $\cos y = \frac{15}{17}$. Calculate the value of $\sin (180 - y)°$.

13. If $\cos x = -\frac{4}{5}$ and $0° < x < 180°$, find $\sin x$.

14. If $\sin y = \frac{4}{5}$ and $90° < y < 180°$, find $\tan y$.

15. If $\cos (180° - A) = -\frac{8}{17}$, and $0° < A < 180°$, find tan A.

The General Angle

The rules stated for obtuse angles can be verified, and extented generally by considering the four quadrants of a circle.

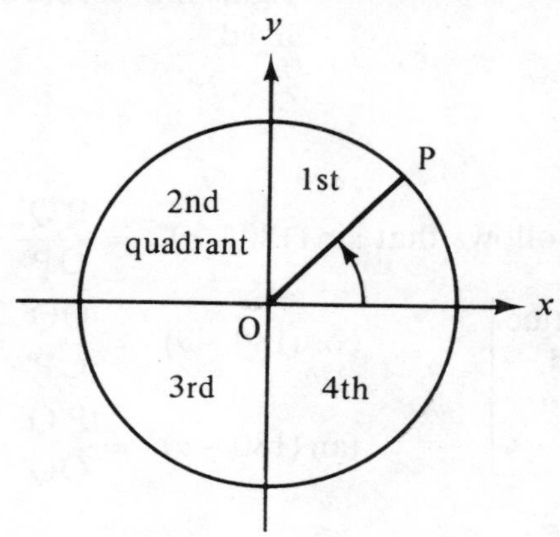

Rotation about the origin of a point P in an *anti-clockwise* direction forms a *positive* angle $x\hat{O}P$. *Clockwise* rotation forms a *negative* angle.

As the radius of the circle OP rotates about the origin it passes through the four quadrants of the circle. These are referred to in the anti-clockwise order as shown in the diagram.

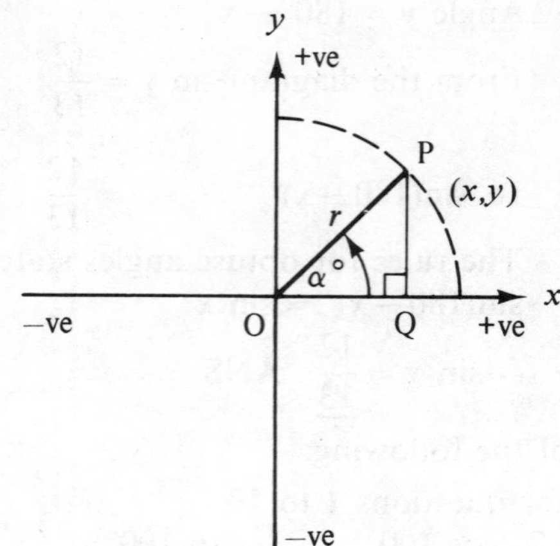

Consider the acute angle $x\hat{O}P = \alpha°$. Let Q be the foot of the perpendicular dropped from P (x, y) onto the x-axis.

Then $\sin \alpha = \dfrac{PQ}{OP}$

$\cos \alpha = \dfrac{OQ}{OP}$

$\tan \alpha = \dfrac{PQ}{OQ}$

[The positive and negative directions of the axes are indicated (+ve, −ve) as a reminder.]

From the diagram OQ represents the x-coordinate of P, and PQ represents the y-coordinate of P.

The radius OP is denoted by r, which is always positive.

Our three trigonometric ratios now become $\sin \alpha = \dfrac{y}{r}$

$\cos \alpha = \dfrac{x}{r}$

and $\tan \alpha = \dfrac{y}{x}$

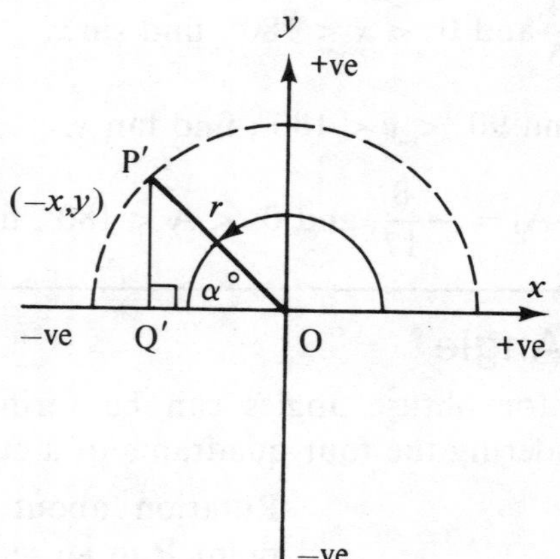

Let the radius OP now rotate about the origin through an angle of $+(180-\alpha)°$ so that its new position (OP′) is a reflection of OP in the y-axis.

i.e. the x-coordinate changes its sign, the y-coordinate is unchanged.

It follows that $\sin (180-\alpha)° = \dfrac{P'Q'}{OP'} = \dfrac{y}{r} = \sin \alpha$

$\cos (180-\alpha)° = \dfrac{OQ'}{OP'} = \dfrac{-x}{r} = -\cos \alpha$

$\tan (180-\alpha)° = \dfrac{P'Q'}{OQ'} = \dfrac{y}{-x} = -\tan \alpha$

[The numerical value of α is unchanged, but the signs must be considered.]

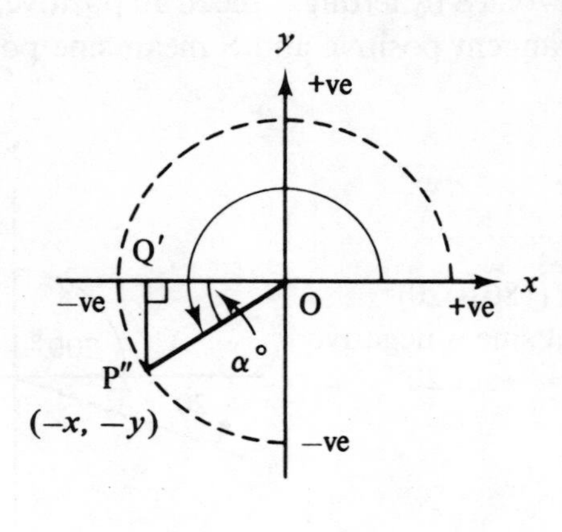

Similarly, for angles in the 3rd quadrant:

$$\sin(180+\alpha)^\circ = \frac{Q'P''}{OP''} = \frac{-y}{r}$$
$$= -\sin\alpha$$
$$\cos(180+\alpha)^\circ = \frac{OQ'}{OP''} = \frac{-x}{r}$$
$$= -\cos\alpha$$
$$\tan(180+\alpha)^\circ = \frac{Q'P''}{OQ'} = \frac{-y}{-x}$$
$$= \frac{y}{x} = \tan\alpha$$

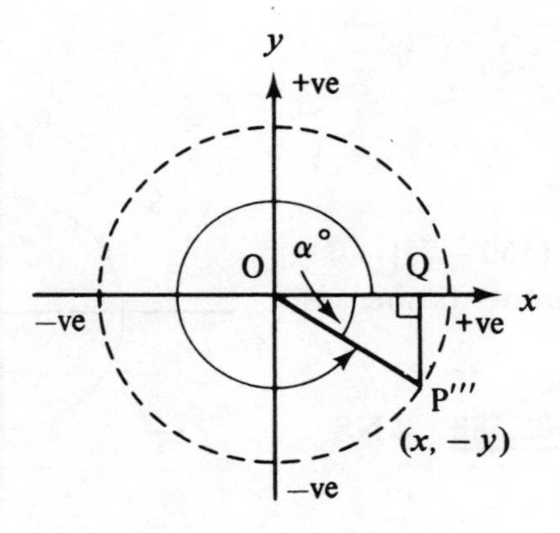

and in the 4th. quadrant:

$$\sin(360-\alpha)^\circ = \frac{QP'''}{OP'''} = \frac{-y}{r}$$
$$= -\sin\alpha$$
$$\cos(360-\alpha)^\circ = \frac{OQ}{OP'''} = \frac{x}{r}$$
$$= \cos\alpha$$
$$\tan(360-\alpha)^\circ = \frac{QP'''}{OQ} = \frac{-y}{x}$$
$$= -\tan\alpha$$

It can be seen that the signs in each quadrant are:

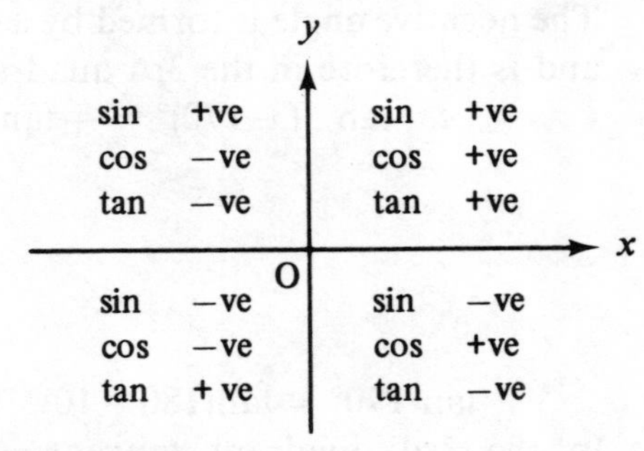

which can be summarised by:

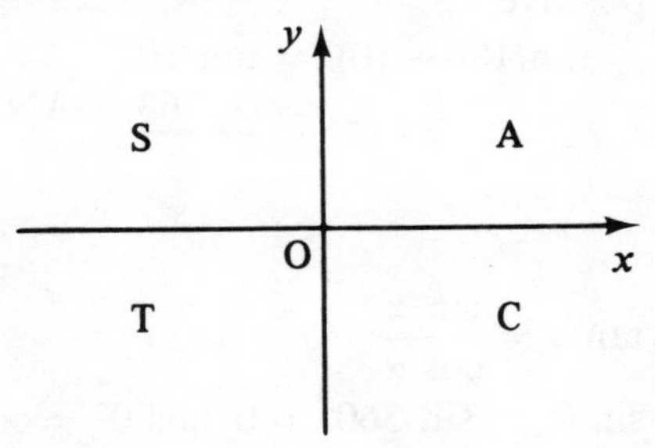

this aid to memory ACTS by letting A mean all positive, C mean cosine positive, T mean tangent positive and S mean sine positive.

Examples

Find the values of

1. $\sin 200°$

$$\sin 200° = \sin(180+20)°$$

In the 3rd quadrant sine is negative

$\therefore \sin(180+20)° = -\sin 20°$

[From tables]

$= \underline{-0.342}$ ANS

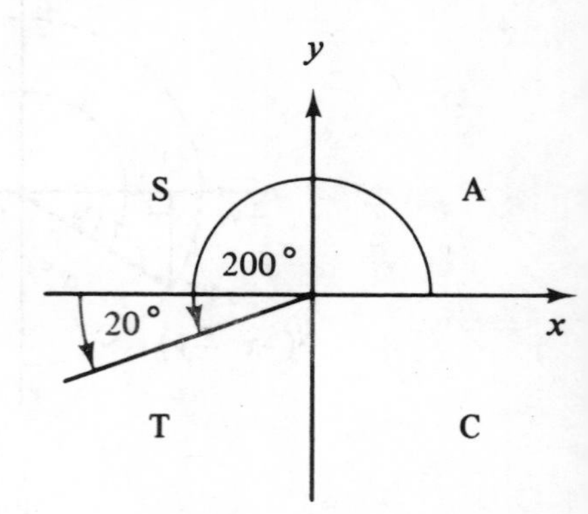

2. $\cos 285°$

$$\cos 285° = \cos(360-75)°$$

In the 4th quadrant cosine is positive

$\therefore \cos(360-75)° = \cos 75°$

$= \underline{0.2588}$ ANS

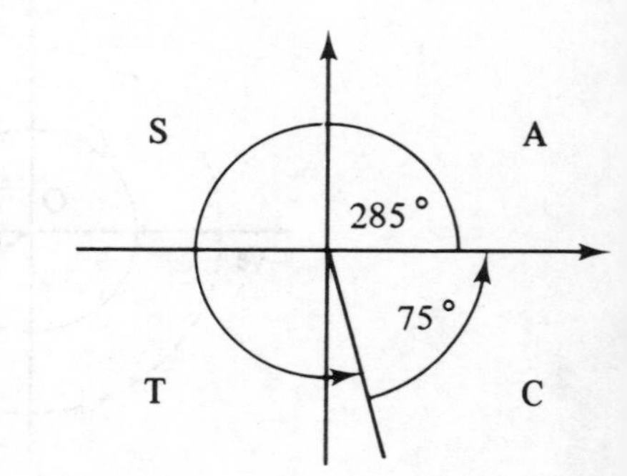

3. $\tan(-170°)$

The negative angle is formed by a clockwise rotation about the origin, and is therefore in the 3rd quadrant.

i.e. $\tan(-170)° = +\tan 190°$ [The corresponding anti-clockwise rotation.]

$$\tan 190° = \tan(180+10)°$$

In the 3rd quadrant tangent is positive

$\therefore \tan(180+10)° = \tan 10°$

$= \underline{0.1763}$ ANS

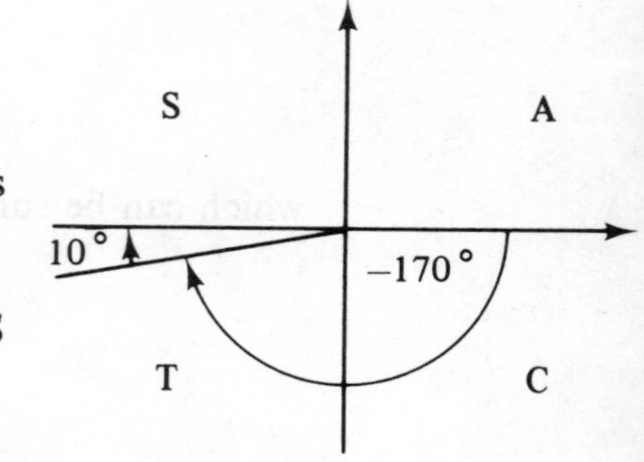

Note

$\tan\alpha = \dfrac{\sin\alpha}{\cos\alpha}$;

$\sin 0° = \sin 360° = 0$; $\cos 0° = \cos 360° = 1$;

$\sin 90° = 1$; $\cos 90° = 0$;
$\sin 180° = 0$; $\cos 180° = -1$;
$\sin 270° = -1$; $\cos 270° = 0$.

Exercise 96 Find the values of:

1. sin 60° **2.** sin 220° **3.** sin 300° **4.** cos 150°
5. cos 210° **6.** cos 350° **7.** tan 112° **8.** tan 258°
9. tan 310° **10.** sin(−150)°.

Polar Coordinates

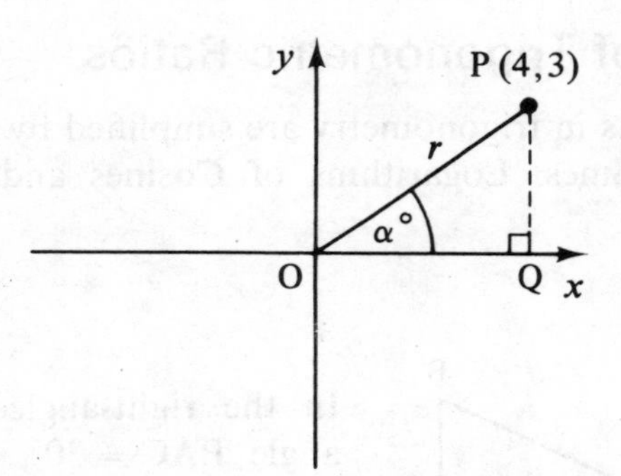

Consider the point P(4, 3). The coordinates of P are given as Cartesian coordinates. An alternative method is to locate P in terms of r (the radius OP) and the angle α.

[Here, OQ = 4, PQ = 3] Using Pythagoras Theorem:

$$OQ^2 + PQ^2 = OP^2$$
$$\text{i.e. } 4^2 + 3^2 = r^2$$
$$\therefore \sqrt{25} = r = \underline{5}$$
$$\text{Using } \tan\alpha = \frac{PQ}{OQ}$$
$$= \frac{3}{4} = 0.75,$$

[From tables] $\alpha = \tan^{-1} 0.75 = \underline{36° 52'}$

∴ P has **polar coordinates** (5, 36° 52′).

To convert from one set of coordinates to the other we use the relationship:

$$\boldsymbol{x = r\cos\alpha; \quad y = r\sin\alpha; \quad \tan\alpha = \frac{y}{x}.}$$

Example Convert the polar coordinates (4, 30°) to Cartesian coordinates.

x-coordinate $= 4\cos 30° = 4 \times 0.866 = \underline{3.464}$
y-coordinate $= 4\sin 30° = 4 \times 0.5 = \underline{2.0}$

∴ The cartesian coordinates are (3.464, 2.0)

Exercise 97 Find (correct to 1 decimal place) the Cartesian coordinates of the points whose polar coordinates are:

1. (3, 30°) **2.** (2, 40°) **3.** (1, 35°)
4. (3, 120°) **5.** (5, 210°)

Find the polar coordinates of the points whose Cartesian coordinates are:

6. (5, 12) **7.** (2, 2) **8.** (4, 1)
9. (– 3, 3) **10.** (4, – 2)

Logarithms of Trigonometric Ratios

Many calculations in trigonometry are simplified by using the tables; Logarithms of Sines, Logarithms of Cosines and Logarithms of Tangents.

Example

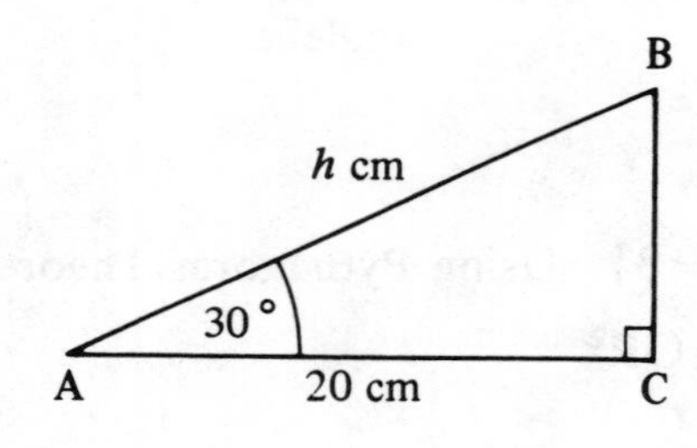

In the right-angled triangle ABC angle BAC = 30°, AB = h cm and AC = 20 cm.
Calculate the value of h correct to two significant figures.

[The sides involved are those adjacent to (20) the angle of 30°, and the hypotenuse (h)]

Using $\cos 30° = \frac{20}{h}$

[Rearranging gives] $h = \frac{20}{\cos 30°}$

[Using logs.]

[The log. of cos 30° is obtained from the log. of cosines page; remember to subtract difference figures.]

No.	Log.
20.0	1.3010
cos 30°	$\bar{1}.9375$ –
	1.3635

[Antilog.] $\therefore h = 23.10$
[Correct to 2 s.f.] $= \underline{23 \text{ cm}}$ ANS

Angles of Elevation and Depression

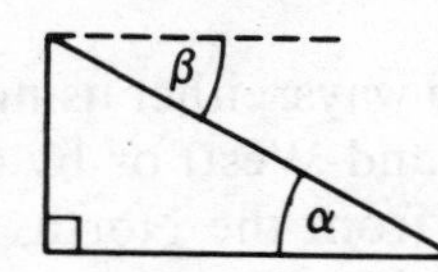

In this diagram α is the angle of elevation, β is the angle of depression.

Note α and β are 'alternate angles' and therefore equal.

Example A point P is 110m vertically above a point Q at sea level. Calculate the angle of elevation of P when viewed from a point at sea 300m from Q.

[The angle of elevation is α]

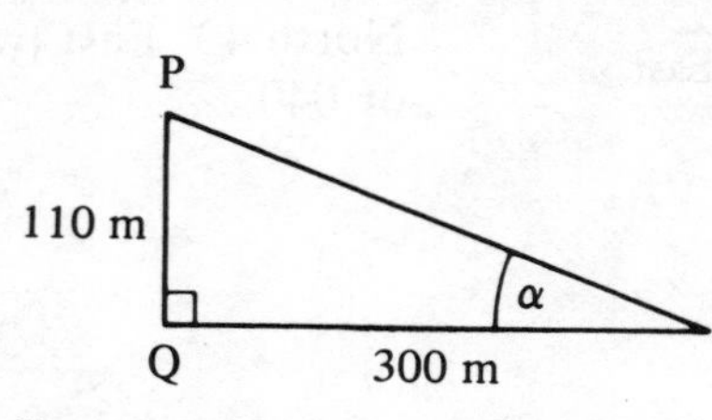

Using $\tan \alpha = \dfrac{110}{300}$

No.	Log.
110	2.0414
300	2.4771 –
	$\bar{1}.5643$

[*Short cut*: look up $\bar{1}.5643$ in the main part of the log. of tangent tables.]

$= \underline{20° 08'}$ ANS

Exercise 98 In questions **1** to **5** calculate the lettered side correct to 2 s.f.:

1.

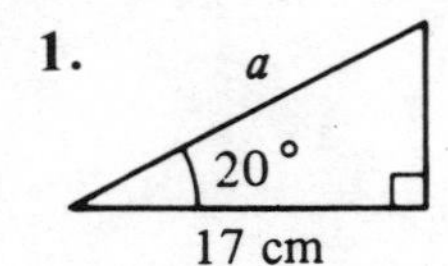

2.

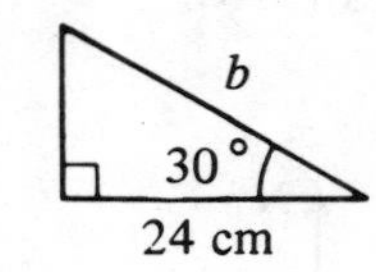

3.

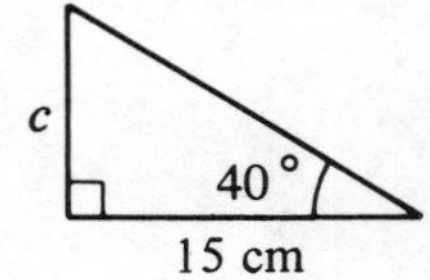

4.

50°
10 cm
d

5.

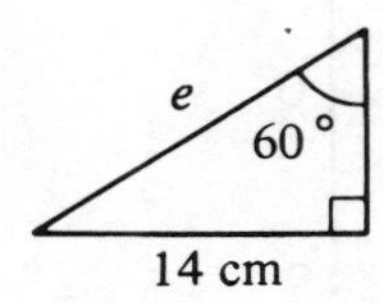

In questions **6** to **10** calculate angle α:

6.

4 cm
α
7 cm

7.

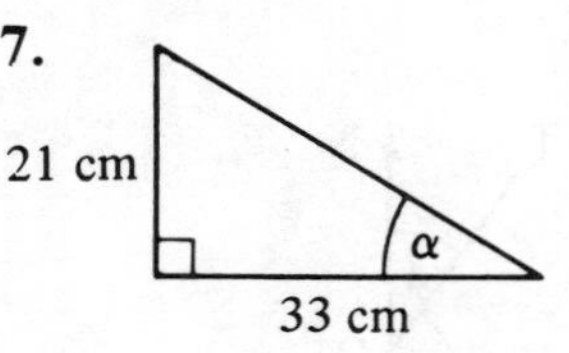

8.

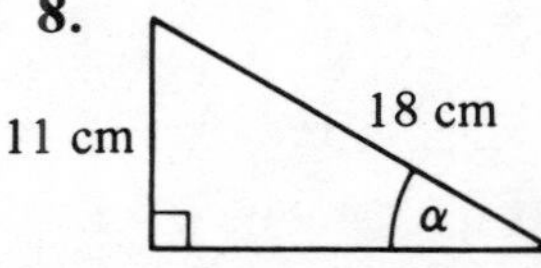

9.

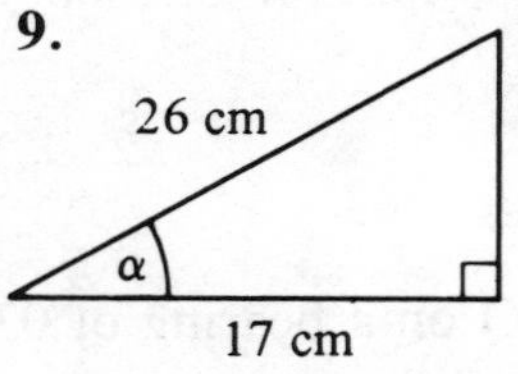

10.

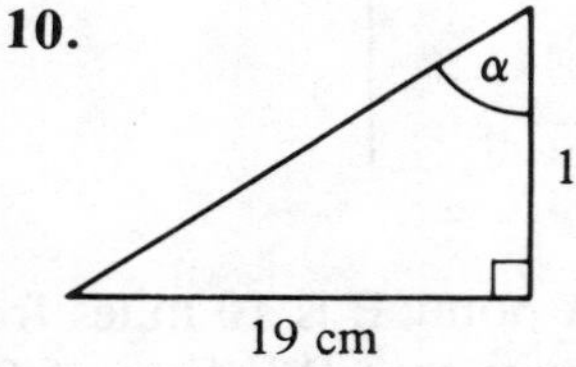

Bearings

A bearing can be given in one of two ways; either using the main points of a compass (North, South, East and West) or by stating the angle measured in a clockwise direction from the North, employing three digits to do so as in the following examples.

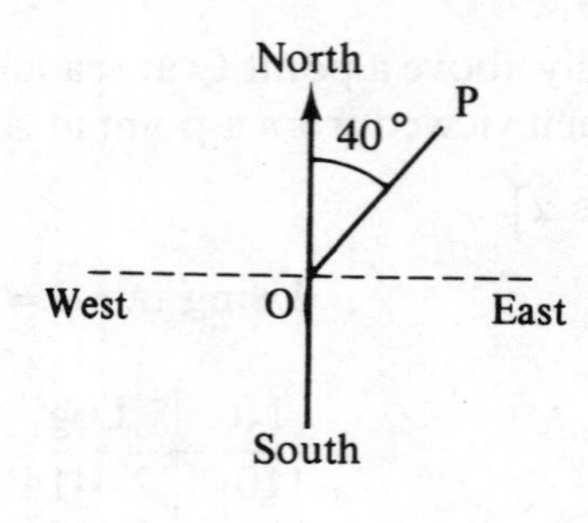

1. The bearing of P from O is North 40° East (written as N40°E) or 040°.

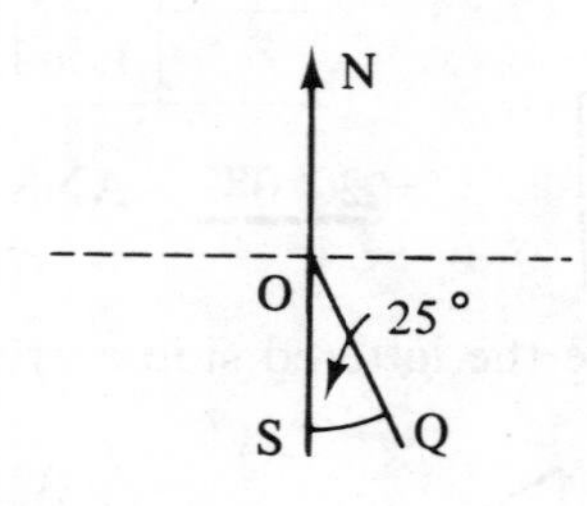

2. The bearing of Q from O is S25°E or (180° − 25°) = 155°.

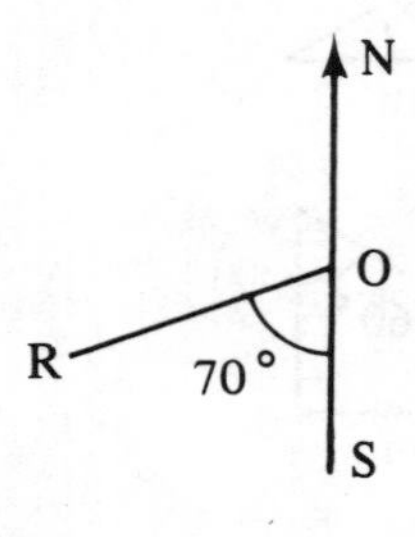

3. The bearing of R from O is S70°W or (180° + 70°) = 250°.

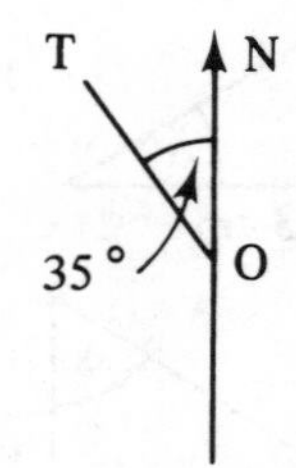

4. The bearing of T from O is N35°W or (360° − 35°) = 325°.

Examples

1. A point P is 10 miles from Q on a bearing of 035°. Calculate the distance that P is East of Q.

[x represents the required distance:]

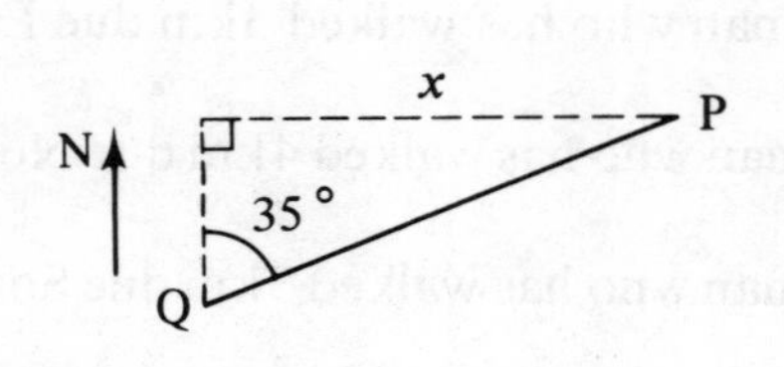

Using $\sin 35° = \frac{x}{10}$

$10 \times \sin 35° = x$

$10 \times 0.5736 = x$

[The sides involved are those opposite (x) the angle of 35°, and the hypotenuse (10).]

$x = \underline{5.736}$ miles ANS

2. A man leaves A and walks 3.4km due West to B. From B he walks 5.7km due South to C. Calculate the bearing of C from A.

[α represents the required angle:]

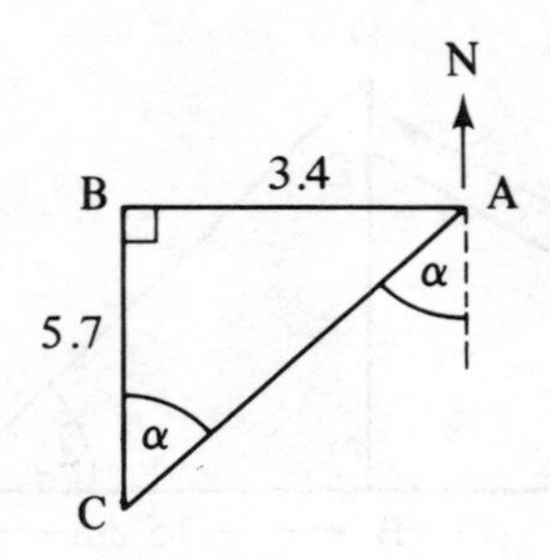

Using $\tan \alpha = \frac{3.4}{5.7}$

No.	Log.
3.4	0.5315
5.7	0.7559 –
	$\bar{1}.7756$

[From Logarithm of Tangent tables] $\alpha = 30°49'$

The bearing of C from A $= 180° + \alpha°$

$= \underline{210°49'}$ ANS

Exercise 99 Draw pencil and ruler sketches to illustrate:

1. A point P 20 miles from Q on a bearing of 049°, how far is P East of Q?
2. A point S 100km from T on a bearing of 107°, how far is S East of T?
3. A point X 55km from Y on a bearing of 237°, how far is X West of Y?
4. A point A 45km from B on a bearing of N35° E, how far is A North of B?
5. A point C 110km from D on a bearing of S50° W, how far is C South of D?

6. The position of a man who has walked 2km due West followed by 3km due South.
7. The position of a man who has walked 3km due East followed by 3km due South.
8. The position of a man who has walked 4km due North followed by 3km due East.
9. The position of a man who has walked 3km due South followed by 4km due West.
10. The position of a man who has walked 2.5km in a direction N45° E followed by 3.5km in a direction S45° E.

Examination Standard Questions

Worked examples

1. In the figure, DB is perpendicular to the line ABC. Calculate the length of DE.

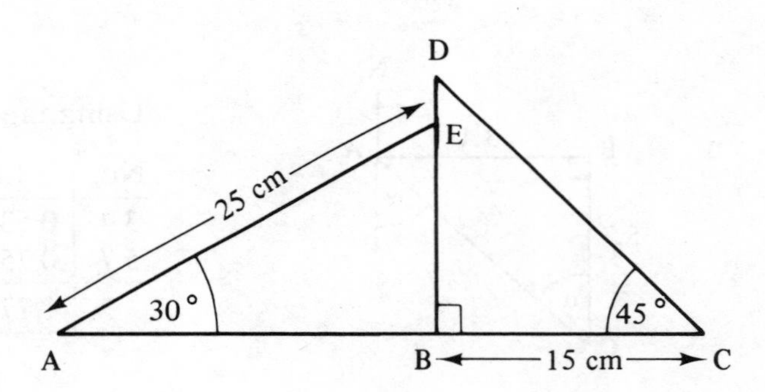

To find length EB:

[Using the larger triangle AEB the sides involved are those opposite (EB) the angle 30° and the hypotenuse (25).]

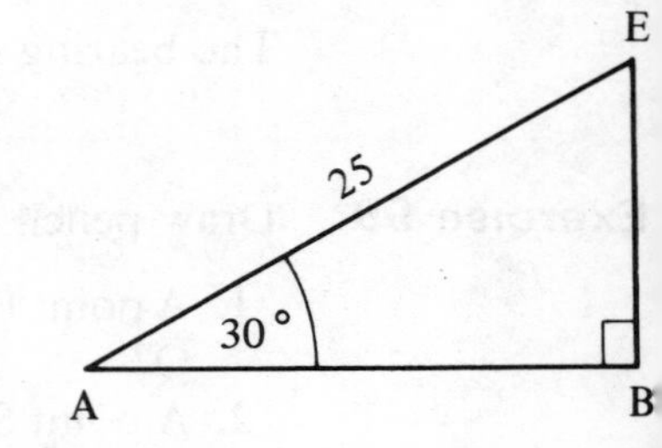

Using $\sin 30° = \dfrac{EB}{25}$

[Multiply both sides by 25] $\therefore 25 \times \sin 30° = EB$

[Using Natural Sine tables] $\therefore 25 \times 0.5 = EB$

$\therefore EB = \underline{12.5\text{cm}}$

To find the length DB:

[Using the smaller triangle BDC the sides involved are those opposite to (DB) and adjacent to (15) the angle of 45°.]

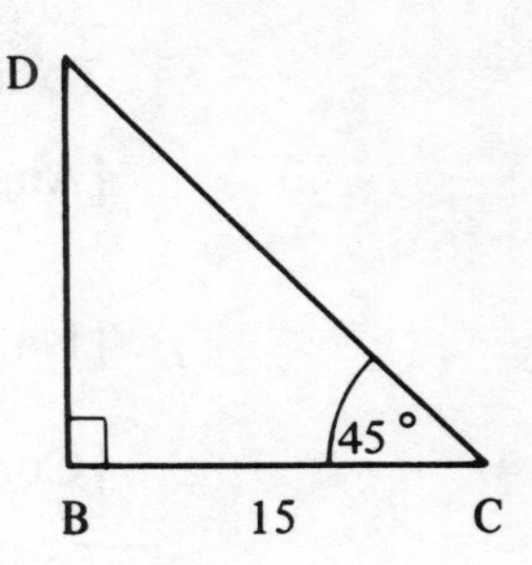

$$\text{Using } \tan 45° = \frac{DB}{15}$$

[Multiply both sides by 15] $\therefore 15 \times \tan 45° = DB$

[Using Natural Tangent tables] $\therefore 15 \times 1.0 = DB$

$\therefore DB = \underline{15\text{cm}}$

$DE = DB - EB = 15 - 12.5 = \underline{\underline{2.5\text{cm}}}$ ANS

2. Find the two possible values of the angle $\alpha°$ such that $0 < \alpha < 360$ and $\sin \alpha = 0.788$.

Sine is positive in the 1st. and 2nd. quadrants.

[From tables] In the 1st. quadrant $\alpha = \sin\ ^{1}0.788 = 52°$

In the 2nd. quadrant $\sin \alpha = \sin (180 - \alpha)°$

$= 180° - 52°$

$= 128°$

The two values of α are $\underline{52°}$ and $\underline{128°}$ ANS

3. Three points A, B, C, on level ground are such that A and B are 50m apart, B is due North of A, C is on a bearing 250° from B, and angle ACB = 90°. Calculate the distances AC and BC correct to three significant figures.

[See diagram]

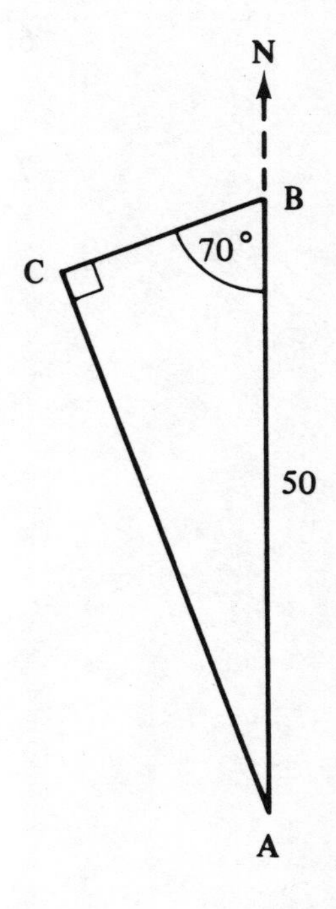

To find length AC:

[Angle ABC = 250° − 180°] Using $\sin 70° = \dfrac{AC}{50}$

$\therefore 50 \times \sin 70° = AC$

[Natural Sine tables] $\therefore 50 \times 0.9397 = AC$

$\therefore AC = 46.985$

[Correct to 3 s.f.] $= \underline{47.0\text{m}}$ ANS

To find length BC:

Using $\cos 70° = \dfrac{BC}{50}$

$\therefore 50 \times \cos 70° = BC$

$\therefore 50 \times 0.342 = BC$

$\therefore BC = \underline{17.1\,\text{m}}$ ANS

Exercise 100

1. Use tables to find the value of sin 113° 31′. (WJEC)
2. Write down the value of cos 130° correct to two decimal places. (JMB)
3. The angle A is acute and $\sin A = \dfrac{8}{17}$. Without using tables, calculate the value of cos(180° − A). (JMB)
4. Find the obtuse angle A such that sin A = sin 50°. (JMB)
5. Given that $x°$ is an acute angle and $\tan x° = \dfrac{12}{5}$, find, without using tables, leaving your answers as fractions, (a) $\sin x°$, (b) $\cos(180° - x°)$. (L)
6. Find the value of cos 108° + sin 108°, correct to 2 decimal places. (O)
7. Y is a point vertically above P, and X is a point 32 m vertically above Q, where PQ is horizontal and of length 20 m. Given that the angle of elevation of X from Y is 42°, calculate the distance YP. (L)

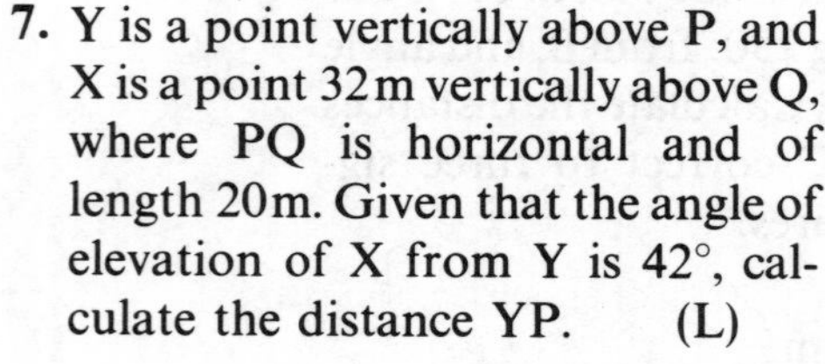

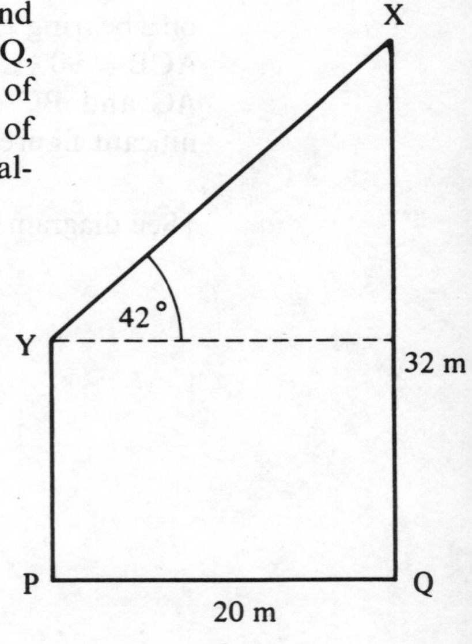

8. Find the two possible values of the angle $\theta°$ such that $0 < \theta < 360°$ and $\cos\theta = 0.423$. (O)

9. (a) If $\sin x° = \sin 27°$ and $90 < x < 180$, write down the value of x.
(b) If $\sin y° = -\sin 27°$ and $180 < y < 270$, write down the value of y. (C)

10.

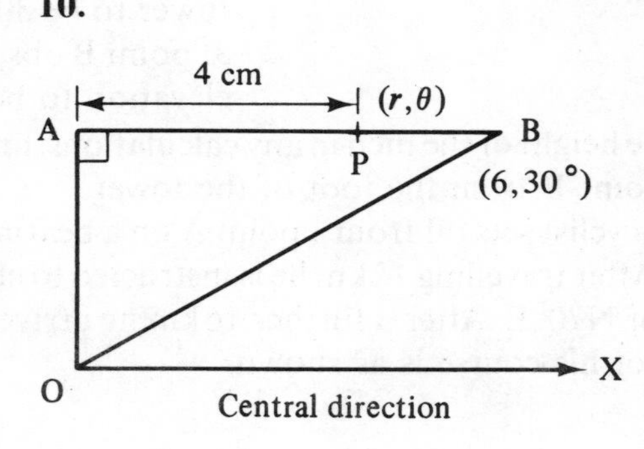

In the diagram, with origin O, the polar coordinates of the point B are (6, 30°), APB is a straight line parallel to the central direction OX, and OA is perpendicular to OX. (All distances are measured in centimetres.)
(a) What are the polar coordinates of A?
(b) If AP = 4 cm, what are the values of r and θ? (AEB)

11. An isosceles triangle ABC has a base AC 5cm long and a height of 6cm. Calculate the vertical angle $\hat{B}$. (JMB)

12. From a lighthouse balcony 20 metres above the sea, a buoy is seen at an angle of depression of 10°. Calculate the distance of the buoy from the foot of the lighthouse which is at sea level. (JMB)

13. A man observes the angle of elevation of the top of a tower to be 26°. The height of the tower is known to be 45m. Neglecting the height of the man, how far is he from the foot of the tower? (L)

14.

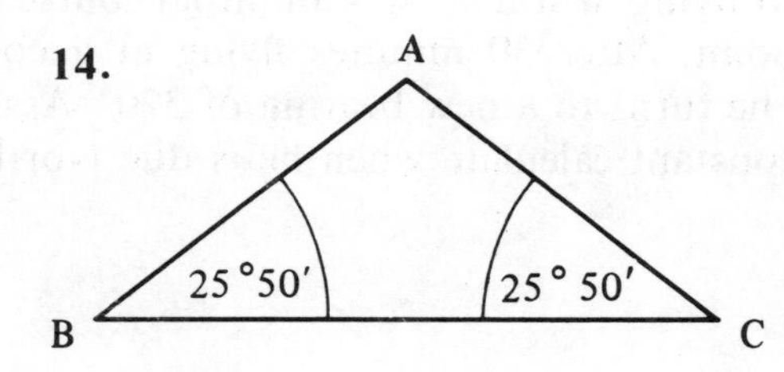

The diagram shows part of a roof support in the form of an isosceles triangle ABC. Given that the beam BC is of length 7.2 m, calculate the length of the beam AB. (C)

15. A point P is 10km from O on a bearing of 175°. Calculate, correct to three significant figures, the distance that P is South of O.

16. From a point A 120m South-West of a lighthouse, a small boat sails in a straight line to a point B, 200m South-East of the lighthouse. Calculate the bearing on which the boat sails. (AEB Part question)

17. Three points P, Q and R are on level ground. The point R is due North of Q and 3750 metres due East of P. On a map of scale 1 in 25000, QR is represented by a line of length 15cm. Find (a) the actual distance QR in metres, (b) the bearing of Q from P. (C)

18. Two men stand so that a vertical tower is directly between them as shown. The man at point A, 50m from the base of the tower, observes the angle of elevation of the top of the tower to be 30°, the other man at point B observes the angle of elevation to be 50°.

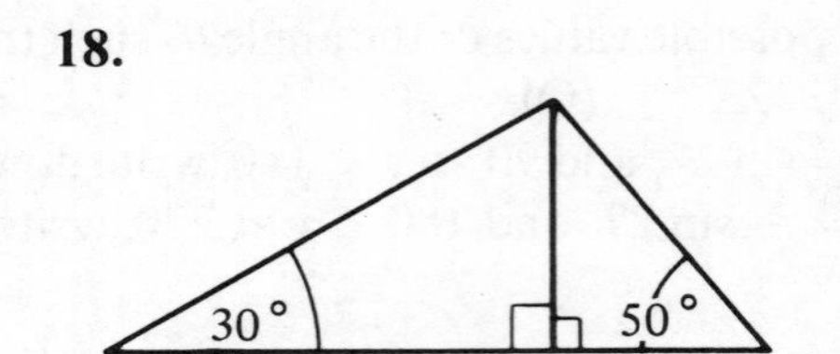

Neglecting the height of the men in any calculations, find the distance of the man at point B from the foot of the tower.

19. A motor-cyclist sets off from a point A on a bearing of N30° E on a test run. After travelling 12km he is instructed to change course to a bearing of N70° E. After a further 16km he arrives at a point C. A diagram of his course is as shown.

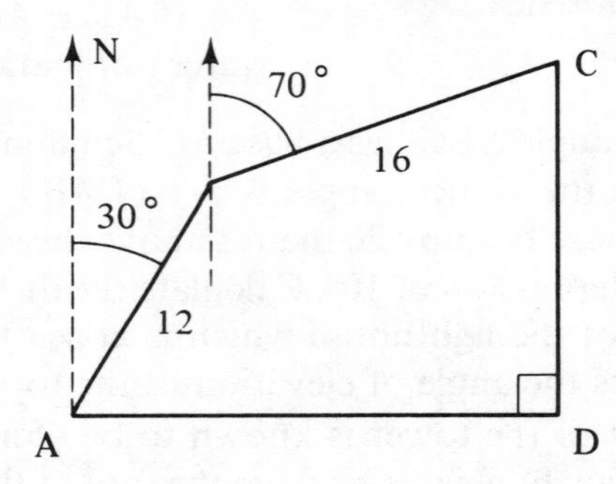

Point D is due South of C and due East of A. Calculate the distances CD and AD correct to the nearest kilometre.

20. An aircraft flying in still air sets an initial course of 050° from his starting point. After 30 minutes flying at a constant speed of 300km/h he turns to a new bearing of 320°. Assuming his speed remains constant calculate when he is due North of his starting point.

14

Further Trigonometry Including Three Dimensional Problems

The angle between a line and a plane

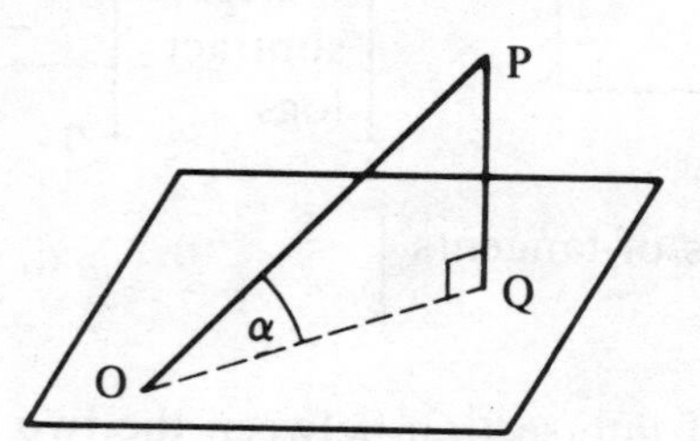

The angle POQ is the angle between the line OP and the plane, where PQ is perpendicular to this plane.

The angle between two planes

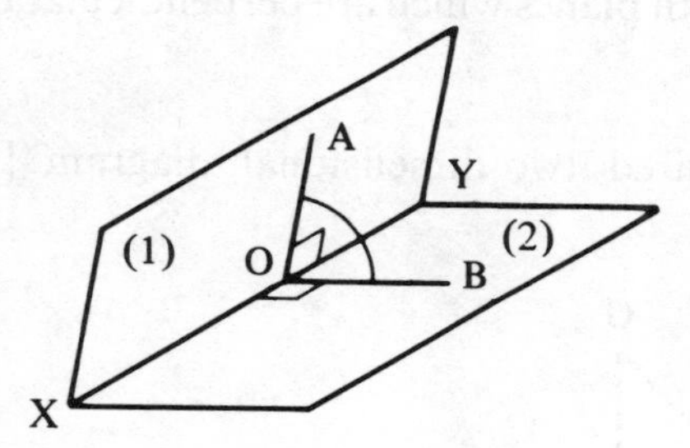

XY is the line of intersection of the two planes. OA is a straight line drawn in plane (1) at right-angles to XY. OB is a straight line drawn in plane (2) at right-angles to XY. The angle between the two planes is angle AOB.

Examples

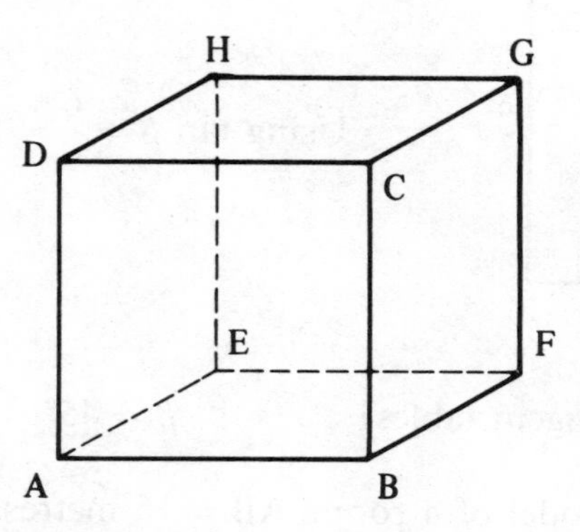

In a cube ABCDEFGH of side 6cm, find (a) the length AF, (b) the angle the line AG makes with the base (AEFB), (c) the angle which the plane HABG makes with the base.

(a) Forming the required two-dimensional diagram:

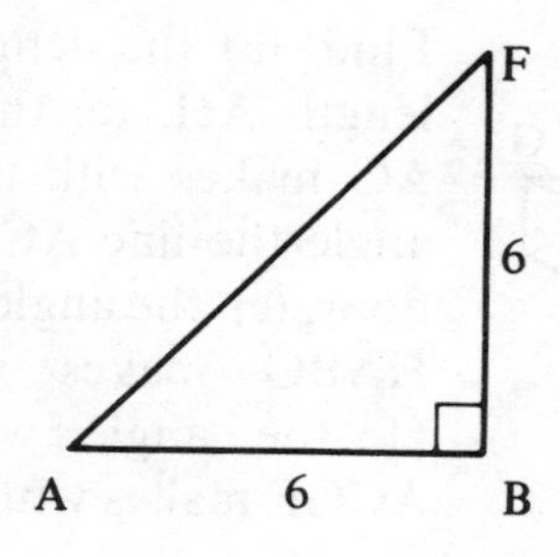

Using Pythagoras theorem gives:

$$AB^2 + BF^2 = AF^2$$

i.e. $6^2 + 6^2 = AF^2$

$$\therefore\ 72 = AF^2$$

$$\therefore\ AF = \sqrt{72}$$

$$= \underline{8.485} \quad \text{ANS}$$

(b) Forming the required two-dimensional diagram [where α is the required angle]:

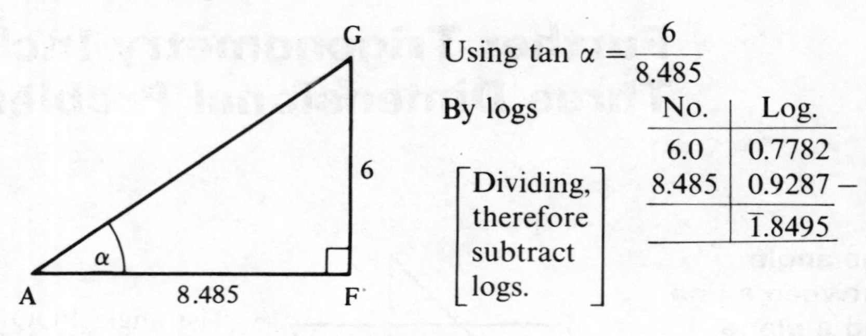

Using $\tan \alpha = \dfrac{6}{8.485}$

By logs

No.	Log.
6.0	0.7782
8.485	0.9287 –
	$\bar{1}.8495$

[Dividing, therefore subtract logs.]

[Using Logarithms of tangents table gives] $\therefore \alpha = \underline{35^\circ 16'}$ ANS

(c) AB is the line of intersection between the two planes, ABFE and ABGH. Lines in both planes which are perpendicular to AB are BF and BG.

Forming the required two-dimensional diagram [where β is the required angle].

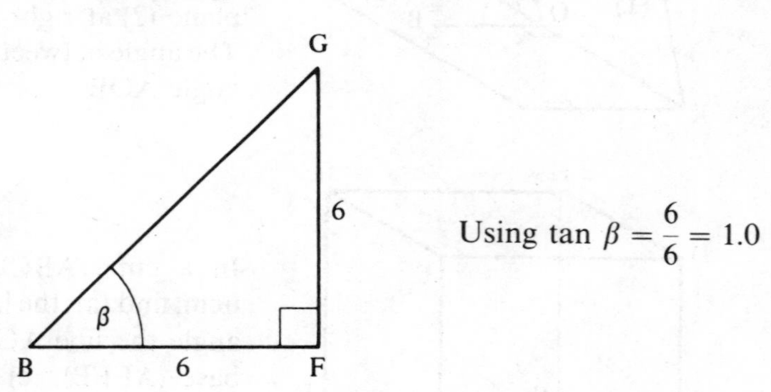

Using $\tan \beta = \dfrac{6}{6} = 1.0$

[From Natural Tangent tables] $\therefore \beta = \underline{45^\circ}$ ANS

Exercise 101 ABCDEFG is a model of a room. AB = 15 metres, AD = 3 metres, BF = 10 metres.

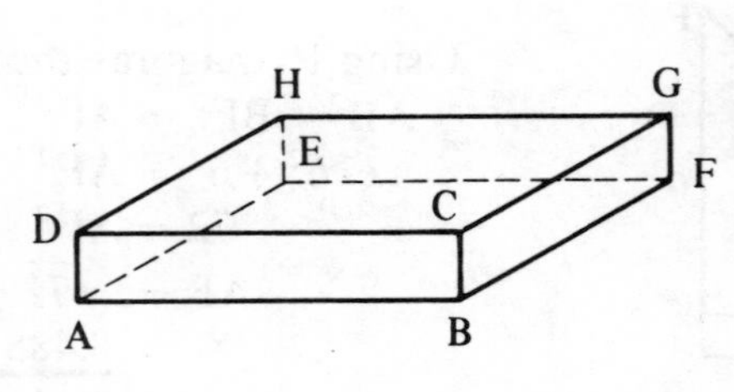

Find (a) the length AF, (b) the length AH, (c) the angle the line AG makes with the floor, (d) the angle the line AC makes with the floor, (e) the angle which the plane HABG makes with the floor, (f) the angle which the plane ACGE makes with the wall BCGF.

Example

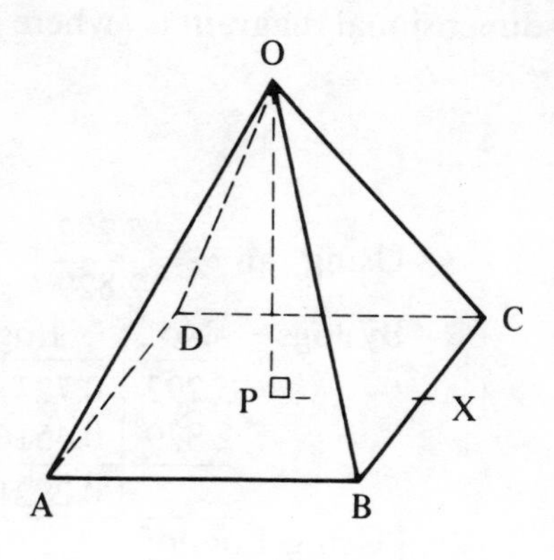

A pyramid on a square base ABCD, where P is the mid-point of the base and O is the vertex of the pyramid, has lengths $AB = BC = CD = DA = 4$cm, and $OA = OB = OC = OD = 6$cm. Calculate (a) the distance AP, correct to 1 d.p., (b) the vertical height, OP, of the pyramid, correct to 1 d.p., (c) the angle OC makes with the base, (d) the angle which the plane OBC makes with the base.

(a) To find AP $(= \frac{1}{2}AC)$

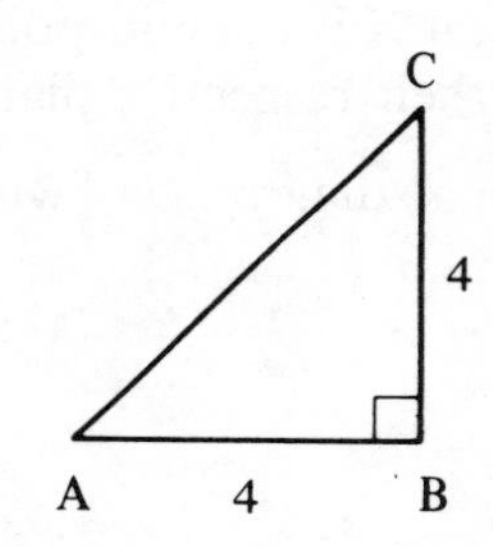

Using Pythagoras theorem gives:

$$AC^2 = AB^2 + BC^2$$
$$= 4^2 + 4^2$$
$$= 32$$
$$\therefore AC = \sqrt{32} = \underline{5.657\text{cm}}$$
$$AP = \tfrac{1}{2}AC = 2.829$$
$$= \underline{2.8\text{cm}} \quad \text{ANS}$$

(correct to 1 d.p.)

(b) Forming a suitable two-dimensional diagram:

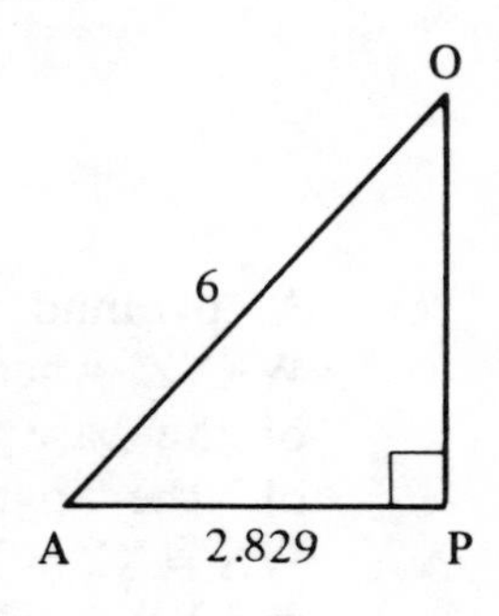

Using Pythagoras theorem gives:

$$AO^2 = AP^2 + OP^2$$
$$\therefore AO^2 - AP^2 = OP^2$$
$$\therefore 6^2 - 2.829^2 = OP^2$$

[Using tables of squares]

$$\therefore 36 - 8.003 = OP^2$$
$$\therefore 27.997 = OP^2$$

[Using square root tables]

$$\therefore \sqrt{27.997} = OP$$
$$\therefore OP = 5.292$$

[Correct to 1 d.p.]

$$= \underline{5.3\text{cm}} \quad \text{ANS}$$

(c) The required two-dimensional diagram is [where α is the required angle]:

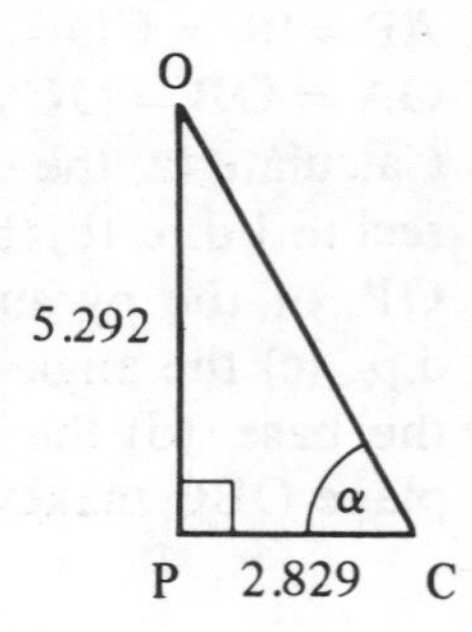

Using $\tan \alpha = \dfrac{5.292}{2.829}$

By logs

No.	Log.
5.292	0.7237
2.829	0.4516 –
	0.2721

[Using Log. of Tangents] $\alpha = \underline{61^\circ 53'}$ ANS

(d) On the original diagram, if X is the mid-point of BC, XP and XO are perpendicular to BC in their respective planes.

The required two-dimensional diagram is [where β is the required angle]:

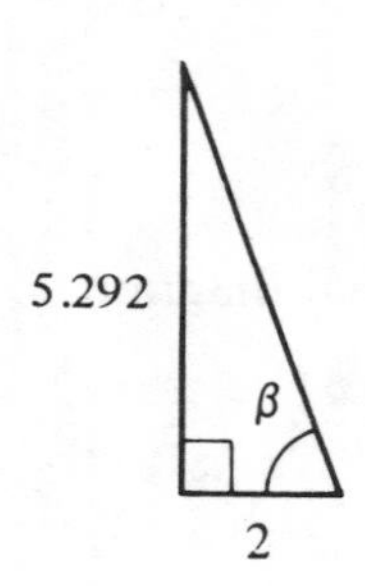

Using $\tan \beta = \dfrac{5.292}{2}$

$\therefore \beta = \tan^{-1} 2.646$

$= \underline{69^\circ 18'}$ ANS

Exercise 102

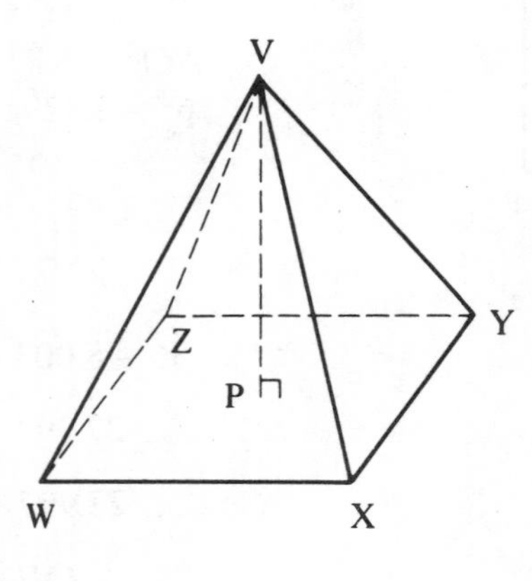

A pyramid on a square base WXYZ, where P is the mid-point of the base and V is the vertex of the pyramid, has lengths WX = XY = YZ = ZW = 10cm, and VW = VX = VY = VZ = 15cm. Calculate (a) the distance XP, correct to 1 d.p. (b) the vertical height VP, of the pyramid, correct to 1 d.p. (c) the angle VX makes with the base, (d) the angle which the plane VWZ makes with the base.

Sine & Cosine Formulae

In Chapter 13 the sin, cos and tan ratios were used in the solution of right-angled triangles. Consider the acute and obtuse angled triangles shown below, we would be unable to use the three ratios as there are no right angles. Two new formula are now given;

$$\frac{a}{\sin A} = \frac{b}{\sin B} = \frac{c}{\sin C},$$ known as the **sine formula**

and $c^2 = a^2 + b^2 - 2ab \cos C$, which rearranged becomes

$$\cos C = \frac{a^2 + b^2 - c^2}{2ab},$$ known as the **cosine formula**.

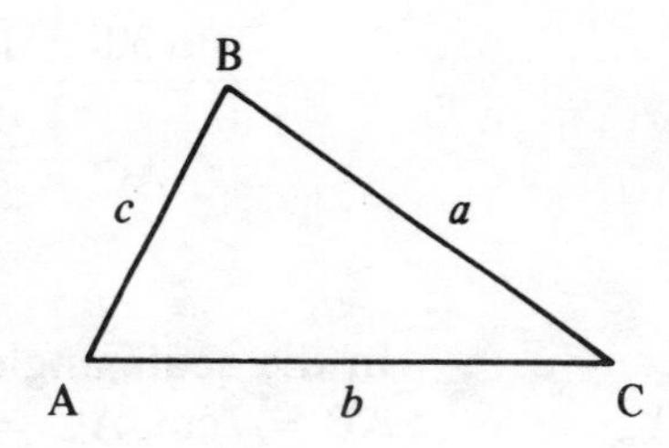

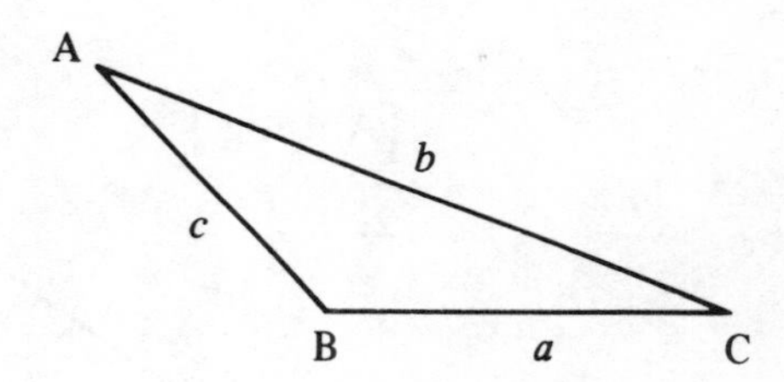

These formulae apply to any triangle, but should be avoided if the problem can be 'broken up' into right-angled triangles.

Examples

1. In the triangle ABC, BC = 5cm, $B\hat{A}C = 30°$, $A\hat{B}C = 130°$. Calculate the lengths AB and AC.

$$\hat{C} = 180° - (130° + 30°) = 20°$$

Using the sine formula $\quad \frac{a}{\sin A} = \frac{c}{\sin C}$

[Where a is the side opposite $\hat{A}$, c is the side opposite $\hat{C}$.] $\quad \therefore \frac{5}{\sin 30°} = \frac{c}{\sin 20°}$

[Multiply both sides by sin 20°] $\quad \therefore \frac{5 \sin 20°}{\sin 30°} = c$

	No.	Log.	
	5.0	0.6990	
[From Log. of Sine tables]	sin 20°	$\bar{1}.5341$	[add]
		0.2331	
	sin 30°	$\bar{1}.6990$	[subtract]
		0.5341	

[Antilog.] $\quad \therefore$ AB = 3.421 cm $\quad$ ANS

Using the sine formula $\dfrac{a}{\sin A} = \dfrac{b}{\sin B}$

[Where b is the side opposite $\hat{B}$] $\quad \therefore \dfrac{5}{\sin 30^\circ} = \dfrac{b}{\sin 130^\circ}$

[Multiply both sides by sin 130°] $\quad \therefore \dfrac{5 \sin 130^\circ}{\sin 30^\circ} = b$

$\left[\begin{array}{l} \sin 130^\circ = \sin (180 - 130)^\circ \\ \qquad\quad\; = \sin 50^\circ \end{array}\right]$ $\quad$ i.e. $\dfrac{5 \sin 50^\circ}{\sin 30^\circ} = b$

No.	Log.	
5.0	0.6990	
sin 50°	$\bar{1}.8843$	[add]
	0.5833	
sin 30°	$\bar{1}.6990$	[subtrac
	0.8843	

[From Log. of Sine tables]

[Antilog.] $\quad \therefore$ AC = 7.661 cm $\quad$ ANS

2.

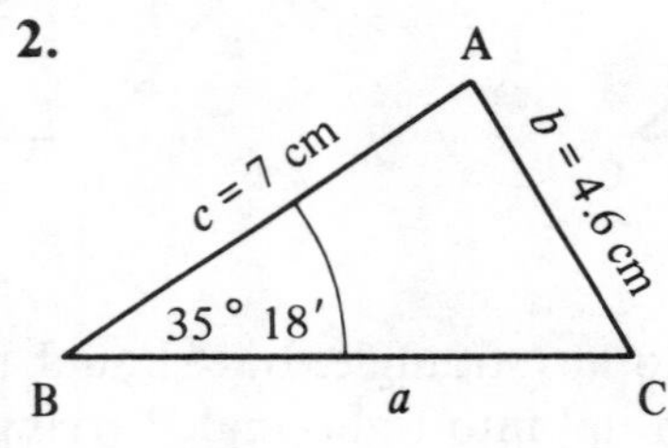

In the acute angled triangle ABC AB = 7cm, AC = 4.6cm and $A\hat{B}C = 35^\circ 18'$. Calculate $B\hat{C}A$.

Let $B\hat{C}A = \alpha^\circ$

Using the sine formula $\dfrac{b}{\sin B} = \dfrac{c}{\sin C}$

i.e. $\quad \dfrac{4.6}{\sin 35^\circ 18'} = \dfrac{7}{\sin \alpha}$

[Multiply both sides by sin α] $\quad \therefore \dfrac{4.6 \sin \alpha}{\sin 35^\circ 18'} = 7$

[Cross multiply] $\quad \therefore 4.6 \sin \alpha = 7 \sin 35^\circ 18'$

[Divide both sides by 4.6] $\quad \therefore \sin \alpha = \dfrac{7 \sin 35^\circ 18'}{4.6}$

No.	Log.	
7.0	0.8451	
sin 35° 18′	$\bar{1}.7618$	[add]
	0.6069	
4.6	0.6628	[subtract]
	$\bar{1}.9441$	

[i.e. log sin $\alpha = \bar{1}.9441$, therefore use log. sine tables to find angle.] $\quad \therefore \alpha = 61^\circ 33'$ $\quad$ ANS

ote

The previous question stipulated that ABC was an acute angled triangle. Using the information given in the question it is also possible to construct an obtuse angled triangle. As $\sin 61°33' = \sin(180° - 61°33')$ angle α *could* be $118°27'$. This is known as the ambiguous case.

3.

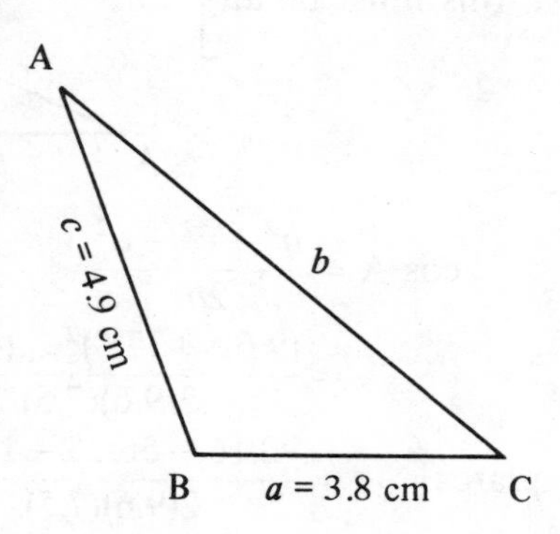

In triangle ABC, AB = 4.9cm, BC = 3.8cm and $A\hat{B}C = 110°$. Calculate the length AC.

As we do not know the angles opposite the given sides we cannot use the sine formula. The previously noted cosine formula, however, can be rearranged to give:

$$b^2 = a^2 + c^2 - 2ac\cos B$$

[$\cos 110° = -\cos(180° - 110°) = -\cos 70°$]

$$= (3.8)^2 + (4.9)^2 - 2(3.8)(4.9)(\cos 110°)$$
$$= (3.8)^2 + (4.9)^2 - 2(3.8)(4.9)(-\cos 70°)$$

[Note the change of sign]

$$= (3.8)^2 + (4.9)^2 + 2(3.8)(4.9)(\cos 70°)$$

[From table of squares]

$$(3.8)^2 = 14.44$$
$$(4.9)^2 = 24.01 +$$
$$38.45$$

[Use logs to calculate $2ac\cos\hat{B}$]

No.	Log.
2.0	0.3010
3.8	0.5798
4.9	0.6902
cos 70°	$\bar{1}.5341$
	1.1051

[Antilog] $= 12.74$

$$b^2 = 38.45$$
$$12.74 +$$
$$51.19$$

$$\therefore b = \sqrt{51.19}$$

$\therefore$ AC = 7.154cm ANS

Note If it was required to calculate a second angle of the triangle it would be easier to use the sine formula in the further calculation.

4. In the triangle ABC, AB = 7.5cm, BC = 4.2cm and AC = 9.6cm. Calculate all three angles.

[Calculate the angle opposite the shortest side first, this must be an acute angle.]

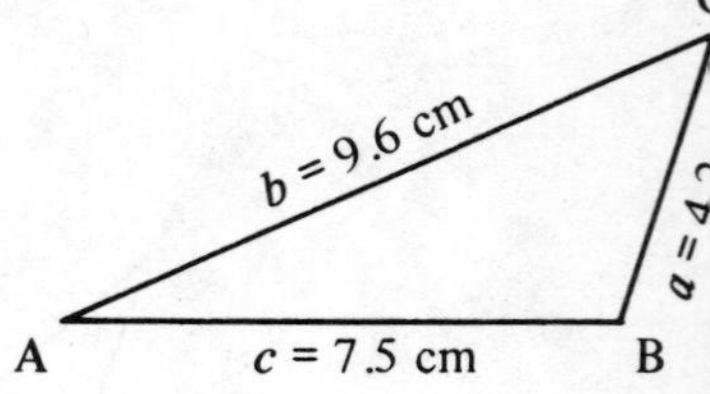

Using $\cos A = \dfrac{b^2 + c^2 - a^2}{2bc}$

$$= \frac{(9.6)^2 + (7.5)^2 - (4.2)^2}{2(9.6)(7.5)}$$

[Using table of squares] $= \dfrac{92.16 + 56.25 - 17.64}{2(9.6)(7.5)}$

$$\cos A = \frac{130.77}{2(9.6)(7.5)}$$

No.	Log.	
2.0	0.3010	
9.6	0.9823	
7.5	0.8751	[add]
	2.1584	
130.8	2.1165	
	2.1584	[subtract]
	$\bar{1}.9581$	

[Using log. cos. tables] $\hat{A} = \underline{24^\circ 46'}$ ANS

Using sine formula $\dfrac{4.2}{\sin 24^\circ 46'} = \dfrac{7.5}{\sin C}$

[The angle opposite the next shortest side must also be acute] $\therefore \sin C = \dfrac{7.5 \sin 24^\circ 46'}{4.2}$

No.	Log.	
7.5	0.8751	
sin 24°46′	$\bar{1}.6221$	[add]
	0.4972	
4.2	0.6232	[subtract]
	$\bar{1}.8740$	

[Using log. sin tables] $\hat{C} = \underline{48^\circ 25'}$ ANS

$$\therefore \hat{B} = 180^\circ - (48^\circ 25' + 24^\circ 46')$$
$$= \underline{106^\circ 49'} \text{ ANS}$$

The three angles are therefore $\underline{\hat{A} = 24°46'}$, $\underline{\hat{B} = 106°49'}$, $\underline{\hat{C} = 48°25'}$ ANS

xercise 103 Each solution should be illustrated with a diagram:

1. Triangle ABC has AB = 7.5cm, $A\hat{B}C = 55°$, $A\hat{C}B = 47°$. Calculate the lengths AC and BC.
2. Triangle ABC has AC = 8.3cm, $A\hat{B}C = 110°50'$, $B\hat{A}C = 43°17'$. Calculate the lengths AB and BC.
3. Triangle ABC has BC = 11.93cm, $B\hat{A}C = 77°07'$, $A\hat{C}B = 31°11'$. Calculate the lengths AB and AC.
4. Triangle ABC has AB = 6.3cm, $A\hat{B}C = 50°10'$, $B\hat{A}C = 46°17'$. Calculate the length BC.
5. Triangle ABC has BC = 4.7cm, and $A\hat{B}C = 23.1°$, $A\hat{C}B = 79.4°$. Calculate the length AC.
6. In triangle ABC; $A\hat{B}C = 109°30'$, AB = 5.3cm, BC = 4.1cm. Calculate the length AC and the angle $B\hat{A}C$.
7. Calculate the size of the largest angle in a triangle whose lengths of sides are 9.1cm, 5.3cm and 6.0cm.
8. A parallelogram has sides of length 4.2cm and 5.9cm, the acute angle being 61°. Calculate the length of the shorter diagonal.
9. Find the length of the shorter diagonal of a rhombus whose sides are of length 2.3cm and has an acute angle of 79°.
10. A kite has sides of length 2cm and 3cm, with an obtuse angle of 110°. Calculate the length of the longer diagonal.

Graphs of Trigonometric Ratios

he Graph of = sin x°

Setting up a table of values at 30° intervals, (correct to 2 d.p.):

x	0°	30°	60°	90°	120°	150°	180°	210°	240°	270°	300°	330°	360°
$y = \sin x°$	0	0.5	0.87	1.0	0.87	0.5	0	−0.5	−0.87	−1.0	−0.87	−0.5	0

The sine curve can now be plotted:

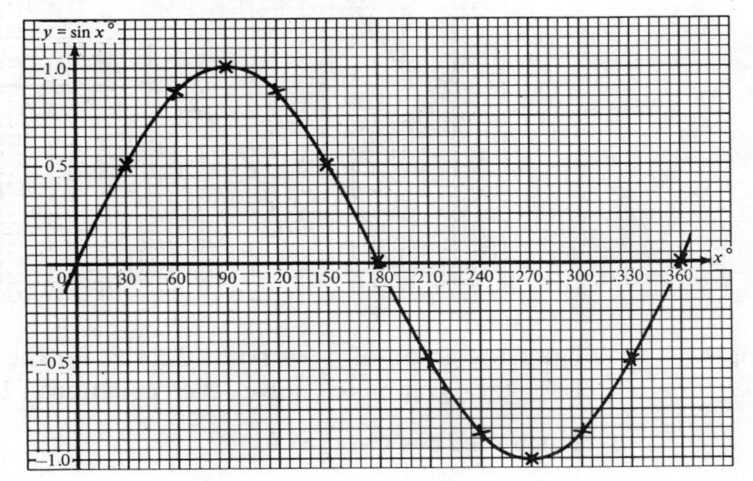

The Graph of $y = \cos x°$

The cosine curve is of the same form as the sine curve, they hav different positions relative to the y-axis.

x	0°	30°	60°	90°	120°	150°	180°	210°	240°	270°	300°	330°	360°
$y = \cos x°$	1.0	0.87	0.5	0	−0.5	−0.87	−1.0	−0.87	−0.5	0	0.5	0.87	1.0

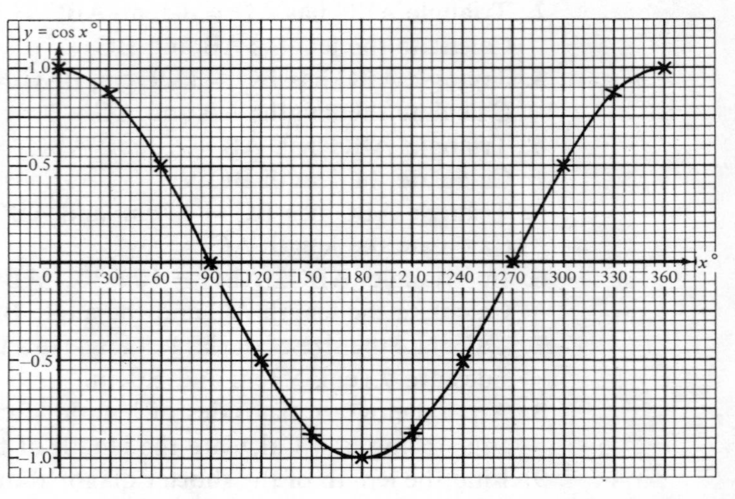

The Graph of $y = \tan x°$

x	0°	30°	60°	90°	120°	150°	180°	210°	240°	270°	300°	330°	360°
$y = \tan x°$	0	0.58	1.73	∞	−1.73	−0.58	0	0.58	1.73	∞	−1.73	−0.58	0

At the points $x = 90°$ and $x = 270°$ the graphs are **discontinuous.**

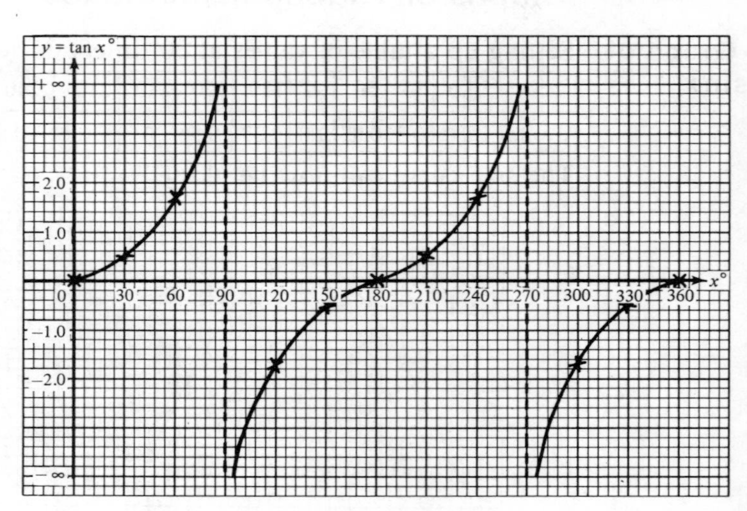

Example

Draw the graph of $y = \sin x° + 2\cos x°$ for the range $0° \leqslant x° \leqslant 180$ Use the graph to solve the equations **1.** $\sin x + 2\cos x = 1.$

2. $\sin x + 2\cos x - \dfrac{x}{15} = 0.$

Setting up a table of values at 15° intervals:

$x°$	0°	15°	30°	45°	60°	75°	90°	105°	120°	135°	150°	165°	180°
$\sin x°$	0	0.26	0.5	0.71	0.87	0.97	1.0	0.97	0.87	0.71	0.5	0.26	0
$2\cos x°$	2.0	1.94	1.74	1.42	1.0	0.52	0	−0.52	−1.0	−1.42	−1.74	−1.94	−2.0
y	2.0	2.20	2.24	2.13	1.87	1.49	1.0	0.45	−0.13	−0.71	−1.24	−1.68	−2.0

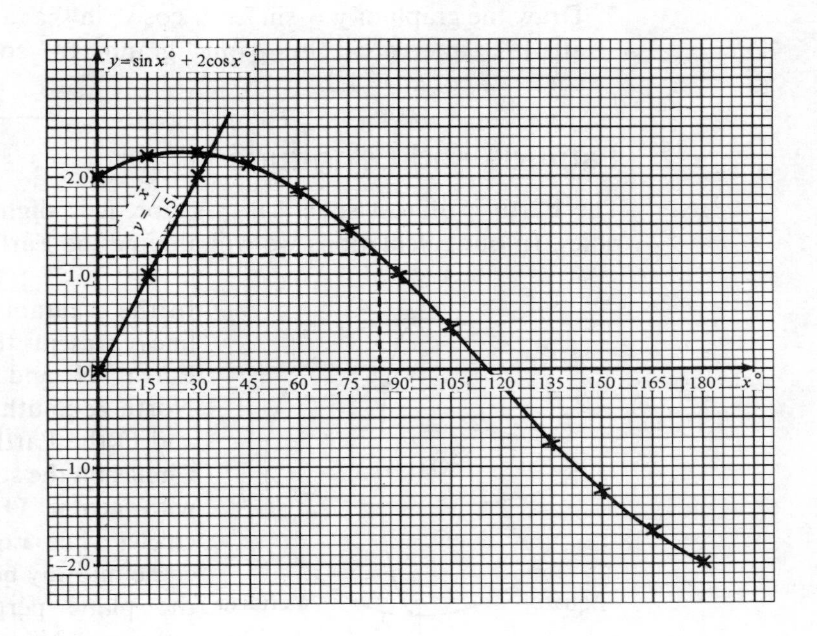

1. When $\sin x + 2\cos x = 1.2$, $y = 1.2$

[Shown by the dotted lines] From the graph $\underline{x = 84°}$ ANS

2. Given $\sin x + 2\cos x - \dfrac{x}{15} = 0$

[Rearranging] $\therefore\ \sin x + 2\cos x = \dfrac{x}{15}$

We therefore need to draw the graph of $y = \dfrac{x}{15}$

When $x = 0$, $y = 0$; when $x = 15$, $y = 1$; when $x = 30$, $y = 2\ldots$

$\therefore$ From the graphs $\underline{x = 33°}$ ANS

xercise 104

1. Draw the graph of $y = \sin x°$ in the range $0° \leqslant x° \leqslant 180°$ at 15° intervals. (a) Use your graph to solve the equation $\sin x° = 0.9$, (b) what is the maximum value of $\sin x°$ in this range?

2. Draw the graph of $y = \cos x°$ in the range $-90° \leqslant x° \leqslant 90°$ at 15° intervals. (a) Use your graph to solve the equation $\cos x° = 0.6$, (b) what is the maximum value of $\cos x°$ in this range?

3. Draw the graph of $y = \tan x^\circ$ in the range $-90^\circ \leqslant x^\circ \leqslant 90^\circ$ at 1[illegible] intervals. Use your graph to solve the equations (a) $\tan x^\circ = 1.$[illegible]
(b) $\tan x^\circ - \dfrac{x}{45} = 0.$
4. Draw the graph of $y = 2\sin x^\circ + \cos x^\circ$ in the range $0^\circ \leqslant x^\circ \leqslant 18$[illegible] at 15° intervals. Use your graph to solve the equation $2 \sin$[illegible] $+ \cos x^\circ = 1.2$.
5. Draw the graph of $y = \sin 2x^\circ + \cos x^\circ$ in the range $0^\circ \leqslant x^\circ \leqslant 18$[illegible] at 15° intervals. Use your graph to solve the equatio[illegible] $\sin 2x^\circ + \cos x^\circ = 1.2$.

Latitude and Longitude

The Earth is almost spherical in shape, it is slightly flattened at th[illegible] poles. In our calculations we will assume the Earth is a sphere.

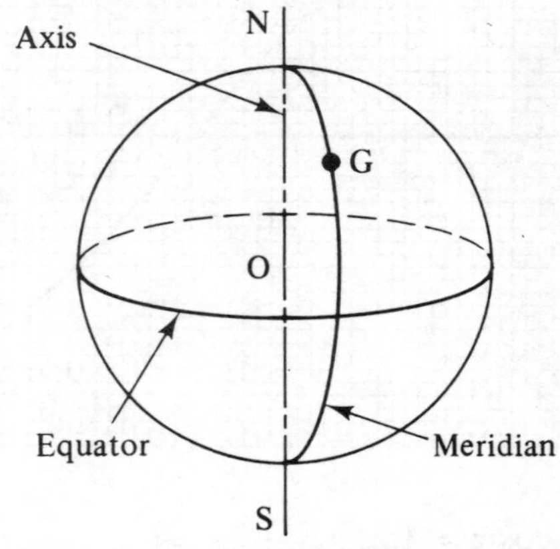

In the diagram the broken centr[illegible] line represents the Polar Axis of th[illegible] Earth, N and S indicating th[illegible] North & South poles respectivel[illegible] and O the Earth's centre.
Circles on the surface of the spher[illegible] with centre O are called **grea[illegible]** **circles**. The **Equator** is the gre[illegible] circle midway between the poles i[illegible] the plane perpendicular to th[illegible] polar axis NS.

The semi-circle NGS is half of a great circle through the poles and [illegible] called a **meridian**. G represents the position of Greenwich, Londo[illegible] and the great circle through G is referred to as the **Greenwich meridia[illegible]**

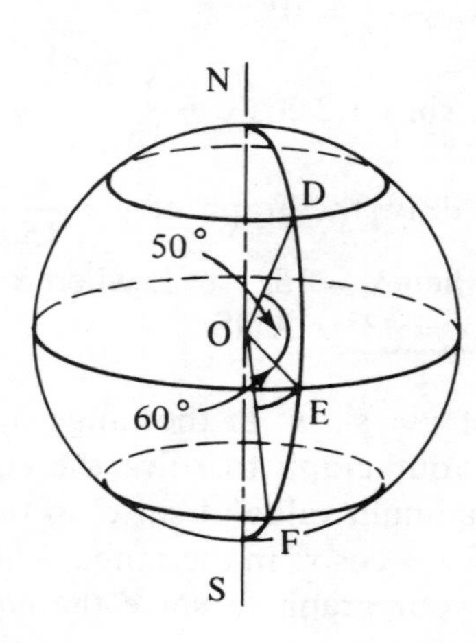

A **small circle** is any circle on th[illegible] surface of the sphere other than [illegible] great circle (i.e. whose centre is n[illegible] O). The small circle through D [illegible] parallel to the plane of the equat[illegible] and is called a **parallel of latitud[illegible]** E is a point on the Equator on th[illegible] same meridian as D, $D\hat{O}E = 50$[illegible] D is said to have latitude 50°N[illegible] Similarly F has a latitude 60°[illegible] Parallels of latitude are in the rang[illegible] 0° to 90° north or south of th[illegible] Equator. Their angles of referenc[illegible] being measured at the centre of th[illegible] Earth (O).

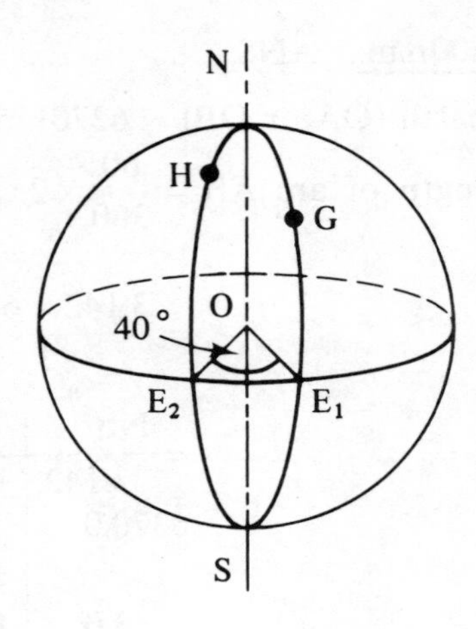

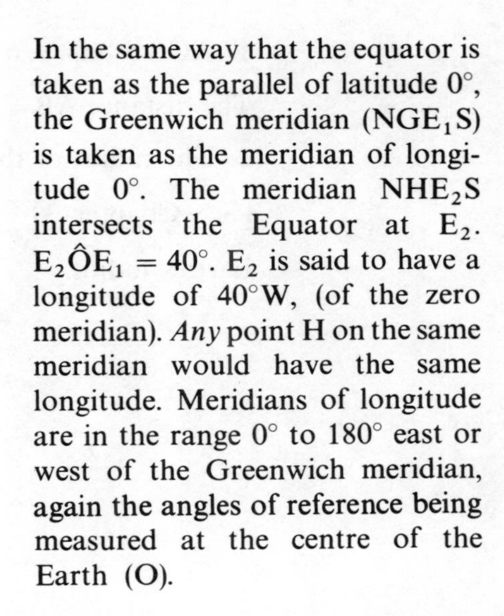

In the same way that the equator is taken as the parallel of latitude 0°, the Greenwich meridian (NGE_1S) is taken as the meridian of longitude 0°. The meridian NHE_2S intersects the Equator at E_2. $E_2\hat{O}E_1 = 40°$. E_2 is said to have a longitude of 40°W, (of the zero meridian). *Any* point H on the same meridian would have the same longitude. Meridians of longitude are in the range 0° to 180° east or west of the Greenwich meridian, again the angles of reference being measured at the centre of the Earth (O).

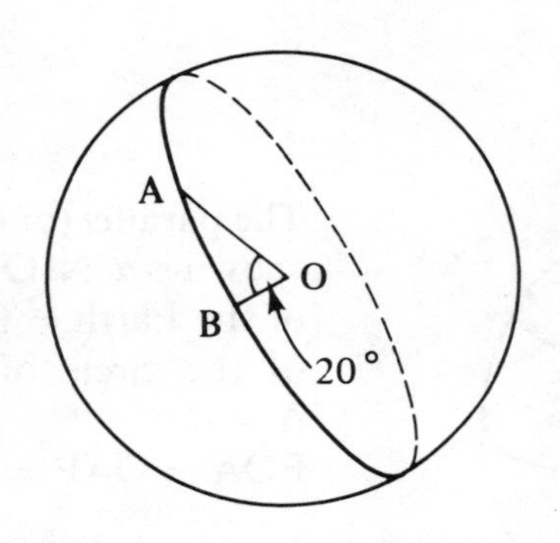

A helicopter pilot wishing to travel from point A to point B on the Earth's surface by the shortest route would move along a great circle. Such distances along great circles can be measured easily in nautical miles. **A nautical mile is that part of the arc of a great circle subtending an angle of 1′ at the Earth's centre.**

$A\hat{O}B = 20° = (20 \times 60)' = 1200'$

$\therefore$ The distance A to B along this great circle = <u>1200 n.m.</u>

ote A speed of 1 nautical mile per hour is called a **knot**.

xample

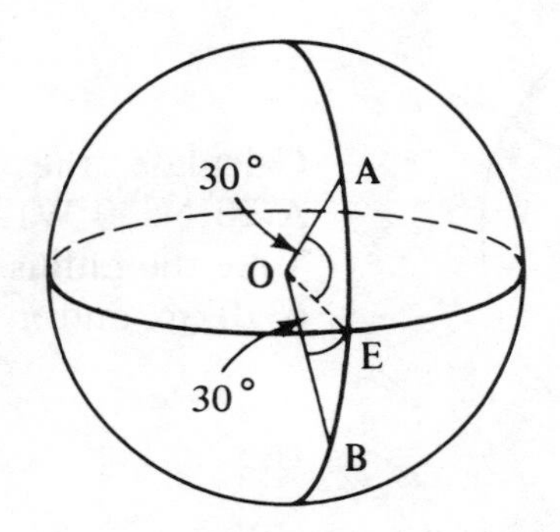

Calculate the shortest distance between A(30°N, 20°E) and B(30°S, 20°E) along the Earth's surface **1.** in nautical miles, **2.** in kilometres, taking the radius of the earth to be 6370km, and $\pi = 3.142$.

Let NAEBS represent the meridian of longitude 20°E.

$A\hat{O}B = 30° + 30° = 60°$.

1. $A\hat{O}B = 60^\circ = (60 \times 60)' = 3600'$

The distance AB = <u>3600n.m.</u> ANS

2. The radius of the earth (OA or OB) = 6370km.

[See Chapter 3] Length of arc $AB = \frac{60}{360} \times 2\pi R$, where R is th

Earth's radius.

$$= \frac{3.142 \times 6370}{3}$$

No	Log	
3.142	0.4972	
6370.0	3.8041	[add]
	4.3013	
3.0	0.4771	[subtract
	3.8242	

[Antilog.] = <u>6671km</u> ANS

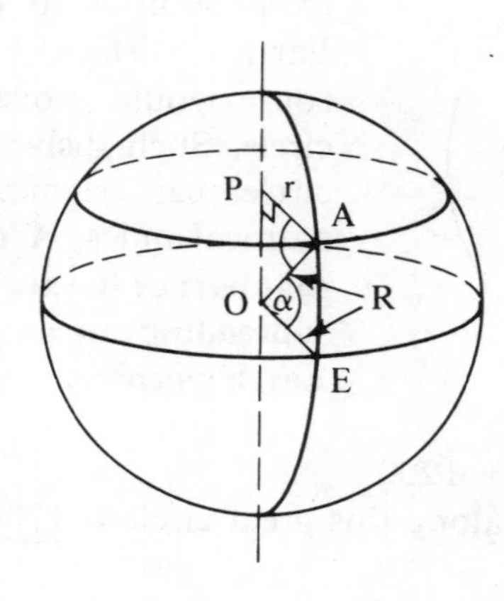

The parallel (or circle) of latitude A shown is α°N. OE and OA are rad of the Earth = R. PA is the radiu of the circle of latitude throug A = r.

$E\hat{O}A = O\hat{A}P = \alpha^\circ$ (alt. $\angle$s)

In triangle OAP, $\cos\alpha^\circ = \frac{AP}{AO} = \frac{r}{R}$

$\therefore r = R\cos\alpha^\circ$

Example

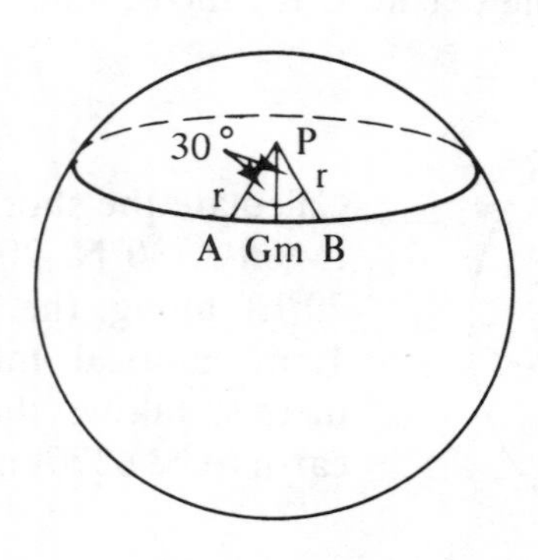

Calculate the distance betwee A(20°N, 30°W) and B(20°N, 30°E Take the radius of the Earth to b 6370km, and $\pi = 3.142$

Let P represent the centre of the small circle of latitude 20°N, and G, the point of intersection of the Greenwich meridian with this circle.

$$\therefore A\hat{P}B = 30^\circ + 30^\circ = 60^\circ$$

If r is the radius of the small circle (latitude 20°N) and R the radius of the Earth,

$$\text{then } r = R\cos 20° \qquad (\therefore\ r = 6370\cos 20°)$$

$$\text{Length of arc AB} = \frac{60}{360} \times 2\pi r$$

$$= \frac{60}{360} \times 2\pi \times 6370\cos 20°$$

$$= \frac{3.142 \times 6370\cos 20°}{3}$$

No.	Log.	
3.142	0.4972	
6370.0	3.8041	
cos 20°	$\bar{1}.9730$	[add]
	4.2743	
3.0	0.4771	[subtract]
	3.7972	

[Antilog.] = 6269 km ANS

Exercise 105 Take the radius of the Earth as 6370km, and $\pi = 3.142$.

1. Find, in nautical miles, the shortest distance between A(40°N, 30°E) and B(10°S, 30°E). Hence calculate the time taken by an aircraft flying at a speed of 300 knots in still air to cover the distance.
2. Find, in nautical miles, the shortest distance between P(33.5°N, 17°W) and Q(19.7°N, 17°W). Hence calculate the time taken by an aircraft flying at a speed of 350 knots in still air to cover the distance.
3. Find, in kilometres, the shortest distance between A(30°N, 20°E) and B(10°S, 20°E)
4. Find, in kilometres, the shortest distance between P(47.2°N, 11°W) and Q(12.8°S, 11°W)
5. Find, in kilometres, the distance between A(10°N, 25°W) and B(10°N, 15°E) along their circle of latitude.
6. Find, in kilometres, the distance between P(56.3°N, 47.2°W) and Q(56.3°N, 11.2°W) along their circle of latitude.
7. Calculate the shortest distance between the North and South poles.
8. Calculate the circumference of the Earth at the Equator.
9. If the Earth rotates once every 24 hours calculate the distance travelled by a place on the Earth's surface latitude 40°N during that time.
10. Calculate the distance travelled by a place on the Earth's surface 22°S 15°W in one hour.

Examination Standard Questions

Worked examples

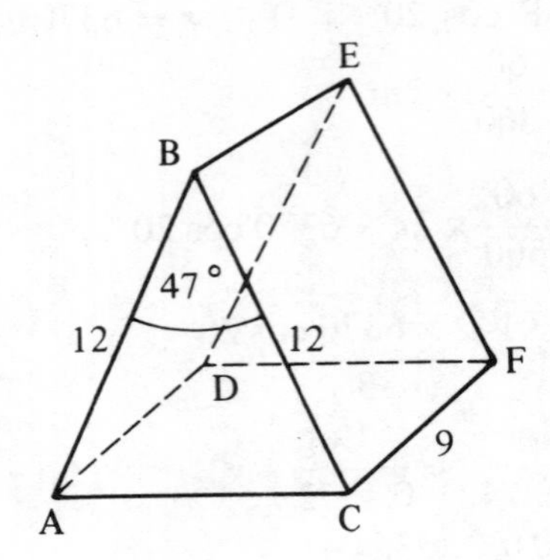

1. The cross-section of a right prism is an isosceles triangle ABC which has AC as base, AB = BC = 12cm, and angle ABC = 47°. The equal edges AD, BE and CF are parallel and each of length 9cm. Calculate (a) the length of AC, (b) the length of AE, (c) the angle between AE and EC, (d) the angle between AE and the face ACFD.

(a) Forming the required two-dimensional diagram:

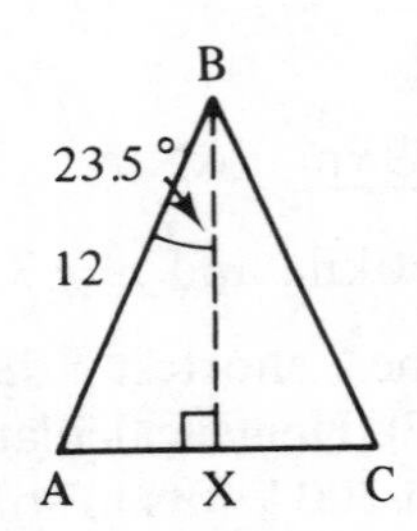

Using $\sin 23.5^\circ = \frac{AX}{12}$

$\therefore 12 \times \sin 23.5^\circ = AX$

$\therefore 12 \times 0.3987 = AX = \underline{4.7844\text{cm}}$

$AC = 2AX = \underline{9.5688\text{cm}}$ ANS

(b) Forming the required two-dimensional diagram.

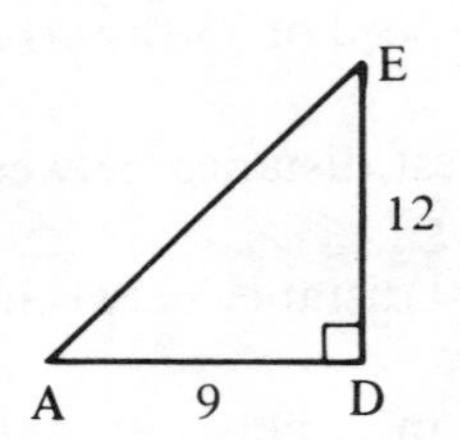

Using Pythagoras theorem:

$AD^2 + DE^2 = AE^2$

$\therefore 9^2 + 12^2 = AE^2$

$\therefore 81 + 144 = AE^2$

$\therefore AE = \sqrt{225} = \underline{15\text{cm}}$ ANS

(c) Forming the required two-dimensional diagram:

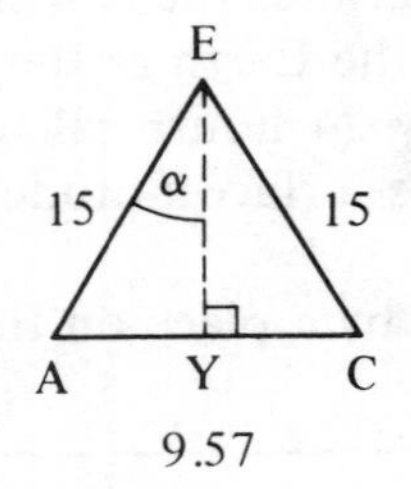

AE = EC = 15cm,
AC = 9.57cm (correct to 2d.p.)
As triangle AEC is isosceles, the line EY bisects the angle AEC.

$[AY = \frac{1}{2} \text{ of } 9.57 = 4.785]$ Using $\sin \alpha = \dfrac{4.785}{15}$

$$\therefore\ \alpha = \sin^{-1} 0.319$$
$$= 18^\circ\, 36'$$
$$\text{Angle } A\hat{E}C = 2\alpha = \underline{37^\circ\, 12'} \quad \text{ANS}$$

(d) This is a case of the angle between a line and a plane:

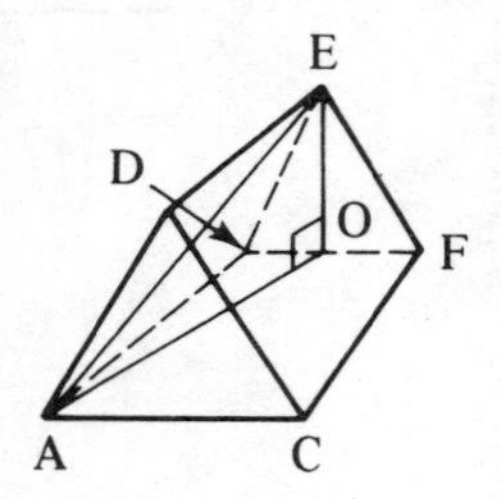

OE is perpendicular to the face ACFD

Forming the required two-dimensional diagram, where β is the required angle:

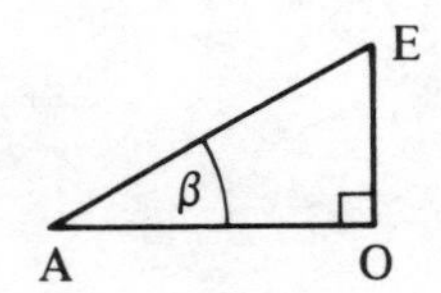

AE = 15cm, OE is the perpendicular height of triangle DEF.

To find the perpendicular height OE:

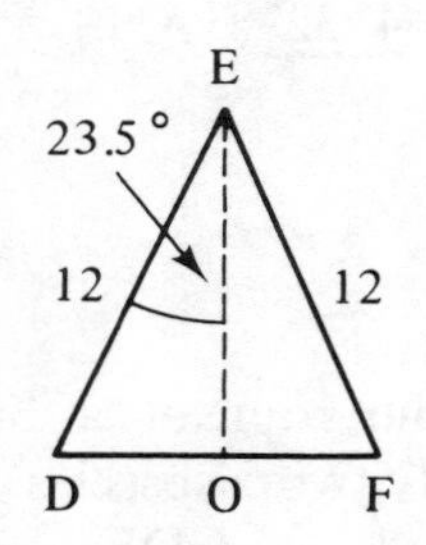

Given that angle $D\hat{E}F = 47^\circ$, and that DE = 12cm.

$$\text{Using } \cos 23.5^\circ = \frac{OE}{12}$$
$$\therefore\ 12 \times \cos 23.5^\circ = OE$$
$$\therefore\ 12 \times 0.9171 = OE = \underline{11.01\,\text{cm}}$$
(correct to 2 d.p.)

Returning to the former diagram:

$$\text{Using } \sin \beta = \frac{11.01}{15}$$
$$\therefore\ \beta = \sin^{-1} 0.7333$$
$$= \underline{47^\circ\, 10'} \quad \text{ANS}$$

2. In making a map, a surveyor notes that the three points A, B and C are at the same level, B is due east of A, and C is to the north of the line AB.
He finds that AB = 10km, BC = 15km and AC = 12km.
Calculate $C\hat{A}B$ and the bearing of C from A.

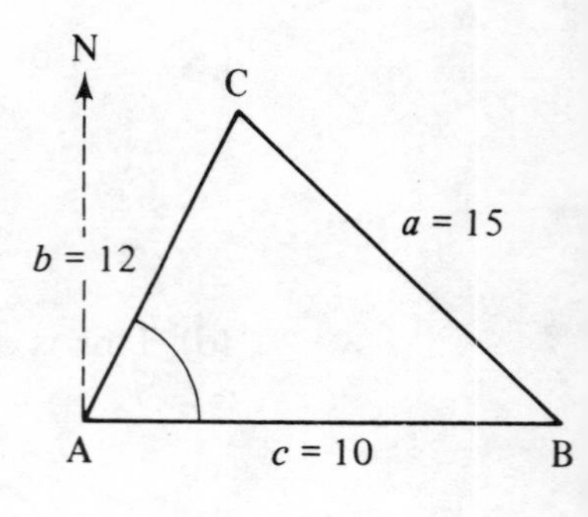

Using the cosine formula

$$\text{i.e. } \cos A = \frac{12^2 + 10^2 - 15^2}{2 \times 12 \times 10}$$

$$= \frac{144 + 100 - 225}{240}$$

$$= \frac{19}{240} = 0.079$$

$$\therefore \quad \hat{A} = \cos^{-1} 0.079 = \underline{85^\circ\, 28'} \quad \text{ANS}$$

The bearing of C from A:

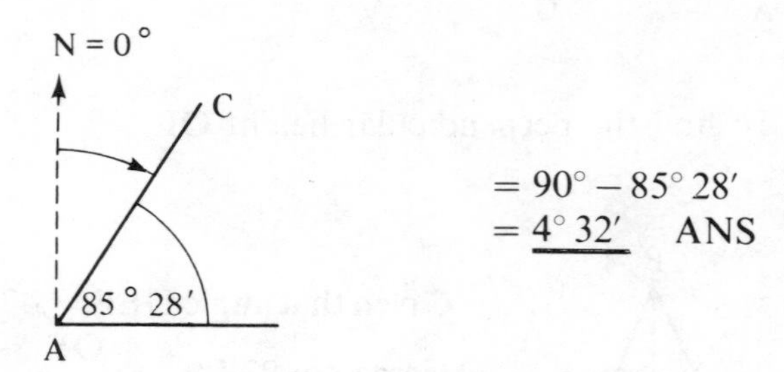

$$= 90^\circ - 85^\circ\, 28'$$
$$= \underline{4^\circ\, 32'} \quad \text{ANS}$$

Exercise 106 **1.**

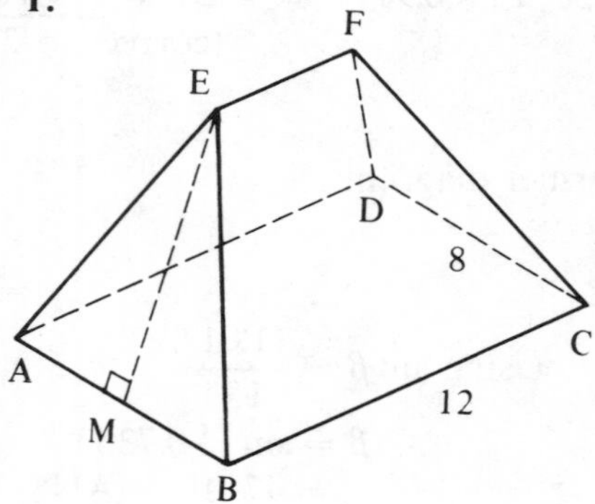

The roof of a building consists of two isosceles triangles ABE and CDF, in which AE = BE = CF = DF, and the two trapezia BCFE and ADFE. All four faces are inclined at the same angle to the horizontal plane of the rectangle ABCD, in which AB = 8m, BC = 12m.
The length of the ridge EF is 4m and EF is 3m vertically above the horizontal plane ABCD.

If M is the foot of the perpendicular from E to AB, show that EM = 5m. Calculate (a) the length of AE, (b) the angle between AE and the plane ABCD, (c) the angle between the plane of triangle AEB and the plane ABCD. (L)

2.

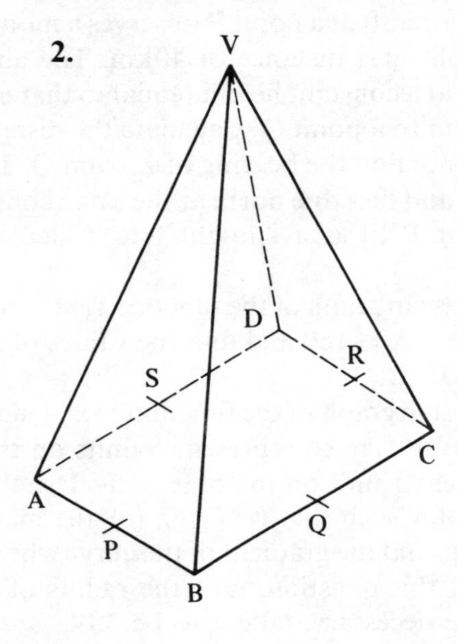

The diagram shows a pyramid with a horizontal rectangular base ABCD and with its vertex V vertically above the centre of the base. P, Q, R and S are the mid-points of the sides of the base.

(a) Using letters given in the figure, name the angle between (i) the edge VA and the base, (ii) the face VAB and the base.

(b) Given that AB = 8cm, BC = 24cm and BV = 20cm, calculate cos $V\hat{Q}S$. (C)

3.

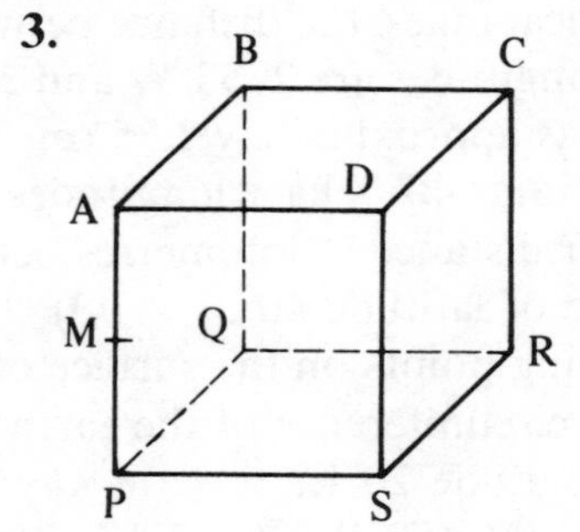

The cube shown in the figure has edges of length 20cm. The mid-point of AP is M.

Calcuate (a) the length of CM, (b) the angle CMR, (c) the angle between MD and the plane PQRS. (C)

4.

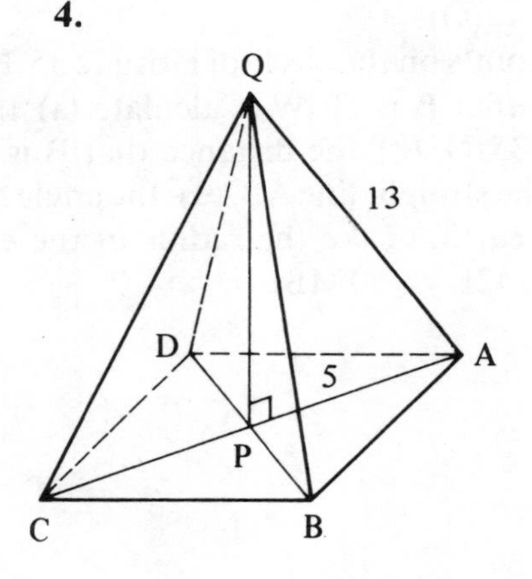

PQ represents a vertical flagpole supported on horizontal ground by four equal wires, AQ, BQ, CQ and DQ. The wires are pegged to the ground at A, B, C and D and ABCD is a square.

Given that each wire is 13m long and each peg is 5m from the base of the pole, P, calculate (a) the height of the pole, (b) the angle between a wire and the ground, (c) the distance between two adjacent pegs, (d) the angle between two adjacent wires. (C)

5. A triangle has sides of lengths 5cm, 8cm, 11cm. Calculate the largest angle. (JMB)

6. In a triangle ABC, AB = 9.40cm, AC = 5.63cm and angle ACB = $113^\circ\,27'$. Calculate angle ABC. (JMB)

7. The pilot of an aircraft at a point P observes a mountain peak X on a bearing of 038° at a distance of 40km. The aircraft flies on a bearing of 070° at a constant height (equal to that of the peak) for a distance of 90km to a point Q. Calculate the distance XQ and the angle XQP. Hence find the bearing of X from Q. The aircraft then changes course and flies due north at the same constant height to a point R so that PXR is a straight line. Calculate the distance RQ. (L)

8. Sketch the Cartesian graph of the function $f(x) = |\cos x^\circ|$ for real x in the domain $0 \leqslant x \leqslant 180$ and find the values of x such that $f(x) = 0.8616$. (JMB)

9. Draw the Cartesian graph of the function $f: x \to \sin 2x^\circ - \sin x^\circ$ for $0 \leqslant x \leqslant 90$, taking 1cm to represent 5 units on the axis of x and 10cm to represent 1 unit on the other axis. Use the graph to find (a) the values of x such that $f(x) = 0$, (b) the maximum value of $f(x)$. By drawing, find the gradient of the curve when $x = 15$. (JMB)

10. In each part of this question, take the radius of the earth to be 6370km. Where necessary, take π to be 3.14.
(a) (i) Find, in nautical miles, the distance between two points on the equator whose longitudes are $2^\circ\,58'$ W and $29^\circ\,31'$ W, (ii) show that 1 nautical mile is approximately 1.85km.
(b) Two points in latitude 40°N have longitudes of 23°W and 67°E. Calculate the shorter distance in kilometres between them, measured along the circle of latitude 40°. (O)

11. Which of the following points on the surface of the earth are at a distance of $\frac{1}{12}$ of the circumference of the earth from the point on the equator with longitude 20°E?
(a) 0°N, 10°W; (b) 0°N, 170°W; (c) 30°S, 50°E; (d) 0°N, 50°E; (e) 30°N, 20°E. (O)

12. A and B are two points on the circle of latitude 35°N. The longitude of A is 25°E and that B is 50°W. Calculate (a) the radius of the circle of latitude 35°N, (b) the distance that B is due West of A, (c) the length of the straight line AB, (d) the angle AOB, where O is the centre of the earth. (Take the radius of the earth to be 3960 miles and π as 3.142). (JMB).

15

Constructions and Loci

Constructions based on the rhombus

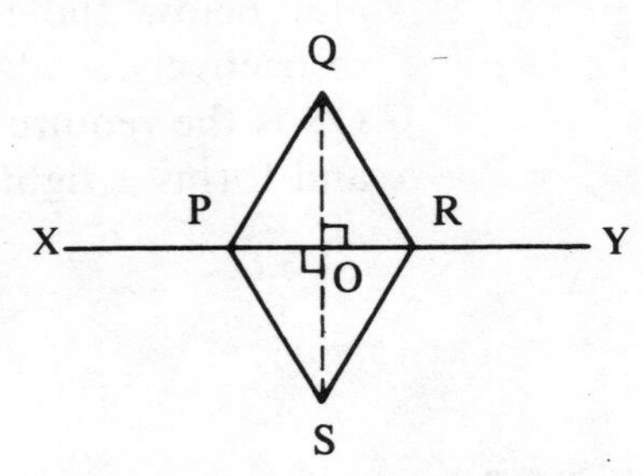

Consider the rhombus PQRS symmetrical about the axis XY. At O its diagonals PR and QS bisect each other at right angles. Opposite angles are equal and bisected by a diagonal.

To construct a perpendicular from a point O on XY to the line XY.

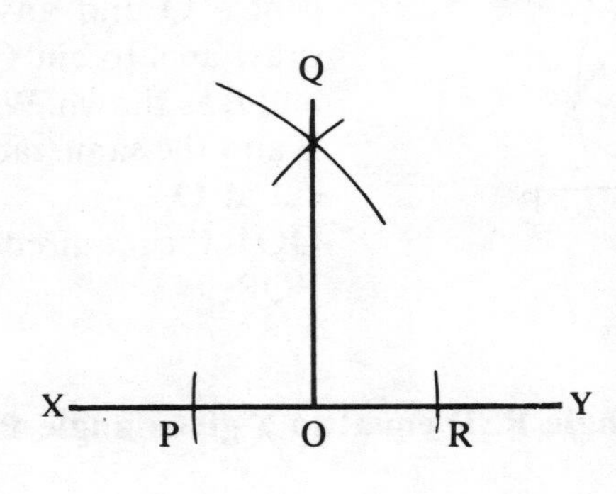

Given: line XY and point O. With centre O and compasses set to any suitable radius draw arcs of a circle to cut the line XY at P and R. With centres P and R, and compasses set to a radius greater than before, draw arcs that cut above the line at Q.
OQ is the required perpendicular from O.

To construct a perpendicular from a point Q, not on the line, to the line XY.

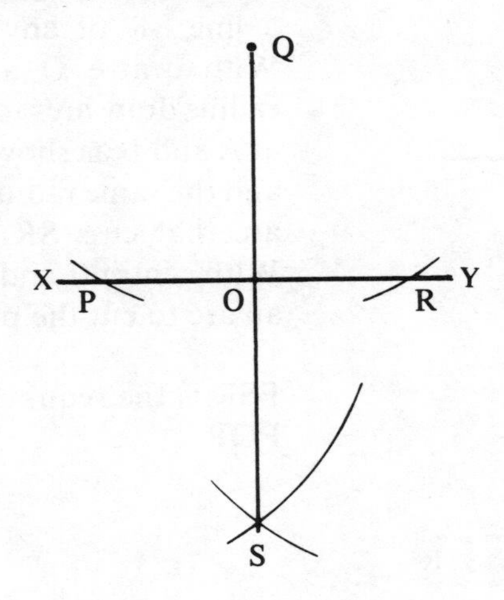

Given: line XY and point Q. With centre Q and any suitable radius draw arcs to cut the line XY at P and R. With centres P and R, keeping the same radius, draw arcs to cut the other side of the line at S. QO is the required perpendicular from Q.

To bisect a given straight line PR.

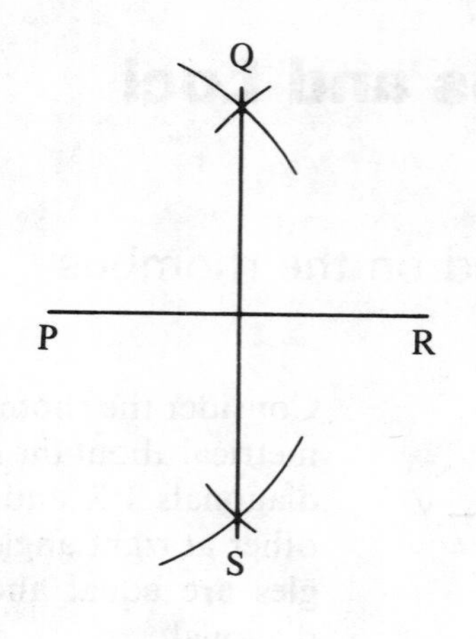

Given: line PR. With centres P and R, using a suitable radius greater than $\frac{1}{2}$PR, draw arcs that cut above and below the line at Q and S respectively.
QS is the required bisector of PR, and forms a right angle.

To bisect a given angle PQR.

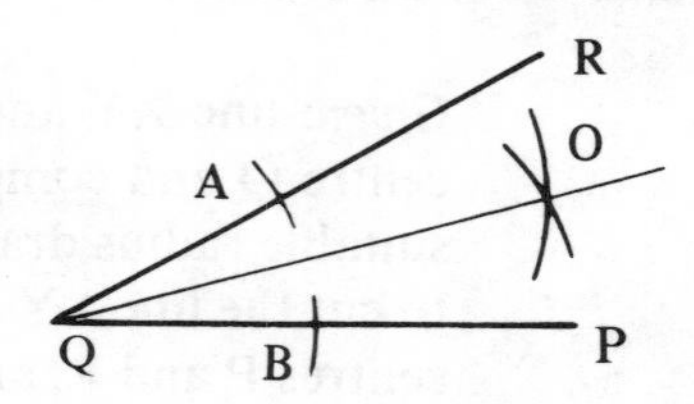

Given: angle PQR as shown. With centre Q and any suitable radius draw arcs to cut QR and QP at A and B as shown. With centres A and B and the same radius draw arcs to cut at O.
QO is the required bisector of angle PQR.

To construct an angle RSP equal to a given angle PQR.

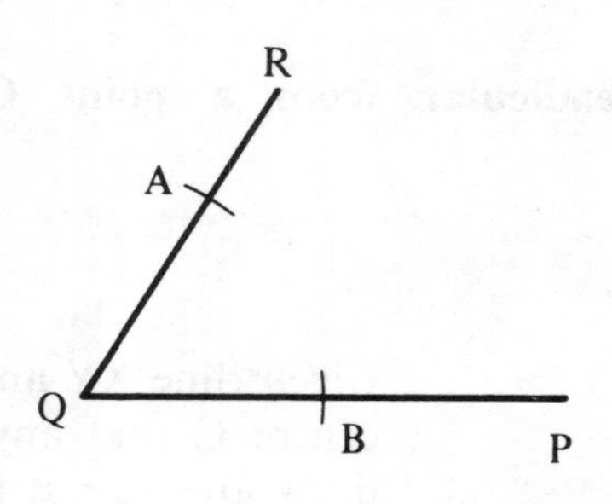

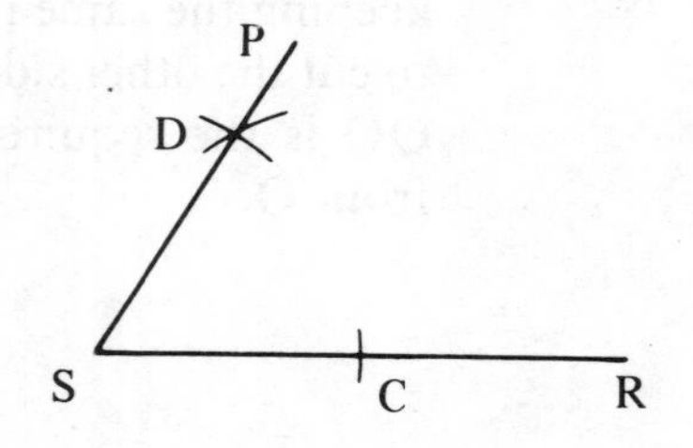

Given: angle PQR as shown. Draw a line SR of any suitable length. With centre Q and any suitable radius draw arcs to cut QR and QP at A and B as shown. With centre S and the same radius draw a suitable arc that cuts SR at C as shown. With centre C and radius AB draw an arc to cut the previous arc at D.

PSR is the required angle equal to PQR.

To construct through a given point Q a line parallel to a given straight line PS.

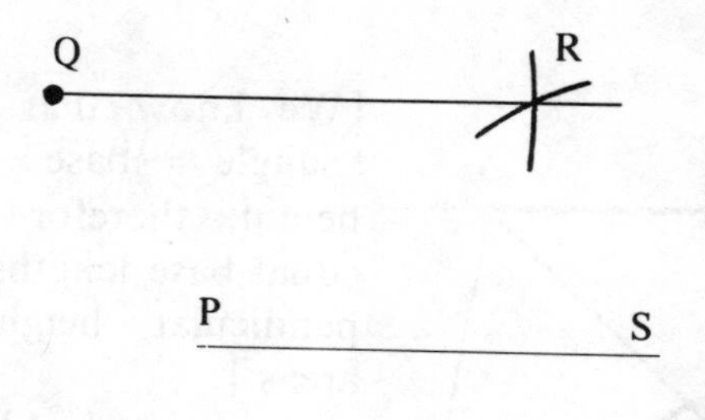

Given: line PS and point Q. With centre Q and radius PS draw a suitable arc as shown. With centre S and radius PQ draw an arc to cut the previous arc at R.
QR is the required line parallel to PS. (QR could have been constructed below PS by the same method if required.)

Constructions Based on Other Plane Figures

To construct an angle of 60°.

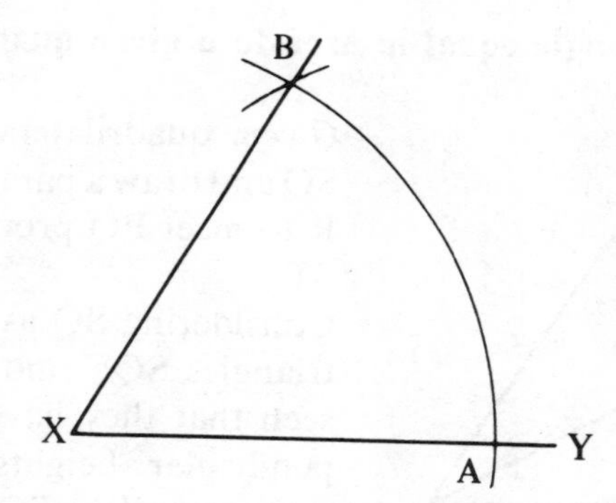

[We need to construct an equilateral triangle.]
Draw any suitable line XY. With centre X and any suitable radius draw an arc to cut XY at A. With centre A and radius the same draw an arc to cut the previous arc at B.
$B\hat{X}A$ is the required angle of 60°

To divide a straight line XY into a given number of equal parts.

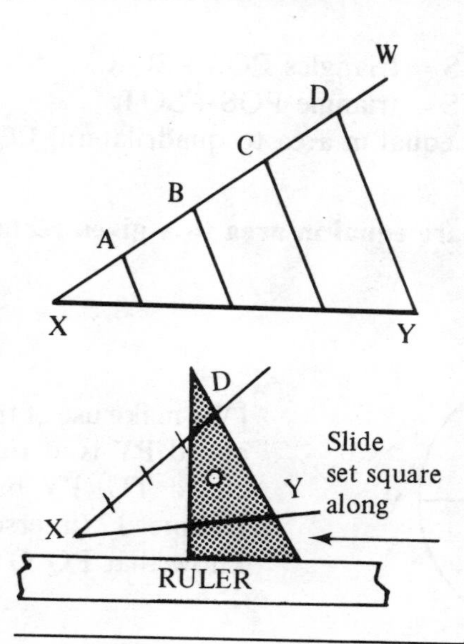

[We need to construct a nest of similar triangles.]
Given: the line XY and the number of parts (4 in this example). Draw a line XW at any suitable angle as shown. With compasses mark off four equal steps XA, AB, BC, CD along the line XW. These steps can be of any suitable length. Join the last mark (D) to Y. Draw lines parallel to DY from C, B and A onto the line XY using a set square and ruler as shown.

These parallel lines divide XY into four equal parts.

Construction of Equal Areas

To construct a triangle equal in area to another triangle on the same base.

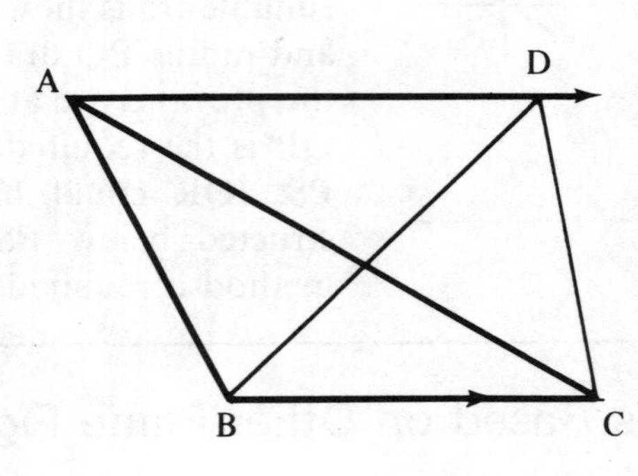

[We know that the area of a triangle = $\frac{1}{2}$base × perpendicular height, therefore triangles with equal base lengths and equal perpendicular heights have equal areas.]

Given: triangle ABC.

Through A draw a line parallel to BC. Any point D on this new line will form with B and C a triangle equal in area to ABC.

To construct a triangle equal in area to a given quadrilateral.

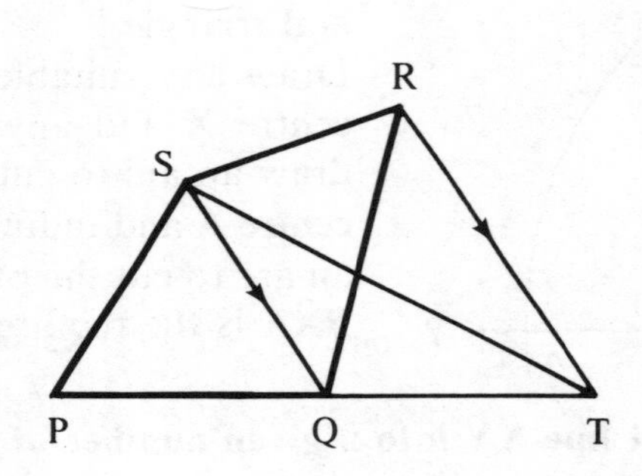

Given: quadrilateral PQRS. Join SQ and draw a parallel line through R to meet PQ produced at T. Join ST.

Considering SQ as the base of the triangles SQT and SQR it can be seen that they have the same perpendicular heights (between the same parallels SQ and RT) and therefore the same area.

Quadrilateral PQRS = triangles PQS + SQR

Triangle PTS = triangle PQS + SQT

∴ Triangle PTS is equal in area to quadrilateral PQRS.

To construct a square equal in area to a given rectangle.

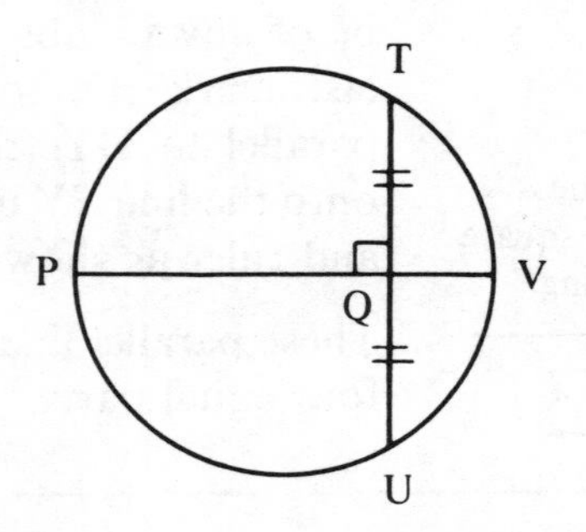

[We make use of the fact that as the chord PV is at right angles to the chord TU, PV bisects TU. From Chapter 12 (intersecting chords) we know that $PQ.QV = QT^2$.]

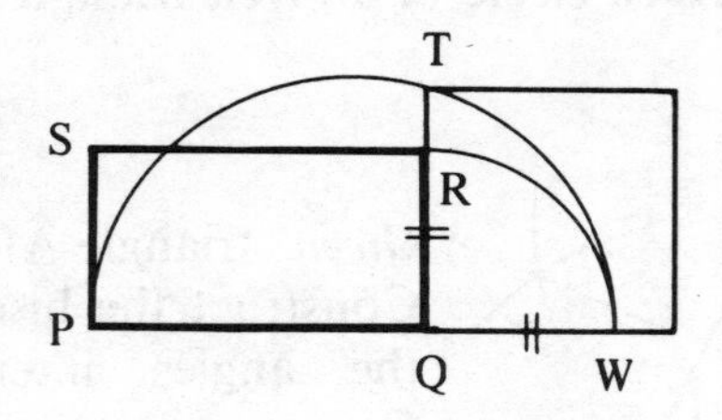

Given: rectangle PQRS. Produce PQ to W so that QR = QW. Find the mid-point of PW and draw a semi-circle with PW as diameter. Produce QR if necessary to meet the semi-circle at T.
QT is a side of the required square as shown.

Further Constructions Involving Circles

To construct a circle to pass through three given points.

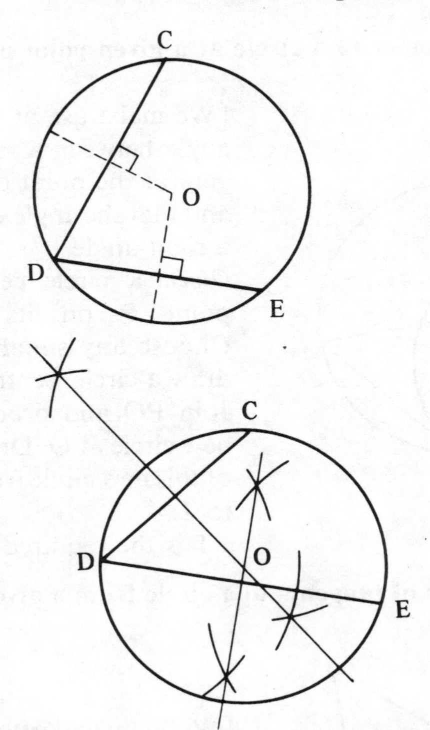

[We make use of the fact that a line drawn from the centre of a circle to the mid-point of a chord is at right angles to that chord.]

Given: three points C, D and E. Join CD and DE. Construct their perpendicular bisectors intersecting at O. O is the centre of the required circle.

To construct the circumcircle of a given triangle.

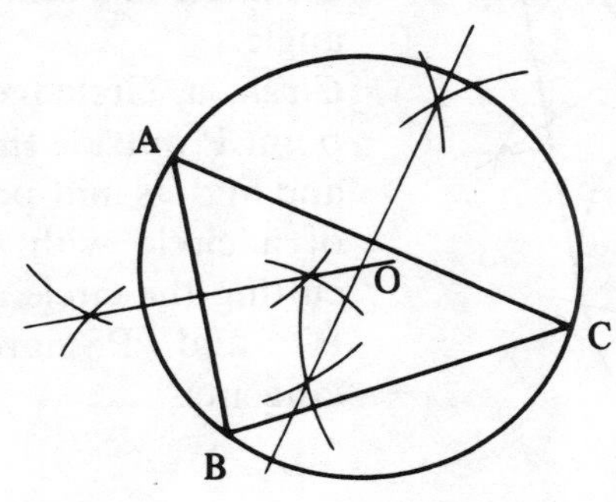

[The method is as for the previous example.]
Given: triangle ABC.
Construct the perpendicular bisectors of two of the sides, the bisectors intersecting at O.
O is the centre of the required circle.

To construct the inscribed circle of a given triangle.

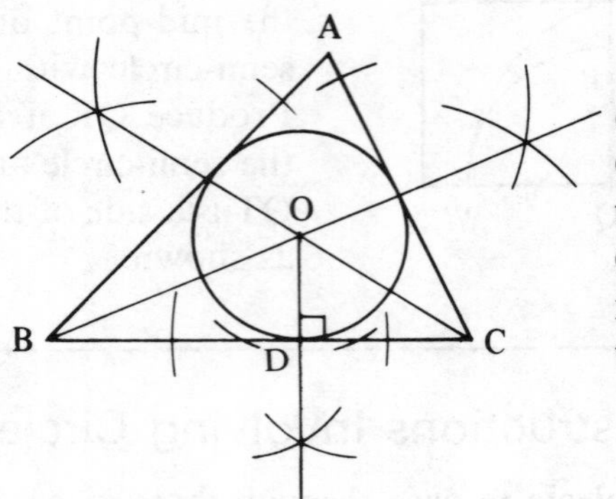

Given: triangle ABC.
Construct the bisectors of two of the angles intersecting at O. Construct a perpendicular OD from O onto one of the sides. OD is the radius of the circle centre O.

To construct a tangent to a circle at a given point on that circle.

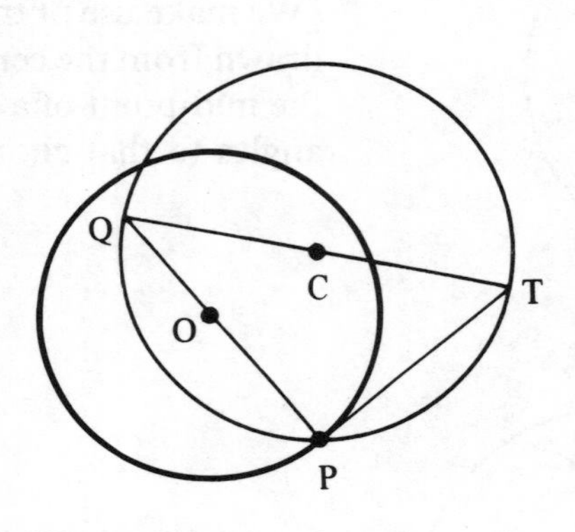

[We make use of the fact that the angle between a radius and a tangent at the point of contact is 90°, and that the angle in a semi-circle is a right-angle.]
Given: a circle centre O with a point P on its circumference. Choose any suitable point C and draw a circle, centre C, through P. Join PO and produce to cut this new circle at Q. Draw the diameter of this new circle from Q through C to T.
PT is the required tangent.

To construct a pair of tangents to a circle from a given point outside the circle.

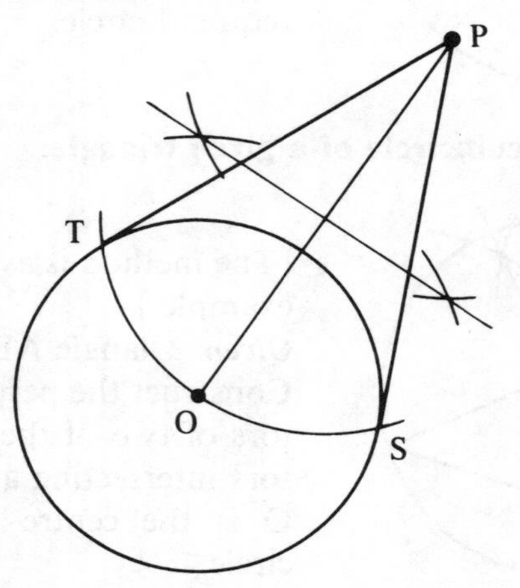

[We again make use of the fact that the angle in a semi-circle is a right-angle.]
Given: a circle centre O with a point P outside the circle. Join OP and find its mid-point. Draw an arc of a circle with OP as diameter cutting the circle at T and S.
PT and PS are the required tangents.

To construct a segment of a circle that contains a given angle.

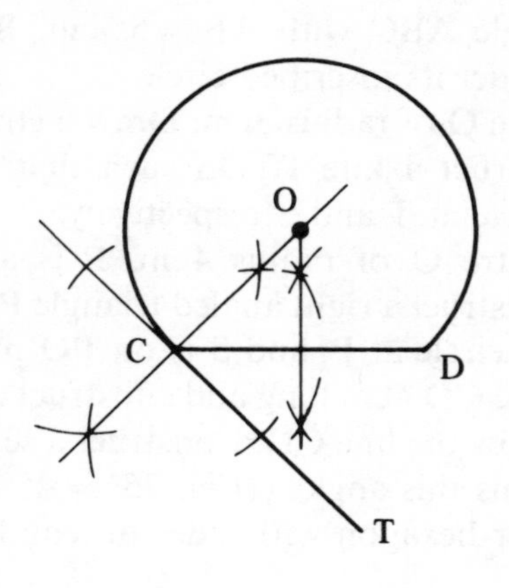

[We make use of the fact that the angle a chord makes with a tangent at the point of contact is equal to the angle in the alternate segment.]
Given: angle DCT.
Let CD be a chord of the circle and draw in the given angle DCT. Construct the perpendicular bisector of CD and a perpendicular to CT from C intersecting at O. With centre O and radius OC draw the major arc of the circle from C to D. This is the required segment.

Note When answering questions involving ruler and compass construction all such work should be clearly shown and not erased. Marks will be lost if essential construction is not clearly shown.

Exercise 107 Use ruler, compasses and pencil only:

1. Construct an isosceles triangle of base length 6cm and perpendicular height 4cm.
2. Construct a right-angled triangle whose shorter sides are of lengths 3cm and 4cm.
3. Construct a triangle ABC with $A\hat{B}C = 45°$ and AB = 4cm, BC = 5cm.
4. Construct an equilateral triangle with sides of length 4cm.
5. Construct a triangle ABC with $A\hat{B}C = 30°$ and AB = 3cm, BC = 5cm.
6. Construct a rhombus whose diagonals are of lengths 5cm and 7cm.
7. Construct a kite whose diagonals are of lengths 6cm and 4cm, the shorter diagonal intersecting the longer 2cm from one end of the longer diagonal.
8. Given that P is any point approximately 4cm above a straight line QR, construct a trapezium PQRS where PS and QR are parallel sides.
9. Construct a triangle equal in area to the trapezium in question **8.**
10. Construct a rectangle having sides of lengths 4cm and 6cm.
11. Construct a square equal in area to the rectangle in question **10.**
12. Construct a parallelogram PQRS with $P\hat{Q}R = 45°$ and PQ = 3cm, QR = 6cm.
13. Construct a triangle ABC, with BC as the base, given that AB = 6cm, BC = 4cm and AC = 5cm.
14. Use part of the line BA in question **13** to divide the line BC into three equal parts.

15. Construct a triangle ABC with AB = 4.2cm, BC = 4.4cm and AC = 2.6cm. Construct its circumcircle.
16. Construct a triangle ABC with AB = 5.5cm, BC = 6.3cm and AC = 5cm. Construct its inscribed circle.
17. Draw a circle centre O of radius 3cm. Draw a straight line OP of length 7cm. Construct a kite PTOS such that PT and PS are tangents to the circle at T and S respectively.
18. Draw a circle centre O of radius 4cm. P is any point on its circumference. Construct a right angled triangle PTS such that PT is a tangent to the circle at P, and S is on PO produced.
19. Draw a straight line CD 4cm long and construct an angle DCE of 75°, where E is below the line CD. Construct a segment of a circle on CD that contains this angle. (*Hint*: $75° = 45° + 30°$)
20. Construct a regular hexagon with sides of length 3cm.

Loci

A **locus** (plural **loci**) is a set of points satisfying specific conditions. It can also be described as the path of a moving point obeying a given law.

Examples

1.

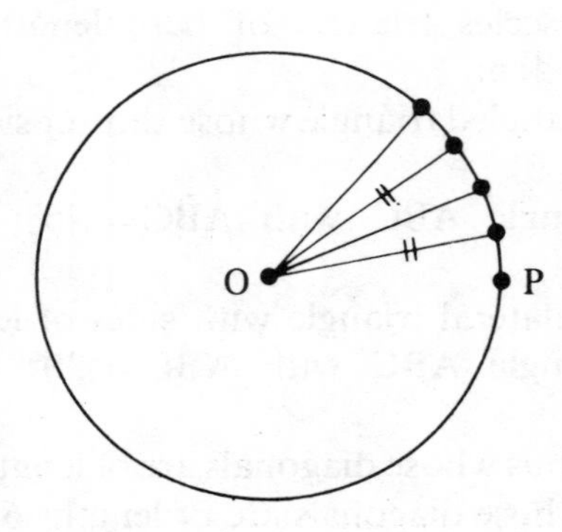

A point P that moves in a plane so that it is always a fixed distance from a point O in the plane traces out a circle.

2.

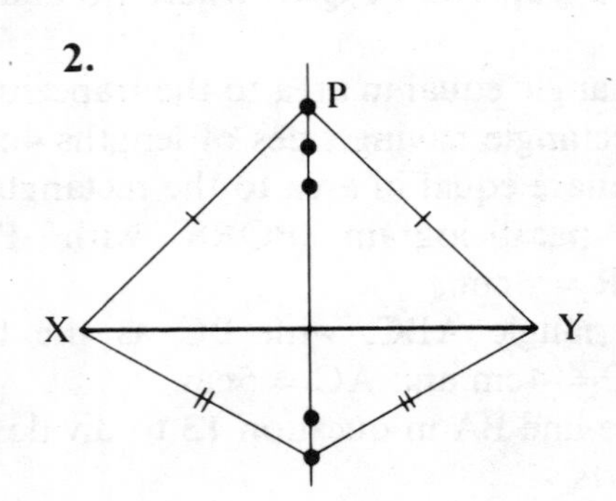

The locus of a point P that moves so that it is equidistant from two fixed points X and Y is the perpendicular bisector of the line XY.

3.

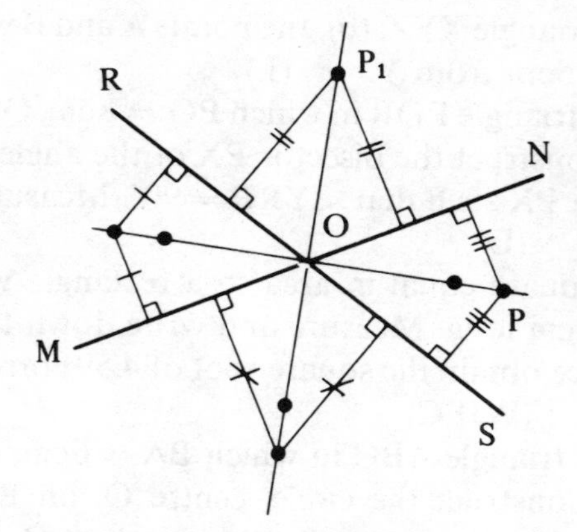

The locus of a point P that moves so that it is equidistant from two straight lines MN and RS that intersect at O is the pair of lines that bisect the angles between MN and RS.

Exercise 108 Use the methods of construction practised earlier in this chapter:

1. Draw a straight line XY 5cm in length. Construct the locus of a point P that moves such that it is always 4cm above XY.
2. An angle AOB = 60° is formed by the intersection of two straight lines OA and OB. Construct this angle and then the locus of a point P that is always the same distance from OA and OB.
3. PQR is a triangle with base PQ = 4cm. Construct that part of the locus of R that is above PQ given that the area of the triangle is $4cm^2$.
4. Construct a square ABCD of side length 3cm with AB as the base. Construct the locus of the point D as the square is pivoted first about B so that BC becomes the base, then about C so that CD becomes the base.
5. Construct a circle of radius 3cm, centre O. T is a fixed point 6cm from O and joined to any point P on the circumference of the circle. M is the mid-point of TP found by construction. Construct the locus of M as the point P moves around the circumference of the circle.

Examination Standard Questions

Exercise 109 Use ruler, compasses and pencil only for the construction:

1. Construct the rectangle ABCD given that AB = DC = 9cm, AD = BC = 5cm. On AB as base construct an isosceles triangle ABE equal in area to ABCD and having AE = EB.
   On the same diagram, and without calculation construct a square equal in area to ABCD. (JMB)
2. Construct a triangle ABC in which AB = 9cm, AC = 7cm, BC = 10cm. Construct the bisector of angle BAC to meet BC at D. Measure AD. Construct the point P on AD such that ∠BPC is a right angle. Measure AP. (JMB)

3. Construct a triangle XYZ in which XY = 7cm, YZ = 8cm and ZX = 9cm. Construct (a) the circle which touches internally all three sides of triangle XYZ, (b) the points A and B which lie on this circle and are 5cm from Y. (L)

4. Construct the triangle PQR in which PQ = 8cm, QR = 8.5cm and RP = 5cm. Construct the bisector PX of the angle QPR and find the point Y on PX such that $\angle YRP = 90°$. Measure the length of YR and PY. (L)

5. Construct a square equal in area to a rectangle whose sides are 3.4cm and 5.4cm long. Measure and write down the length of its sides, and hence obtain the square root of 4.59 correct to one place of decimals. (WJEC)

6. Construct the triangle ABC in which BA = 6cm, AC = 8cm and $B\hat{A}C = 90°$. Construct the circle, centre O, on BA as diameter. Bisect $B\hat{A}C$ and let the bisector cut the circle at D. Produce OD to meet BC at E. Measure BE. (WJEC)

7. Construct two lines AB and BC so that AB = 7cm, BC = 9cm and $\angle ABC = 120°$. Now determine a point P within the angle ABC so that $\angle APB = 60°$ and $\angle BPC = 45°$. Measure BP.
Describe briefly your construction. (O)

8. Draw a straight line AB and on it mark a point X. By a geometrical construction, showing clearly all construction lines, find and mark all points which are 4cm from X and 3cm from the line AB. (L)

9. On a straight line AB of length 6cm construct (a) a segment of a circle containing an angle of 30°, (b) a triangle ABC of area $21\,cm^2$ in which $\angle ACB = 30°$.
Find by measurement the lengths of AC and BC. (L)

10. (a) Construct in the plane of the paper the locus of points equidistant from the two fixed points A and B.

A ——— 5 cm ——— B

(b) On the same figure, construct the locus of points P such that the triangle ABP has area $5cm^2$.
(c) How many isosceles triangles ABP (with AP = PB) are there with area $5cm^2$? (O)

11. Draw a line AB 6cm long and approximately 10cm from the top of the page.
In a single diagram draw accurately the locus of (a) the point P which moves so that AP = PB, (b) the point Q which moves so that $A\hat{Q}B = 45°$, (c) the point R which moves so that the area of triangle ARB is $12cm^2$.
Label each locus clearly.
Mark on your figure a point X such that $AX < XB$, $A\hat{X}B = 45°$ and the area of triangle AXB = $12cm^2$. Measure the lengths AX and BX. (C)

12. In a single diagram construct (a) a triangle PQR in which base QR = 7cm, PQ = 11cm and $P\hat{Q}R = 30°$, (b) the locus of points equidistant from P and R, (c) the circle which passes through P and R and has QR as the tangent at R.
Measure the radius of this circle. (C)

13. Construct in a single diagram (a) a circle, radius 5cm, with diameter AC, (b) a point B on the circumference of the circle such that AB = BC, (c) the point D on the circumference of the circle, but on the side of AC opposite to B, such that $C\hat{A}D = 60°$, (d) the locus of points equidistant from AB and AC. Given that the locus cuts the circle again at P, measure PD. (C)

14. (a) Construct a quadrilateral ABCD with AB = 7cm, AD = 5cm, BC = DC = 10cm and the angle BAD = 90°. Construct the perpendicular from B to DC. Measure and state the length of this perpendicular. Hence calculate the area of the quadrilateral ABCD.
(b) Three marker buoys P, Q and R are such that Q is 90m due East of P, and R is 60m on a bearing from P of 030° (N30°E). Draw an accurate scale diagram to show the relative positions of the marker buoys. Show the position of a fourth buoy S which is equidistant from P and Q and is also equidistant from QR and QP. Measure and state the distance QS. (AEB)

15. (a) Draw a line AB 10cm long and another line AC such that the angle BAC = 50°. Construct a circle on the chord AB such that AB subtends an angle of 50° at the circumference of the major arc of the circle. Measure and state the radius of this circle.
(b) Construct a triangle PQR with PQ = 12cm, PR = 8cm and the angle RPQ = 60°. Show the positions of the points S and T which are both equidistant from the lines RP and RQ, and are also 7.5cm from Q. Measure and state the length ST. (AEB)

16
Probability and Statistics

Bar Charts

A group of students were asked the question 'which television channel do you watch the most, BBC 1, BBC 2 or ITV?' The following 720 replies were recorded:

BBC 1	BBC 2	ITV
300	90	330

A **bar chart** can be drawn to represent this information. The bars must be the same width, their lengths representing the number of students choosing each station. The bars may be drawn vertically or horizontally as shown.

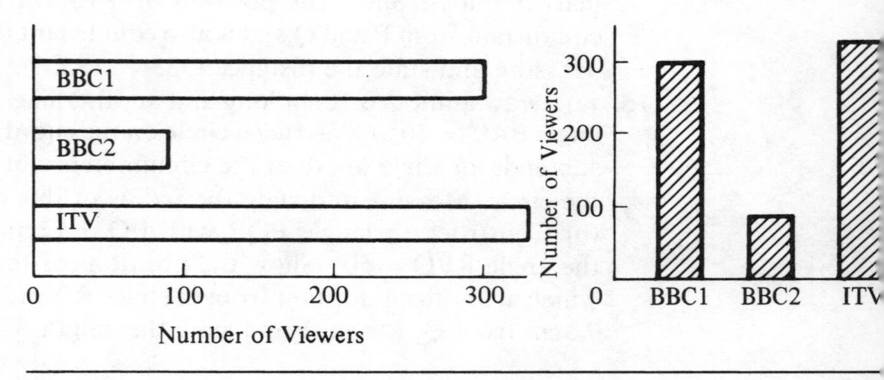

Pie Charts

The same information could be shown in the form of a **pie chart.** The circle represents all the people questioned, each group of people occupying a sector of the circle.

300 out of 720 students chose BBC 1, i.e. $\frac{300}{720}$

To convert this to an angle of a sector we multiply by 360°;

$$\text{i.e.} \quad \frac{300}{720} \times 360° = \underline{150°}$$

BBC 2's viewers are represented by $\frac{90}{720} \times 360° = \underline{45°}$

ITV's viewers ae represented by $\frac{330}{720} \times 360° \quad = \underline{165°}$

The pie chart can now be drawn:

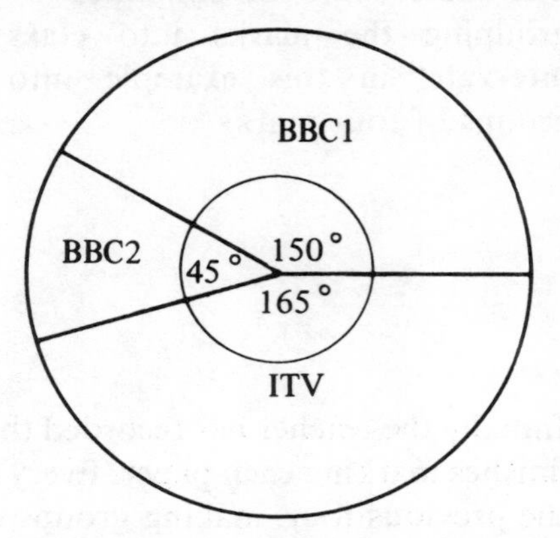

Frequency Distributions

Consider a form of 29 pupils who scored the following marks out of 20 in a test:

15, 9, 9, 17, 12, 8, 9, 17, 12, 9, 10, 14, 20, 7, 19, 8, 14, 14, 10, 14, 14, 9, 16, 14, 6, 5, 14, 7, 16.

The marks could be recorded on a **frequency table** as shown below.

Mark	Tally	Frequency
1		0
2		0
3		0
4		0
5	/	1
6	/	1
7	//	2
8	//	2
9	~~////~~	5
10	//	2
11		0
12	//	2
13		0
14	~~////~~ //	7
15	/	1
16	//	2
17	//	2
18		0
19	/	1
20	/	1
		29 scores

The table could be simplified by grouping the marks into **class intervals**, in this example into groups of four marks.

Marks	Frequency
1–4	0
5–8	6
9–12	9
13–16	10
17–20	4
	29 scores

Initially the teacher has recorded the information as 'tally marks' as he finishes marking each paper. Every fifth mark is used to cross through the previous four, making groups of five for easier counting.

Histograms

Such frequency distributions can be shown in the form of **histograms**. Each score or each class is represented by the area of a rectangle. When the class intervals are equal, as in the example above, the rectangles on the histogram will be of equal width.

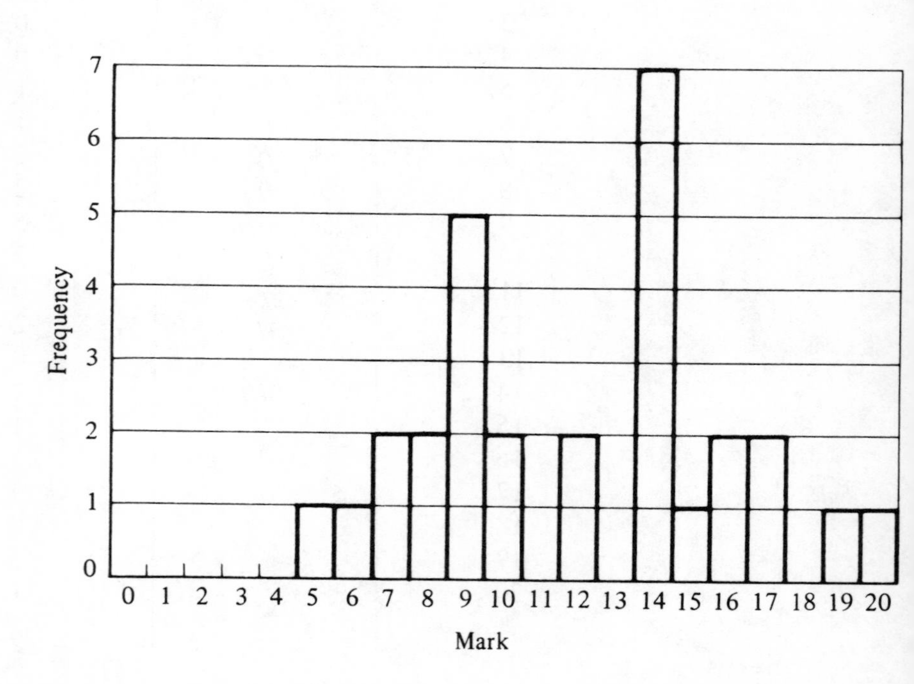

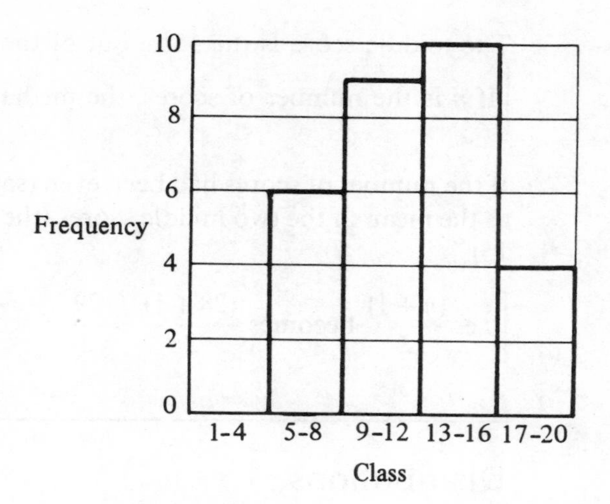

If we join the mid-points of the top end of each rectangle we obtain a **frequency polygon**.

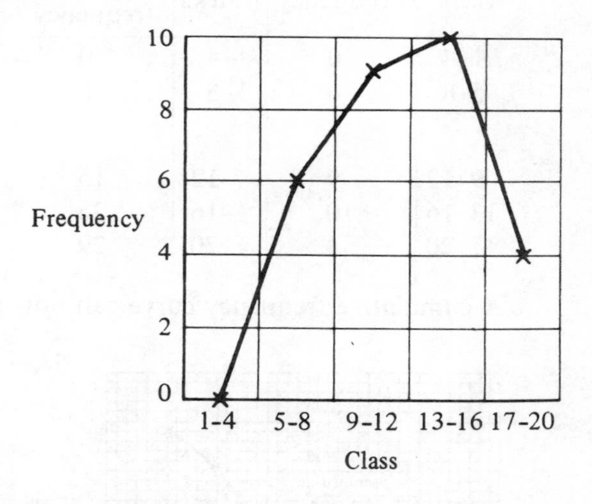

Averages

There are three types of average we must consider.
The **mean**, the **median** and the **mode**.
Using the set of test marks listed earlier;

the **mean** $= \dfrac{\text{the total of all marks scored}}{\text{the number of scores}} = \dfrac{348}{29} = \underline{12 \text{ marks}}$

the **mode** is the most frequent score $= \underline{14 \text{ marks}}$

the **median** is the middle score when the scores are arranged in order, i.e. the 29 scores in ascending order are:

5, 6, 7, 7, 8, 8, 9, 9, 9, 9, 9, 10, 10, 12, 12, 14, 14, 14, 14, 14, 14, 14, 15, 16, 16, 17, 17, 19, 20.

The middle score is the 15th. out of the list of 29, = 12 marks

[If n is the number of scores, the median is the $\frac{(n+1)}{2}$ score.]

Note

If the number of scores had been even (say 28), the median is calculated as the mean of the two middle scores (the scores at the positions 14 and 15).

[i.e. $\frac{(n+1)}{2}$ becomes $\frac{(28+1)}{2} = \frac{29}{2}$ = score at position 14.5.]

Distributions

A **cumulative frequency** distribution table for the set of marks can be constructed as shown:

Marks	Frequency	Marks	Cumulative frequency
1–4	0	1–4	0
5–8	6	1–8	6
9–12	9	1–12	15
13–16	10	1–16	25
17–20	4	1–20	29

[i.e. 6 pupils scored 8 marks or less]

[i.e. 15 pupils scored 12 marks or less]

A cumulative frequency curve can now be drawn:

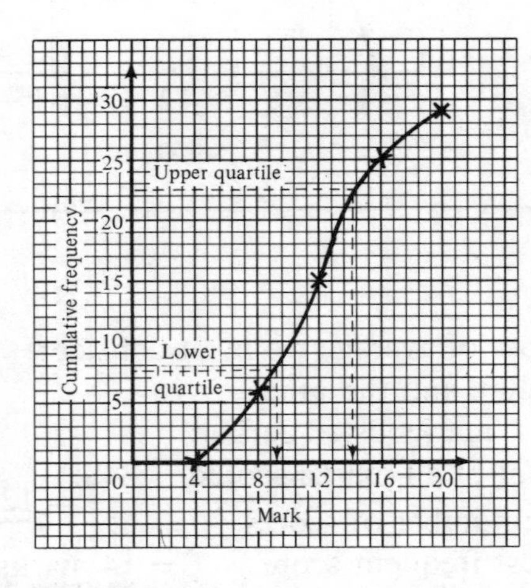

[The cumulative frequencies are plotted at the upper end of the class intervals,
e.g. against 8 in the 5–8 interval.]

The median divides the list of scores into two halves. The list may be further divided into quarters known as **quartiles.** The first quartile called the **lower quartile** has one-quarter of the scores below it. The second quartile is the median, and has half of the scores below it. The third quartile, called the **upper quartile**, has three-quarters of the scores below it.

The lower and upper quartiles can be *estimated* from the graph by taking readings from the cumulative frequency curve for positions $\frac{(29+1)}{4}$ and $\frac{3(29+1)}{4}$ on the vertical axis.

[i.e. Scores 7.5 and 22.5 = 9 marks and 14 marks

The **range** of marks for the test was from 5 marks to 20 marks = 20 − 5 = 15 marks.

This does not tell us how evenly the marks were spread. For this we find the **interquartile range**, which is the difference between the lower and upper quartiles. If this interquartile range is large the marks are well spread, if it is small the scores were close together indicating that the test paper failed to highlight the relative abilities of the pupils in the way the examiner would wish.

In our example; upper quartile − lower quartile = 14 − 9 = 5.

A comparison using the **semi-interquartile range** is sometimes required, this is one-half of the interquartile range.

Exercise 110

1. A group of housewives was asked which of four washing powders they preferred. Draw a bar chart to represent the following results.

Powder	Sudso	Brita	Pow	Wow
No. of housewives	73	30	46	31

2. Draw a pie chart to represent the information given in question **1**.
3. Construct a frequency table for the following set of 35 homework marks. The work was marked out of 15.

   7, 11, 10, 6, 9, 7, 6, 11, 14, 9, 7, 4, 13, 11, 14, 12, 9,
   10, 6, 11, 9, 12, 7, 6, 4, 9, 10, 15, 10, 9, 11, 7, 7, 9, 3,

4. Construct another frequency table for the marks in question **3** using equal class intervals of three marks.
5. Construct a histogram using the frequency table from question **4**.
6. Construct a frequency polygon using the table from question **4**.
7. Find the mean, median and mode of the set of marks in question **3**.
8. Construct a cumulative frequency distribution table from the table used in question **4**.
9. Draw a frequency curve from the table constructed in question **8**.
10. (a) Using the graph from question **9**, or otherwise, determine the lower and upper quartiles of the set of marks given in question **3**.
    (b) Calculate (i) the range of these marks, (ii) the interquartile range of these marks, (iii) the semi-interquartile range of these marks.

Probability

If we toss a coin it is expected to land in one of two ways, 'head' or 'tail' uppermost. We therefore say that the **probability** of a successful call of 'heads' is one chance in two, i.e. $p = \frac{1}{2}$, where p represents the probability of that event occurring.

The probability of selecting the King of Diamonds from a pack of cards is $p(\text{KD}) = \frac{1}{52}$, as there is only one such card in the pack. However, the probability of selecting any King is $p(\text{K}) = \frac{4}{52} = \frac{1}{13}$.

Consider one throw of a fair die, that is one which is unbiased. It is equally likely to land in any of six ways. Therefore the probability of throwing a 3 is one chance in six, i.e. $p(3) = \frac{1}{6}$.

It is obvious that the probability of *not* throwing the 3, and therefore getting a 1, 2, 4, 5 or 6 is $\frac{5}{6}$, which is $1 - p(3)$.

As the probability of throwing a 3 is $\frac{1}{6}$, the probability of throwing a 4 on the next occasion will also be $\frac{1}{6}$. The probability of throwing *either* a 3 *or* a 4 will be two chances in six.

$$\text{i.e. } p(3 \textit{ or } 4) = p(3) + p(4) = \frac{1}{6} + \frac{1}{6} = \frac{1}{3}.$$

These two events are said to be **mutually exclusive**, i.e. they cannot occur together.

What are the chances of throwing a 3 *then* a 4 in consecutive throws? There are six ways that the dice can fall on each throw. We might throw a 1 followed by a 1, i.e. (1, 1); or 1 followed by a 2, i.e. (1, 2). Complete the list of possible results started below, sometimes referred to as the **sample space** of possible outcomes. (1, 1), (1, 2), (1, 3), ...

You should find thirty-six different possible results from the two throws. The pairing (3, 4) appears only once. Therefore there is one chance in thirty-six of throwing a 3 *then* a 4 in consecutive throws.

$$\text{i.e. } p(3 \textit{ then } 4) = p(3) \times p(4) = \frac{1}{6} \times \frac{1}{6} = \frac{1}{36}.$$

These two events are said to be **independent**.

Earlier we found that the probability of selecting a King from a pack of cards is $\frac{1}{13}$. What is the probability of drawing a King again in consecutive draws if the original King is replaced?

$$p(\text{K } \textit{then} \text{ K}) = p(\text{K}) \times p(\text{K}) = \frac{1}{13} \times \frac{1}{13} = \frac{1}{169}.$$

What is the probability of drawing two consecutive Kings if the first King is *not* replaced?
If the first King is not replaced, there are only 3 Kings left in a pack now containing 51 cards. The chances are therefore $\frac{3}{51}$.

$$\therefore p(\text{K then K } \textit{without} \text{ replacement}) = \frac{1}{13} \times \frac{3}{51} = \frac{1}{221}$$

These events are said to be **dependent.**

Examples

1. What is the probability of a total score of 8 or more when two fair dice are thrown?

The ways of achieving scores of 8, 9, 10, 11, 12 are:
(2, 6), (3, 6), (4, 6), (5, 6), (6, 6),
(3, 5), (4, 5), (5, 5), (6, 5),
(4, 4), (5, 4), (6, 4),
(5, 3), (6, 3),
(6, 2). = 15 ways

$$\therefore p(8 \text{ or more}) = \frac{15}{36} = \underline{\frac{5}{12}} \quad \text{ANS}$$

2. A coin is spun three times, what is the probability of there being 3 tails?
The ways the 3 coins can fall are:
(H, H, H), (H, H, T), (H, T, H), (T, H, H), (H, T, T), (T, H, T), (T, T, H), (T, T, T).
= 8 ways, only one of which is (T, T, T).

$$\therefore p(\text{TTT}) = \underline{\frac{1}{8}} \quad \text{ANS}$$

3. The probability of getting 'heads' on one spin of a coin is $\frac{1}{2}$. What is the probability of getting 'heads' each time on four spins of the coin?

$$p(\text{H}) = \frac{1}{2}$$

$$\therefore p(\text{HHHH}) = \frac{1}{2} \times \frac{1}{2} \times \frac{1}{2} \times \frac{1}{2} = \underline{\frac{1}{16}} \quad \text{ANS}$$

4. A bag contains 3 red discs, 2 blue discs and 1 green disc. What is the probability of drawing, without replacement, from the bag (a) two blue discs in two draws, (b) one of each colour in three draws?

(a) There are 2 chances in 6 of drawing a blue disc first time, then only 1 blue disc remains out of 5 discs

i.e. $p(2B) = \frac{2}{6} \times \frac{1}{5} = \frac{2}{30} = \underline{\frac{1}{15}}$ ANS

(b) First consider the probability that we get the order (R, B, G).

i.e. $p(R, B, G) = \frac{3}{6} \times \frac{2}{5} \times \frac{1}{4} = \frac{6}{120} = \frac{1}{20}$

now $p(R, G, B) = \frac{3}{6} \times \frac{1}{5} \times \frac{2}{4} = \frac{6}{120} = \frac{1}{20}$

and $p(B, R, G) = \frac{2}{6} \times \frac{3}{5} \times \frac{1}{4} = \frac{6}{120} = \frac{1}{20}$

and $p(B, G, R) = \frac{1}{20}$

and $p(G, B, R,) = \frac{1}{20}$

and $p(G, R, B) = \frac{1}{20}$

$\therefore$ $p(1 \text{ of each}) = 6 \times \frac{1}{20} = \frac{6}{20} = \underline{\frac{3}{10}}$ ANS

Exercise 111

1. If the probability of hitting the target in a fairground rifle range is $\frac{1}{5}$, what is the probability of not hitting the target?
2. What is the chance of selecting the two-of-diamonds from an ordinary pack of 52 cards?
3. What is the chance of selecting any diamond from a full pack of cards?
4. What is the probability of selecting an Ace followed by an Ace again in two draws from a pack of cards if the first card is replaced?
5. What is the probability of selecting two consecutive Aces from a pack of cards if the first Ace is not replaced?
6. What is the probability of throwing a double-six with a pair of fair dice?
7. What is the probability of throwing a four followed by an odd number with two throws of an unbiased die?
8. What is the probability of a total score of 9 or more when two fair dice are thrown?

9. A coin is spun twice. What is the chance of getting a 'head' followed by a 'tail'?
10. A coin is spun five times. What is the chance of all five spins resulting in 'tails'?
11. A bag contains 8 marbles; 4 are blue, 3 are green and 1 is red. What are the probabilities of (a) drawing a blue marble at the first attempt, (b) drawing two blue marbles in two attempts, (c) drawing a blue followed by a green in two attempts, (d) drawing one of each colour in three attempts?
12. A group of pupils standing in the playground consists of 6 boys and 3 girls. If two are selected at random to help a teacher, what is the chance that both are boys?
13. What is the probability of selecting three consecutive diamonds from a pack of cards if none are replaced?
14. If a pair of fair dice are thrown what is the probability of the total score being seven?
15. A coin is biased so that the 'heads' comes down twice as often as the 'tails'. If the coin is spun three times what is the probability of (a) they're all 'heads', (b) they're all 'tails'?

Examination Standard Questions

Worked examples

1. A fair coin is spun four times: (a) show the sample space of possible outcomes, (b) find the probability that at least two tails occur.

2. Two coins are spun simultaneously, one is fair the other biased so that the probability it comes down tails is $\frac{3}{4}$. Find the probabilities that

(a) both coins show heads,
(b) both coins show the same.

1. (a) The sample space is

(H, H, H, H), (H, H, H, T), (H, H, T, H), (H, T, H, H,)
(T, H, H, H), (H, H, T, T), (H, T, T, H), (T, T, H, H),
(T, H, H, T), (T, H, T, H), (H, T, H, T), (H, T, T, T),
(T, H, T, T), (T, T, H, T), (T, T, T, H), (T, T, T, T).

(b) Eleven of the sixteen possible outcomes contain at least two tails,

$$\therefore\ p(\text{at least 2 tails}) = \underline{\frac{11}{16}} \quad \text{ANS}$$

2. (a) $p(\text{H on the fair coin}) = \frac{1}{2}$

$p(\text{H on the unbiased coin}) = 1 - \frac{3}{4} = \frac{1}{4}$

$\therefore\ p(\text{H, H}) = \frac{1}{2} \times \frac{1}{4} = \underline{\frac{1}{8}}$ ANS

(b) Possible results are (H, H) and (T, T).

We know that $p(\text{H, H}) = \frac{1}{8}$,

$p(\text{T on the fair coin}) = \frac{1}{2}$, $p(\text{T on the biased coin}) = \frac{3}{4}$,

$\therefore\ p(\text{T, T}) = \frac{1}{2} \times \frac{3}{4} = \frac{3}{8}$

$\therefore\ p(\text{H, H or T, T}) = \frac{1}{8} + \frac{3}{8} = \underline{\frac{1}{2}}$ ANS

Exercise 112

1. (a) The mean of the five numbers x, 2, 3, 5 and 9 is 3. Find x.
(b) A pie chart is drawn to represent three commodities. The angles of two of the sectors are 107° and 208°. Express the third sector as a percentage of the whole pie chart. (C)

2. In an examination taken by 100 candidates, the marks scored were as shown in the following table:

Range	No. of candidates
0–9	6
10–19	8
20–29	9
30–39	10
40–49	20
50–59	15
60–69	11
70–79	9
80–89	7
90–99	5

(i.e. 6 candidates scored a mark in the range 0 to 9 inclusive, and so on).
(a) State the modal class.
(b) Estimate the median.
(c) Estimate the lower quartile. (C)

3. Use the given cumulative frequency curve to estimate (a) the lower quartile, (b) the inter-quartile range. (C)

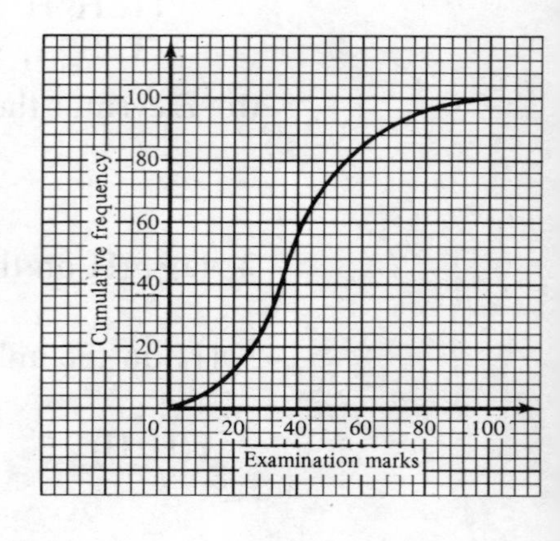

4. The members of a class of 30 children were asked to record their weekly pocket money in pence. The results were:

15 30 35 25 25 36
40 15 35 48 40 20
30 30 35 40 40 30
50 35 30 25 15 40
20 20 25 36 35 30

(a) Tabulate these results in a frequency table.

(b) Determine (i) the mean, (ii) the mode, (iii) the median of these figures.

(c) Prepare a second frequency table using three classes: those receiving less than 25p; those receiving at least 25p but less than 45p; those receiving 45p or more. Illustrate this distribution by a pie chart using a circle of radius 5cm. (O)

5. The following *cumulative* frequency table shows the distribution of heights of a certain group of 100 people.

Height (cm)	≤165	≤169	≤173	≤177	≤181	≤185	≤189	≤193
Cumulative frequency	0	9	31	60	85	94	99	100

(a) Find the median height for this group of people.

(b) Use the given figures to form the frequency distribution, and hence calculate an estimate of the mean height. (O)

6. The table below gives information concerning the masses at birth of the first 100 babies born in a maternity home in 1975.

Mass in kg		No. of babies
Exceeding	Not exceeding	
1.5	2.5	2
2.5	3.0	4
3.0	3.5	18
3.5	4.0	38
4.0	4.5	28
4.5	5.0	8
5.0	6.0	2

Represent this information by (a) a histogram, (b) a cumulative frequency polygon.

Find (c) the median of the masses at birth of these 100 babies, (d) the number of babies whose masses at birth were within 1kg of this median. (JMB)

7. A merchant buys rods each with a specified length of 1 m. The rods in two samples A and B each containing 100 rods are measured to the nearest cm and the following distributions found.

Length (cm)	98	99	100	101	102
Sample A	5	25	35	25	10
Sample B	5	20	50	20	5

Estimate the mean length of the rods in sample A and state whether this mean length is greater than or less than the mean length of the rods in sample B.
Using the same scales and axes, draw a cumulative frequency diagram for each distribution. From the graph estimate the semi-interquartile range for each sample. (JMB)

8. The following scores were obtained by 40 children in a multiple choice test:

30 31 32 33 35 35 35 37 40 41
43 44 44 45 47 49 50 50 50 51
53 54 54 54 54 55 57 57 58 58
58 59 59 59 60 61 61 62 64 64

(a) Write down the median and the mode.
(b) Prepare a frequency table using class intervals of 5 units (taking 30–34 as the first class). Using your table estimate the arithmetic mean.
(c) Prepare a second frequency table using class intervals of 7 units (taking 30–36 as the first class) and again estimate the arithmetic mean.
Explain briefly what assumption in your calculations may account for the difference in the answers to (b) and (c). (O)

9. An experiment was performed in which 4 unbiased pennies were tossed simultaneously 500 times and the number of heads noted each time. The following frequency distribution was obtained.

Number of heads	0	1	2	3	4
Frequency	27	130	198	110	35

From the distribution, calculate (a) the arithmetic mean of the number of heads, (b) the relative frequency of (i) no heads, (ii) one head.
By listing all the possible outcomes of the experiment, or otherwise, calculate the probability of getting (c) no heads, and (d) one head. (JMB)

10. One hundred pods of a new variety of garden pea were opened to

find the number of peas in each pod. The frequency distribution is given below.

Number of peas per pod	1	2	3	4	5	6
Number of pods	2	15	30	25	20	8

Find the values of the mode and the median of this distribution, labelling each answer, and calculate the mean number of peas per pod.
If two pods are selected at random, find the probability that they both contain 4 peas. (C)

11. A special pack of 24 playing cards contains the two, three, four, five, six and seven of each of the four suits, spades, hearts, diamonds and clubs.
(a) What are the probabilities that if a single card is turned up from a shuffled pack it is (i) a two, (ii) a spade?
(b) Three cards are turned up in succession. What are the probabilities that they (i) all show the number two, (ii) all show the same number, (iii) are all spades, (iv) are all of the same suit? (O)

12. Two ordinary dice are thrown together. Find the probability that the sum of the two scores is 4. (JMB)

13. Two boys, Alan and Bob, attempt independently to solve a problem. The probability of Alan solving it is $\frac{4}{5}$ and the probability of Bob solving it is $\frac{3}{4}$. Find the probability (a) that they both solve it, (b) that Alan solves it and Bob does not. (JMB)

14. I select two *different* digits at random from the set $\{2, 3, 4, 5, 6\}$.
(a) How many different pairs are there? (Count the pair 2, 3 and the pair 3, 2, for example, as the same pair.)
What is the probability that (b) one of the numbers of the pair is a 6; (c) neither of the numbers in the pair is even; (d) the sum of the numbers in the pair is divisible by 3? (O)

15. If I pick at random an integer between, but not including, 10 and 20, what is the probability that it is prime? (O)

16. The probability of an event A happening is $\frac{1}{5}$ and the probability of an event B happening is $\frac{1}{4}$. Given that A and B are independent, calculate the probability that (a) neither event happens, (b) just one of the two events happen. (C)

17. Bag A contains 10 balls, of which 3 are red and 7 are blue. Bag B contains 10 balls of which 4 are red and 6 are blue. A ball is drawn at random from each bag.

Find the probability that (a) both are red, (b) at least one is blue. (C)

18. (a) A marksman hits a circular target every time he fires a shot. In the middle of the target is a circular bulls-eye with radius equal to one-third of the radius of the target. A shot which hits the bulls-eye scores 10 points. A hit elsewhere on the target scores 1 point. Given that a shot is equally likely to hit any part of the target, calculate the probability that the marksman will score (i) 10 points, (ii) 1 point.

Given that the marksman fired 90 shots at the target, find his probable score.

(b) (i) A card is drawn from a pack of 52 playing cards and is then replaced. A second draw is made. Find the probability that the same card is drawn each time.

(ii) If the first card is not replaced in the pack, find the probability that the second card is of the same suit as the first card. (C)

19. A die, with its faces numbered 1 to 6, is biased so that the probabilities of throwing 1, 2, 3, 4, 5 are $\frac{1}{12}, \frac{1}{8}, \frac{1}{8}, \frac{1}{6}, \frac{1}{4}$ respectively.

(a) Show that the probability of throwing 6 is $\frac{1}{4}$.

(b) Which is the greater, the probability of throwing an even number or the probability of throwing an odd number?

(c) The die is thrown twice. Draw a diagram showing the sample space of possible outcomes.

(d) If the die is thrown twice, what is the probability that the sum of the two scores is 4? (JMB)

20. Three former school friends, Jones, Smith and Thomas, unexpectedly met in a cafe. They agreed to meet again at the same cafe a month later. The probabilities that they would remember, and be able to keep, the appointment were $\frac{9}{10}$ for Jones, $\frac{5}{6}$ for Smith and $\frac{1}{10}$ for Thomas. Show that the probability that Jones and Smith would be there but not Thomas is $\frac{27}{40}$.

Calculate, similarly, the probabilities that Jones and Thomas alone meet and that Smith and Thomas meet alone. Hence show that the probability that just two of them meet is slightly less than 0.7. If, in fact, Thomas went what was the probability that he met at least one of the others? (O)

Summary of Essential Facts and Formulae

Area of triangle $= \frac{1}{2}$base $\times$ perpendicular height
Area of parallelogram $=$ base $\times$ perpendicular height
Area of trapezium $= \frac{1}{2}$(sum of parallel sides) $\times$ distance between
Circumference of circle $= 2\pi r$ or πD
Area of circle $= \pi r^2$

Length of circular arc $= \dfrac{\alpha}{360} \times 2\pi r$

Area of circular sector $= \dfrac{\alpha}{360} \times \pi r^2$

Total surface area of cylinder $= 2\pi r(h+r)$
Volume of uniform solid $=$ base area $\times$ height
Volume of cone or pyramid $= \frac{1}{3}$base area $\times$ perpendicular height
Curved surface area of cone $= \pi rs$ (where s is the slant height)
Surface area of sphere $= 4\pi r^2$
Volume of sphere $= \frac{4}{3}\pi r^3$
1 are(a) $= 100\text{m}^2$
1 hectare (ha) $= 10\,000\text{m}^2$
1 litre $= 1000\text{cm}^3$
$(a+b)^2 = a^2 + 2ab + b^2$
$(a-b)^2 = a^2 - 2ab + b^2$
$(a+b)(a-b) = a^2 - b^2$

If $ax^2 + bx + c = 0$, then $x = \dfrac{-b \pm \sqrt{b^2 - 4ac}}{2a}$

$x^p \times x^q = x^{p+q}$
$x^p \div x^q = x^{p-q}$
$(x^p)^q = x^{pq}$
$x^{-p} = \dfrac{1}{x^p}$

$x^0 = 1$
$x^{\frac{p}{q}} = (\sqrt[q]{x})^p$
$\text{f}: x \to x^2$ means a function f maps x onto x^2.
gf(x) means do f first then g
If $\text{f}: x \to 2x$ then $\text{f}^{-1}: x \to \frac{1}{2}x$

If $y = ax^n$ then $\frac{dy}{dx} = anx^{n-1}$

and $\int y\,dx = \frac{ax^{n+1}}{n+1} + c$

$v = \frac{dx}{dt}$, $\int v\,dt = x + c$

$a = \frac{dv}{dt}$, $\int a\,dt = v + c$

Area under curve $= \int y\,dx$

Volume of revolution $= \pi \int_a^b y^2\,dx$, (where a and b are the limits)

$x \geqslant y$ means x is greater than or equal to y

$x \leqslant y$ means x is smaller than or equal to y

$a \in A$ means that a is a member of the set A

$b \notin A$ means that b is not a member of the set A

$A \subset B$ means that A is a subset of the set B

$\{\ \}$ or ϕ means that the set is empty

$A \cup B$ means the union of the sets A and B

$A \cap B$ means the intersection of the sets A and B

Matrix multiplication $\begin{pmatrix} a & b \\ c & d \end{pmatrix}\begin{pmatrix} e & f \\ g & h \end{pmatrix} = \begin{pmatrix} ae+bg & af+bh \\ ce+dg & cf+dh \end{pmatrix}$

Sum of interior angles of convex polygon $= (2n-4) \times 90°$ (where n is number of sides)

Length of arc $= r\theta^c$

Area of sector $= \frac{1}{2}r^2\theta^c$

In a right-angled triangle:

$a^2 + b^2 = c^2$ (where c is the hypotenuse)

$\tan A = \frac{\text{opposite side}}{\text{adjacent side}}$

$\sin A = \frac{\text{opposite side}}{\text{hypotenuse}}$

$\cos A = \frac{\text{adjacent side}}{\text{hypotenuse}}$

$\sin(180° - \alpha) = \sin\alpha$

$\cos(180° - \alpha) = -\cos\alpha$

$\tan(180° - \alpha) = -\tan\alpha$

In any triangle:

$\frac{a}{\sin A} = \frac{b}{\sin B} = \frac{c}{\sin C}$

$c^2 = a^2 + b^2 - 2ab\cos C$ or $\cos C = \frac{a^2 + b^2 - c^2}{2ab}$

$r = R\cos\alpha°$ (where R is the radius of the Earth, r is the radius of the circle of latitude, $\alpha°$ is the latitude)

Answers

Chapter 1

Page 3 Exercise 1

1. $\frac{6}{16}$ **2.** $\frac{4}{6}$ **3.** $\frac{12}{27}$ **4.** $\frac{4}{20}$ **5.** $\frac{2}{3}$ **6.** $\frac{1}{2}$ **7.** $\frac{5}{6}$
8. $\frac{2}{3}$ **9.** $\frac{9}{51}$ **10.** $\frac{1}{17}$ **11.** $1\frac{3}{8}$ **12.** $\frac{7}{8}$ **13.** $\frac{1}{2}$
14. $\frac{11}{12}$ **15.** $\frac{2}{7}$ **16.** $1\frac{1}{4}$ **17.** $4\frac{5}{8}$ **18.** $\frac{13}{20}$
19. $5\frac{7}{16}$ **20.** $2\frac{1}{4}$

Page 4 Exercise 2

1. $\frac{1}{4}$ **2.** $\frac{2}{5}$ **3.** $\frac{2}{3}$ **4.** $1\frac{3}{4}$ **5.** 3 **6.** 4 **7.** $1\frac{1}{2}$
8. $\frac{3}{4}$ **9.** $1\frac{5}{8}$ **10.** $11\frac{1}{4}$

Page 5 Exercise 3

1. 4.34 **2.** 43.34 **3.** 0.357 **4.** 0.5 **5.** 11.047
6. 10.01 **7.** 18.0 **8.** 0.00054 **9.** 0.33 **10.** 45.3

Page 5 Exercise 4

1. 45.4 **2.** 4540 **3.** 5896.4 **4.** 1189.04 **5.** 49.1
6. 468 **7.** 0.0047 **8.** 20.0 **9.** 41.3 **10.** 1.0

Page 6 Exercise 5

1. 228.78 **2.** 108.846 **3.** 531.3025 **4.** 8.6 **5.** 111.35

Page 6 Exercise 6

1. 50 % **2.** 75 % **3.** $62\frac{1}{2}$ % **4.** $33\frac{1}{3}$ %
5. $41\frac{2}{3}$ % **6.** 40 % **7.** 4 % **8.** 45 % **9.** 66.6 %
10. 175 % **11.** $\frac{7}{10}$ **12.** $\frac{1}{10}$ **13.** $\frac{1}{8}$ **14.** $\frac{19}{20}$
15. $\frac{99}{100}$

Page 7 Exercise 7

1. 3×10^2 **2.** 4.96×10^2 **3.** 2.5×10 **4.** 4.006×10^3
5. 2.3109×10^4 **6.** 3×10^{-1} **7.** 3×10^{-3} **8.** 1.67×10^{-3}
9. 4.96143×10^3 **10.** 3.00006×10^2

Page 8 Exercise 8

1. 1 : 2 **2.** 1 : 4 **3.** 1 : 20 **4.** 1 : 20 000 **5.** 1 : 50
6. $\frac{1}{10}$ **7.** $\frac{1}{100}$ **8.** $\frac{4}{25}$ **9.** $\frac{1}{40}$ **10.** $\frac{1}{400}$

Page 9 Exercise 9

1. 42_{10} **2.** 101_2 **3.** 1100_2 **4.** 42_6 **5.** 114_7
6. 101_4 **7.** 9_{10} **8.** 10_2 **9.** 101_2 **10.** 14_5
11. 16_8 **12.** 2_3 **13.** 275_{10} **14.** 1001_2
15. 10100_2 **16.** 413_5 **17.** 1736_8 **18.** 9_{10}
19. 110_2 **20.** 4_5

Page 11 Exercise 10

1. $\frac{10}{13}$ **2.** $1\frac{5}{14}$ **3.** $2\frac{3}{10}$ **4.** 0.208 **5.** 0.1981
6. 33p **7.** 56% **8.** 10% **9.** 27.2 cm **10.** 6:5
11. £598 **12.** 101100 **13.** -2 **14.(a)** 11110
(b) 1010 **15.** 1001001 **16.** 123 **17.(a)** 325
(b) 53 **18.(a)** True **(b)** True **(c)** False
(d) True

Chapter 2

Page 14 Exercise 11

1. 1.4624 **2.** 0.4624 **3.** 2.4624 **4.** 2.7559
5. $\bar{1}.7559$ **6.** 0.9445 **7.** 2.9445 **8.** 3.0 **9.** $\bar{3}.0$
10. 4.9956

Page 15 Exercise 12

1. 1.3711 **2.** 0.6294 **3.** 4.9238 **4.** $\bar{1}.6154$
5. 1.4685 **6.** 2.5997 **7.** 3.9241 **8.** 5.6210
9. $\bar{3}.5228$ **10.** $\bar{1}.4777$

Page 16 Exercise 13

1. 1.5856 **2.** 6.3328 **3.** 5.7996 **4.** 2.1611
5. $\bar{5}.1145$ **6.** 1.7643 **7.** 3.417 **8.** 1.229
9. 0.7184 **10.** $\bar{3}.822$

Page 17 Exercise 14

1. 85.01 **2.** 1 356 000 **3.** 14.65 **4.** 0.000 165 7
5. 257.2

Page 17 Exercise 15

1. 18.95 **2.** 9600 **3.** 76.88 **4.** 116.1 **5.** 0.044 68

Page 18 Exercise 16

1. 576 **2.** 594.4 **3.** 1.22×10^{11} **4.** 2.459×10^{-6}
5. 6.41×10^{-5}

Page 19 Exercise 17

1. 7 **2.** 23.84 **3.** 0.4117 **4.** 0.2783 **5.** 0.3797

Page 20 Exercise 18

1. 0.1057 **2.** 34.8 **3.** 0.0795 **4.** 1.32 **5.** 52.46
6. 15.4 **7.** 2^{32} **8.** No, Yes, Yes, No **9.** (e)
10.(a) 0.3685 **(b)** 0.059 73

Chapter 3

Page 26 Exercise 19

1. 40 cm^2 **2.** 49 cm^2 **3.** 400 cm^2 **4.** 12 cm **5.** 20 cm^2
6. 31.5 m^2 **7.** 74 cm^2 **8.** 9 cm **9.** 30 cm^2
10. 14 cm^2 **11.** 8000 cm^2 **12.** 5 cm **13.** 44 cm, 154 cm^2 **14.** 66 cm, 346.5 cm^2 **15.** 62.8 cm, 314 cm^2
16. 2.5 cm, 19.625 cm^2 **17.** $\frac{\pi r}{2}, \frac{\pi r^2}{4}$ **18.** $\frac{2\pi r}{3}, \frac{\pi r^2}{3}$
19. $\left(\frac{540}{\pi}\right)^\circ, \frac{3r^2}{2}$ **20.** $72^\circ, \frac{2\pi r}{5}$ **21.** 1056 cm^2, 2618 cm^3
22. 4620 cm^2, 19404 cm^3 **23.** 251.2 cm^2, 301.44 cm^3
24. 282.6 cm^2, 349.7 cm^3 **25.** 50 cm^3, $16\frac{2}{3}$ cm^3
26. 105 cm^3, 35 cm^3 **27.** $19\frac{1}{4}$ cm^3, $6\frac{5}{12}$ (6.417) cm^3
28. 10.35 cm^3, 3.45 cm^3 **29.** 616 cm^2, $1437\frac{1}{3}$ cm^3
30. 1017.4 cm^2, 3052 cm^3

Page 28 Exercise 20

1.(a) 2:7, **(b)** $\frac{7}{9}$ **2.** 19 cm **3.(a)** 22 cm
(b) 3.5 cm, 13.6 cm **(c)** 10.5 cm **4.** 84° **5.** 88 m^2
6. 7.29×10^5 **7.** F, F **8.** 880 cm^2
9. 1188 cm^3, 11 cm **10.** 7 cm

Chapter 4

Page 31 Exercise 21

1. 25 **2.** -25 **3.** 1 **4.** $3x$ **5.** 0 **6.** $-5x$
7. $-4x$ **8.** $5a$ **9.** $-5a$ **10.** $-4b$ **11.** 2
12. 8 **13.** -8 **14.** x **15.** $-2x$ **16.** $-9x$
17. $-5x$ **18.** $19a$ **19.** $-6a$ **20.** $6a$

Page 33 Exercise 22

1. 4 **2.** 12 **3.** 6 **4.** $\frac{2}{3}$ **5.** 6 **6.** 18
7. 36 **8.** -6 **9.** -6 **10.** 66 **11.** 8 **12.** 1
13. 3 **14.** 9 **15.** 3

Page 33 Exercise 23

1. 35 **2.** $35x$ **3.** $35x^2$ **4.** $35x^2$ **5.** $-35x^2$

6. $-35x^2$ 7. $35x^2$ 8. $35x^2$ 9. $-18x^2$
10. $110x^2$ 11. 5 12. $5x$ 13. $7x$ 14. $5x$
15. $-5x$ 16. $5x$ 17. $-7x$ 18. x 19. $-35x$
20. $35x$

Page 34 Exercise 24

1. 1 2. 4 3. a 4. $4a$ 5. $4a$ 6. $2a+4$
7. $3a+3b$ 8. $a+b$ 9. $6x-3y$ 10. $2x-y$
11. $2z$ 12. $3x^2+5$ 13. $3x^2-x$ 14. $2x^3+y^2+y$
15. $-a^3+3a^2-a-1$

Page 35 Exercise 25

1. $4x+2$ 2. $3x^2+3x-6$ 3. $2a+6b$ 4. a^2+ab
5. $4x^2-3xz+y^2$ 6. $4a+17b-c$ 7. $4a+5b$
8. $pqr+pqs+prs+qrs$ 9. x^2-2x+3
10. $4x^3+6x^2+2x-21$

Page 36 Exercise 26

1. $xy+xz+y^2+yz$ 2. x^2+5x+6 3. x^3+2x^2+x
4. x^3-x-6 5. $x^2+2xy+y^2$ 6. $x^2-2xy+y^2$
7. x^2-y^2 8. $3p^2+6pq+3q^2$ 9. $4p^2-4q^2$
10. $8p^4+16p^2q^2+8q^4$

Page 38 Exercise 27

1. $x(x^2+x+2)$ 2. $4x^2(1+3x)$ 3. $3(a+b)$
4. $a(b^2+c^2)$ 5. $ab(c-d)$ 6. $8xy(3-x)$ 7. $pq(1-r)$
8. $40(2a-b)$ 9. $3x(6y-1)$ 10. $x(x+y)$
11. $(5+a)(x+y)$ 12. $(3p+5)(q+r)$ 13. $(c+2d)(a-b)$
14. $(x-y)(x+z)$ 15. $(4+x)(5+y)$ 16. $(x+4)(x+5)$
17. $(x+11)(x+6)$ 18. $(x+1)^2$ 19. $(x+10)(x+8)$
20. $(x+3)(x+1)$ 21. $(x+6)(x+9)$
22. $(x+3)(x+8)$ 23. $(x+8)(x+12)$ 24. $(x-6)(x-8)$
25. $(x-4)(x-5)$ 26. $(x+3)(x-9)$ 27. $(x+7)(x-10)$
28. $(2x+5)(x+7)$ 29. $(4x+5)(x+2)$ 30. $(7x-2)(3x+1)$
31. $(3x+2)(2x-3)$ 32. $(x+7)(x-7)$ 33. $(a+c)(a-c)$
34. $(b+10)(b-10)$ 35. $(2x+y)(2x-y)$
36. $(3x+4y)(3x-4y)$ 37. $(3+x)(3-x)$ 38. $(6x+5)(6x-5)$
39. $(a+1)(a-1)$ 40. $(17+13)(17-13)$

Page 40 Exercise 28

1. $L=\dfrac{A}{B}$ 2. $h=\dfrac{3V}{A}$ 3. $h=\dfrac{2A}{a+b}$ 4. $B=\dfrac{V}{LW}$
5. $h=\dfrac{A}{2\pi r}-r$ 6. $F=\dfrac{9C}{5}+32$ 7. $T=\dfrac{100I}{PR}$

8. $R = \dfrac{100(A-P)}{PT}$ 9. $a = \dfrac{x-u}{t}$ 10. $u = \sqrt{(v^2 - 2ax)}$

11. $x = \dfrac{v^2 - u^2}{2a}$ 12. $f = \dfrac{uv}{u+v}$

Page 41 Exercise 29

1. a^5 2. a^9 3. a^8 4. $6a^5$ 5. $20a^9$

6. a^3b^3 7. $6a^3b^2$ 8. a^2 9. $3a^4$ 10. $\dfrac{2a^2}{b^2}$

Page 42 Exercise 30

1. $\dfrac{1}{x^2}$ 2. $\dfrac{1}{b^2}$ 3. 1 4. a^2 5. a^6 6. $\dfrac{1}{b^7}$

7. b^3 8. $\dfrac{a^3}{b}$ 9. a^2 10. ab^2 11. a^6 12. a^8

13. $\dfrac{1}{a^4}$ 14. $\dfrac{1}{a^4}$ 15. a^6b^3 16. $\dfrac{a^6}{b^3}$ 17. $\dfrac{8}{a^9}$

18. $\dfrac{9a^4}{b^4}$ 19. $36a^2$ 20. $16a^4$

Page 42 Exercise 31

1. $\sqrt{a^3}$ 2. $\sqrt[4]{b}$ 3. $\sqrt[4]{b^3}$ 4. $\sqrt{x^3}$ 5. $\sqrt[3]{x^{-2}}$
6. 5 7. 3 8. 9 9. $\frac{1}{8}$ 10. 343 11. 27
12. $\frac{1}{8}$ 13. 512 14. $\frac{1}{16}$ 15. 25

Page 44 Exercise 32

1.(a) $a(b-4)$ (b) $(x+4)(x-4)$ 2. $13xy$ 3. $\frac{7}{12}$
4.(a) $2(3x+1)(3x-1)$ (b) $(a-2c)(b-3)$
5.(a) $(a+2c)(2a-3b)$ (b) $(3r-2s)(4r-3s)$
(c) $(p+q-2r)(p+q+2r)$ 6.(a) $5(3x+2)(3x-2)$
(b) $(3e+2f)(2g-h)$ 7. 40 8. $c = \sqrt{\left(\dfrac{2b}{a}+d\right)}$

9. $c = \dfrac{a^2 - b^4 d}{b^4}$ 10. $P = \dfrac{b}{H-a}$ 11.(a) 3 (b) 4

12.(a) 27 (b) $\frac{1}{25}$ 13.(a) $\frac{1}{64}$ (b) $\frac{27}{125}$ 14. 157.5
15. 45 16. 6 17. 30 18. 2 19. $2x(2x-1)$
20. $\frac{9}{4}$

Chapter 5

Page 45 Exercise 33

1. 3 2. 9 3. -9 4. $\frac{1}{4}$ 5. 3 6. -3
7. -5 8. -1 9. 2 10. 4 11. 1 12. 5
13. -5 14. -2 15. 4 16. 5 17. -6
18. 10 19. 1 20. 5

Page 47 Exercise 34

1. $x > 2$ 2. $x > \frac{5}{7}$ 3. $x < 2$ 4. $x > 2$
5. $x > 6$ 6. $x < 9$ 7. $x > 2$ 8. $x > 3$
9. $x > 3$ 10. $x \leqslant 6$ 11. $x \leqslant 4$ 12. $x \leqslant 7$
13. $x > 4$ 14. $x > -3$ 15. $x < -3$

Page 48 Exercise 35

1. $n+2$ 2. $2n+1$ 3. $4n-1$ 4. n^2+1 5. n^3
6. $2n^2+1$ 7. $3n^3$ 8. n^3+4n 9. $(\frac{n}{2})^2$
10. $n(-1)^n$ 11. 4 12. 9 13. 8 14. 32
15. 20 16. $\frac{11}{24}$ 17. 5 18. 0 19. $-35\frac{8}{9}$ 20. $\frac{19}{27}$

Page 49 Exercise 36

1. 2, 228 2. x, $15x^2$ 3. $2xy$, $24x^2y^2$
4. $x^2y^2z^4$, $x^3y^5z^4$ 5. $3(x+2)$, $9(x+2)^2$
6. $(x-2)$, $x(x-2)^3$ 7. $(x+7)$, $x(x+7)(x+2)$
8. $(x+2)$, $(x+1)(x+2)(x+3)$ 9. $(x-4)$, $(x+4)(x-4)^2$
10. $(x-5)$, $x(x-5)(x-1)(x+5)$

Page 50 Exercise 37

1. $\frac{1}{2}$ 2. $2ab$ 3. $\frac{b^3c}{3a}$ 4. $\frac{x+4}{x-3}$ 5. $\frac{x-7}{x+2}$
6. 3 7. $\frac{8}{ab^2}$ 8. $\frac{a^3c}{15}$ 9. $\frac{1}{a}$ 10. $\frac{21xy^4p}{11zq}$

Page 51 Exercise 38

1. $\frac{13}{20}$ 2. $\frac{9a}{20}$ 3. $\frac{22a}{35}$ 4. $\frac{5a+7}{6}$ 5. $\frac{3a-17}{10}$
6. $\frac{7a-10}{(a+2)(a-3)}$ 7. $\frac{3a^2-9a-2}{a^2-6a+9}$ 8. $\frac{5a^2+5a-12}{(a-3)(a+3)^2}$
9. $\frac{3(4a^2+17a+2)}{(a-1)(a+2)(a+4)}$ 10. $\frac{9a^2+80a+121}{(a+11)(a-7)(a+8)}$

Page 51 Exercise 39

1. 18 2. -3 3. $5\frac{1}{4}$ 4. -36 5. $-\frac{1}{3}$

Page 52 Exercise 40

1. $x = 3$ 2. $x = 14$ 3. $y = 18$ 4. $y = \frac{3}{16}$
5. $x = 46\frac{3}{4}$

Page 53 Exercise 41

1. $x = 1\frac{1}{3}$ 2. $x = \frac{6}{7}$ 3. $y = 9$ 4. $y = \frac{1}{6}$ 5. $x = 12$

Page 54 Exercise 42

1. 24 **2.** 1 **3.** 15 **4.** 20 **5.** 14

Page 56 Exercise 43

1. $x = 3$, $y = 1$ **2.** $x = 6$, $y = 4$ **3.** $x = 5$, $y = 4$
4. $x = -3$, $y = +3$ **5.** $x = 4$, $y = -1$ **6.** $x = 10$,
$y = -9$ **7.** $x = 2$, $y = 1$ **8.** $x = -2$, $y = 5$
9. $x = 5$, $y = 3$ **10.** $x = -4$, $y = -6$

Page 57 Exercise 44

1. -2, -1 **2.** -4, -3 **3.** -2, $+3$ **4.** -7, $+9$
5. $-\frac{1}{2}$, -3 **6.** $-\frac{2}{3}$, $-1\frac{1}{2}$ **7.** $-\frac{2}{3}$, $+1\frac{1}{2}$
8. $2\frac{1}{3}$, 4 **9.** $-1\frac{2}{5}$, 4 **10.** $\frac{3}{4}$

Page 58 Exercise 45

1. 3.732, 0.268 **2.** 1.7808, -0.2808 **3.** 2.1197,
-0.7863 **4.** 0.2899, -0.6899 **5.** 0.6666, -3.0
6. 1.425, -0.1754 **7.** -0.4531, -1.261
8. 0.481, -2.081 **9.** 0.8792, -0.3792 **10.** 3.14, -4.14

Page 61 Exercise 46

1. -4 **2.** $x = 13$ **3.** $2 < x < 6$ **4.** $x < -3$
5. $24\frac{2}{3}$ **6.** $79\frac{7}{8}$ **7.** $(-1)^n(2^n + n^{-1})$ **8.** $\dfrac{3}{x+2}$
9. $\dfrac{2a-8b}{15}$ **10.** $\dfrac{3x-11}{10}$ **11.** $\dfrac{11}{6x}$ **12.** $1\frac{1}{2}$
13. $x = -0.7$ **14.** $x = 9$ **15.** 2
16. (i) 100 (ii) 6.25 **17.** $5\frac{5}{7}$ **18.** 225 **19.** 6
20. 1.44 **21.** $y = 8x + \frac{3}{x}$ (i) -25 (ii) 0.25 or 1.5
22. $b = -16$, $a = 100$, $t = \frac{1}{4}$ **23.** 3.75 cm **24.** 12
25. $W = \dfrac{kv^2}{1}$, **(a)** $k = 0.8$, **(b)** $l = 57.6$ **26.(a)** 2
(b) 1 **(c)** -1 **27.** $x = 1\frac{1}{2}$ **28.** $x = 2$ or 8 **29.** $x = 5$,
$y = -2$ **30.** $x = 7$, $y = -2$ **31.** $x = 3$, $y = 1\frac{1}{2}$
32. 2.26, -0.59 **33.** 6.1, 0.4 **34.** 2.19, -3.19
35. $3.5 \leqslant x \leqslant 7.5$, $3 \leqslant y \leqslant 42$, 12.5, 133.5, 1.5, 122.5

Chapter 6

Page 65 Exercise 47

1. 4, 2 **2.** 3, -11, 34, -30, 105, -81
5. $(x-3)$ **6.** $(x-1)(x+1)(x-2)$
7. $(x-5)(x-7)(x+7)$ **8.** $(x+1)(x-3)$

Page 66 Exercise 48

1. $\{5, 6, 7\}$ **2.** $\{-3, -6, -9\}$ **3.** $\{0, 2, 8\}$
4. 1 and 2 are 'one to one', 3 is 'many to one' **5.(a)** -1
(b) 1 **(c)** -4 **(d)** -2 **6. (a)** 2 **(b)** 0
(c) 9 **(d)** 1 **7.(a)** 2 **(b)** 3 **(c)** -2 **(d)** -1
8.(a) -20 **(b)** -8 **(c)** -11 **(d)** -17
9.(a) 16 **(b)** 4 **(c)** 1 **(d)** 7 **10.(a)** $2(x+4)$
(b) $2x+4$, No **11.(a)** $(x-1)^2$ **(b)** x^2-1, No
12.(a) $(x^2)^3$ **(b)** $(x^3)^2$, Yes

Page 68 Exercise 49

1.(a) 2 **(b)** $-3\frac{2}{3}$ **2.(a)** 5 **(b)** -4 **3.(a)** 3
(b) -3 **4.(a)** 4 **(b)** -16 **5.(a)** 27 **(b)** 1
(c) 3 **(d)** 5 **(e)** 6 **(f)** 9 **(g)** 1 **(h)** 5
(i) 32 **(j)** 2 **(k)** 2 **(l)** 12 **6.(a)** No
(b) Yes **(c)** Yes

Page 70 Exercise 50

1. $a = 10$, $3x+5$ **2.** $a = -4$, $b = -20$
3.(a) $\{-3, 7\}$ **(b)** $\{-\frac{1}{2}, -3, -7\}$ **4.(a)** $\{1\}$
(b) $\{1, -3, -\frac{3}{2}\}$ **5.** 15, 28 **6.** b only **7.** $p = \frac{1}{2}$,
$q = -\frac{1}{2}$ **8.** $k = 9$, $x = \frac{1}{2}$ **9.(a)** $x > 1$ **(b)** $a = -5$,
$b = 4$ **(c)** all **10.** $2x - \frac{3}{2}$ **11.(a)** $x \to x+3$
(b) $x \to \pm\sqrt{x}$ **(c)** $x \to (x-3)^2$ **(d)** $x \to (x-3)^2 - 3$
(e) $x \to \pm\sqrt{x}+3$ **12.(a)** -7 **(b)** $x = -4$ **(c)** $9x-10$
13.(a) 81 **(b)** -3 **14.(a)** 35 **(b)** $\frac{5}{6}$

Chapter 7

Page 74 Exercise 51

9. $1, 1, \frac{1}{2}, 2, 2, -\frac{1}{3}, -2, 1\frac{1}{2}$ **10.** Upper left to lower right

Page 75 Exercise 52

1.(a) $\frac{1}{2}$, -2 **(b)** $-2\frac{1}{2}$, 1 **2.(a)** $\frac{2}{3}$, -3
(b) $-\frac{1}{3}$, -2 **3.(a)** $1\frac{1}{3}$, $-1\frac{1}{2}$ **(b)** $\frac{1}{3}$, $-\frac{1}{2}$

Page 77 Exercise 53

1. 4 **2.** 12 **3.(a)** $5x^4$ **(b)** $4x$ **(c)** $12x^2$
4.(a) $4x^3+3x^2+2x$ **(b)** $8x^3+6x+1$ **(c)** $\dfrac{4}{x^3}-\dfrac{12}{x^5}$
5.(a) 11 **(b)** 61 **(c)** $-\frac{1}{8}$

Page 81 Exercise 54

1. $-\frac{1}{3}$ min. **2.** $\frac{1}{2}$ min. **3.** $4\frac{3}{4}$ max.

4. $-6\frac{2}{3}$ min. $2\frac{1}{3}$ max.
5. 0 min. 4 max. **6.** $\frac{4}{9}$ max. 0 min.
7. $-1\frac{14}{27}$ max. -11 min. **8.** 400 m^2 **9.** 800 m^2
10. 36

Page 83 Exercise 55

1.(a) 2 m **(b)** -2 m, (2 m from origin in opposite direction)
(c) -2 m/s **(d)** -6 m/s^2 **(e)** Yes **2.(a)** 8 s
(b) -10 m/s^2 **3.(a)** 34 m/s **(b)** 10 m/s^2 **(c)** 4 m/s
4.(b) 8 m **(c)** -16 m/s **(d)** -16 m/s^2
5.(a) $(4s+1)$ m/s **(b)** 4 m/s^2

Page 85 Exercise 56

1. $\frac{x^3}{x}+x+c$ **2.** $\frac{x^5}{5}+c$ **3.** x^4+c **4.** $\frac{x^3}{6}+c$

5. x^5-x^3+c **6.** $\frac{x^3}{2}+\frac{x^2}{2}-3x+c$ **7.** $\frac{x^3}{4}-x^2+x+c$

8. $\frac{t^3}{3}+c$ **9.** y^4+c **10.** $\frac{v^2}{2}-\frac{v^3}{9}+c$

Page 86 Exercise 57

1. 18 m **2.** $15\frac{2}{3}$ m/s **3.** 2.38 or $2\frac{23}{60}$
4. $x=\frac{3t^5}{5}+\frac{t^4}{4}+t^2+c$ **5.** 32 m **6.** 30 m/s
7. 45 m **8.** 32 m **9.** 0 m/s **10.** 50 m

Page 89 Exercise 58

1. $8\frac{2}{3}$ units^2 **2.** 4 units^2 **3.** $4\frac{1}{2}$ units^2 **4.** $1\frac{1}{3}$ units^2
5. $48\frac{2}{5}\pi$ units^3 **6.** 36π units^3 **7.** 27π units^3
8. $1\frac{1}{15}\pi$ units^3

Page 93 Exercise 59

1. $0.6r$ **2.(a)** 13.5 **(b)** (1, 5), (3, 27)
3. $A=10x^2$; 60 **4.(a)** 6 seconds **(b)** 6 m/s
(c) 2 m/s^2 **(d)** $12\frac{1}{4}$ m/s **5.(a)** 8 m **(b)** 4 m/sec
(c) 3 m **6.** (1, 6), (3, 2) **7.(a)** 0 m/s
(b) -3 m/s^2 **(c)** -4 m/s **8.(a)** 2 cm/s **(b)** $2\frac{1}{2}$ s
(c) -7 cm/s^2 **(d)** $3\frac{1}{8}$ cm/s **9.** $(-1\frac{1}{2}, 0)$ and $(2\frac{1}{2}, 0)$ and
(0, 15) **(a)** $x=\frac{1}{2}$ **(b)** $y=16$ **10.** (1, 0),
(2, -1), $1<x<2$ **11.** P and R
12. 18, 25, 26, 10; -1.19, 2.52; -1.75, 4.75
13.(a) 0.27, 3.73**(b)** -0.65, 4.65, $-1<x<0$, $2<x<3$

14.(a) $v = t^2 - 6t + 8$, $x = \frac{1}{3t^3} - 3t^2 + 8t$ **(b)** $t = 2$ s, $t = 4$ s **15.** $4\frac{1}{2}$ **16.(a)** $4\frac{2}{3}$ **(b)** $5\frac{1}{3}$ **(c)** $\frac{206\pi}{15}$

(d) 50π **(e)** $\frac{544\pi}{15}$ **18.(a)** 6 m/s **(b)** 2, 3

(d) $\frac{t^3}{3} - \frac{5t^2}{2} + 6t$ **(e)** $\frac{1}{6}$ m (towards O) **19.** (2, 2), $\frac{16\pi}{15}$

20.(a) $8 + 2t - t^2$ **(b)** $8t + t^2 - \frac{t^3}{3}$ **(c)** 4 s **(d)** $26\frac{2}{3}$ cm

Chapter 8

Page 100 Exercise 62

1.(a) (0, −4), (−1, −3); **(b)** 12 **2.(a)** $(-1, -\frac{1}{3})$
(b) 9 **3.** $-1\frac{1}{2}$, $9\frac{3}{4}$

Page 103 Exercise 63

1.(a) 12.8 m/s **(b)** 12.7 m/s (±0.2 m/s) **2.(a)** 2 m/s^2
(b) 2 m/s^2 (±0.4 m/s^2) **3.(a)** 1.33 p.m. (±2 mins)
(b) 2.40 p.m. **(c)** 64.2 km/h **4.(a)** 25 m/s
(b) 20 m/s **5.(a)** 100 m/s^2 **(b)** 120 m/s^2
6.(a) 1.28 p.m. (±2 mins) **(b)** 29 km (±1 km)

Page 104 Exercise 64

1. $x < 4$, $y \geqslant 1$, $y \leqslant x - 1$ **2.(a)** A(−3, 0); B(0, 3)
3.(a) $6 > t > 3\frac{1}{2}$ **(b)** (i) 16.8 (ii) 16, 4
4.(a) −1.8, 0, 3.35 **(b)** −1.6, −0.2, 3.4
(c) 3.55 **5.** $1.4 < x < 6.4$, 0.75, 3.0, $\frac{x}{3} + \frac{3}{x} = 5 - x$
6.(a) 2.6 at $x = 3.3$ **(b)** 1.76, 6.24, .55
7. 7.5 kmh^{-1}, 25 kmh^{-1}, 10.57 a.m. 11.18 a.m.
8. 1.76 h **9.** 2.1 h **10. (a)** 28 s **(b)** 37 m/s
11. 1550 m **12.(a)** 29.8 m
(b) line (2, 0) – (6, 7.6); line (2, 0) – (6, 14.9)
13.(a) 0.75 m/s/s **(b)** 23 s **14.(a)** $a = 7.2$, $b = 2.25$
(b) 0.56 kg **(c)** 25%

Chapter 9

Page 108 Exercise 65

1. P = {natural numbers less than 6},
Q = {positive even numbers less than 12},
R = {prime numbers less than 13}
2. 1, 2, 3, 4, 5. **3.(a)** T **(b)** T **(c)** F
(d) F **(e)** T **(f)** F **(g)** F **(h)** F **4.(a)** Yes
(b) No **(c)** Yes **5.** {2, 4, 6, 8}, {2, 6, 8, 10},
{2, 4, 8, 10}, {2, 4, 6, 10}, {4, 6, 8, 10}

Page 109 Exercise 66

1. {1, 3, 5, 6, 7, 8, 9, } **2.** Yes **3.** {1, 2, 3, 4, 5}
4. No **5.** Yes **6.** Yes **7.** Yes **8.** Yes
9. Yes **10.** No

Page 110 Exercise 67

1. {1, 2, 3, 4, 5, 6, 7, 8, 9, 10} **2.** {2, 4, 5, 6, 7, 8, 9, 10}
3. {1, 3, 5, 6, 7, 8, 9} **4.** ϕ **5.** {6, 8} **6.** {5, 7, 9}
7. X, Y **8.** {1, 2, 3, 4, 5, 6, 7, 8, 9, 10} **9.** ϕ
10. {1, 2, 3, 4, 5, 6, 7, 8, 9, 10} **11.** {1, 3, 5, 7, 9}
12. {2, 4, 6, 8, 10} **13.** {1, 2, 3, 4, 10} **14.** ϕ
15. {2, 4, 10} **16.** Identical **17.** Yes **18.** No

Page 111 Exercise 68

4. {1, 2, 3, 4, 5, 6, 7}, {1, 2, 3, 4, 5, 8}, {3. 4, 5, 6, 7, 8}, {4}, {3, 4}, {4, 5}, {4} **5.** {5, 6}, $\mathscr{E}$, $\mathscr{E}$, ϕ, ϕ.

Page 113 Exercise 69

1.(a) 7 **(b)** 14 **(c)** 12 **2.(a)** 43 **(b)** 39
(c) 0 **(d)** 26 **3.(a)** 84 **(b)** 20 **(c)** 14 **4.** 1

Page 115 Exercise 70

1.(a) (i) 7 (ii) 7 Yes **(b)** Yes
(c) (i) {a, b, c, d, e, f, g, h} (ii) {a, b, c, d, e, f, g, h, }
(d) (i) 12 (ii) 12 Yes **(e)** Yes
(f) (i) {c, e}, (ii) {c, e} Yes **2.(a)** (i) 12 (ii) 12 Yes
(b) Yes **(c)** (i) {a, c, e, f, g, h, i, j, k} (ii), (iii) and
(iv) {a, b, c, d, e, f, g, h, i, j, k) Yes **(d)** (i) 60
(d) (ii) 60 Yes **(e)** Yes **(f)** (i) {e, h} (ii) {e} (iii) {e}
(iv) {e} Yes **3.(a)** (i) 27 (ii) 27 Yes **(b)** Yes
(c) (i) {a, c, e} (ii) {a, c, e} Yes **(d)** (i) 23 (ii) 56 No
(e) No **(f)** (i) {a, b, c, d, e, h} (ii) {a, b, c, d, e, h} Yes

Page 116 Exercise 71

1. $\mathscr{E}$ **2.** $\mathscr{E}$ **3.** $\mathscr{E}$ **4.** A′UB **5.** $\mathscr{E}$

Page 119 Exercise 72

1. $\begin{pmatrix}3\\4\end{pmatrix}$ **2.** $\begin{pmatrix}1\\-2\end{pmatrix}$ **3.** $\begin{pmatrix}4\\2\end{pmatrix}$ **4.** $\begin{pmatrix}3\\9\end{pmatrix}$ **5.** 2.2
6. 3.2 **7.** 6.6 **8.** No **9.** Yes **10.** Yes
11. 1.0 **12.** 2.2

Page 123 Exercise 73

1. A is a 2 by 1; B is a 2 by 2; C is a 2 by 2; D is a 2 by 4; E is a 2 by 2 **2.** (**b**) $\begin{pmatrix} 6 & 12 \\ 2 & 1 \end{pmatrix}$ (**d**) $\begin{pmatrix} 5 & 10 \\ 10 & 1 \end{pmatrix}$

(**e**) $\begin{pmatrix} 6 & 12 \\ 2 & 1 \end{pmatrix}$ (**g**) $\begin{pmatrix} 2 & -6 \\ -4 & 3 \end{pmatrix}$ (**h**) $\begin{pmatrix} -2 & 6 \\ 4 & -3 \end{pmatrix}$

(**j**) $\begin{pmatrix} 17 & 33 \\ 4 & -11 \end{pmatrix}$ (**k**) $\begin{pmatrix} 26 & 21 & 29 & 75 \\ 10 & 17 & 0 & 0 \end{pmatrix}$

(**m**) $\begin{pmatrix} -1 & 24 \\ 13 & 7 \end{pmatrix}$ (**o**) $\begin{pmatrix} 3 \\ 9 \end{pmatrix}$ (**p**) $\begin{pmatrix} 6 & 27 \\ 9 & -3 \end{pmatrix}$

(**q**) $\begin{pmatrix} \frac{3}{2} & \frac{1}{2} \\ \frac{7}{2} & 1 \end{pmatrix}$ (**r**) $\begin{pmatrix} 282 & 147 \\ -65 & -34 \end{pmatrix}$ (**s**) $\begin{pmatrix} 282 & 83 \\ -65 & -18 \end{pmatrix}$

(**t**) $\begin{pmatrix} 50 & 43 \\ 15 & -8 \end{pmatrix}$ (**u**) $\begin{pmatrix} 50 & 43 \\ 15 & -8 \end{pmatrix}$ **3.**(**a**) Yes (**b**) No

(**c**) No (**d**) Yes **4.**(**a**) $\begin{pmatrix} 13 & 18 \\ -6 & 1 \end{pmatrix}$ (**b**) $\begin{pmatrix} 31 & 9 \\ 3 & 28 \end{pmatrix}$

(**c**) $\begin{pmatrix} 34 & 75 \\ -25 & -16 \end{pmatrix}$ **5.**(**a**) B (**b**) B (**c**) 0 (**d**) 0

(**e**) C (**f**) C **6.**(**a**) Yes (**b**) Yes (**c**) Yes

Page 125 Exercise 74

1. $\begin{pmatrix} 7 & -2 \\ -3 & 1 \end{pmatrix}$ **2.** $\begin{pmatrix} -2 & 1 \\ 5 & -2 \end{pmatrix}$ **3.** Singular

4. $\begin{pmatrix} 1\frac{1}{2} & -1 \\ -\frac{1}{2} & \frac{3}{4} \end{pmatrix}$ **5.** $\begin{pmatrix} 0.6 & -0.4 \\ -0.2 & 0.3 \end{pmatrix}$

Page 131 Exercise 75

1.(**a**) (1, −2), (3, −1) (**b**) (−1, −2), (−3, −1) (**c**) (2, 1), (1, 3) (**d**) (2, 4), (6, 2) (**e**) (1, −2), (3, −1) **2.** 3 units^2 **3.** (−6, 3), (−4, 3)

Page 133 Exercise 76

1. 6 **2.** $n = 3$ **3.** (i) $\frac{1}{3}x - \frac{1}{6}y$; $y - x$; $x - \frac{3}{4}y$ (iii) 3:1 **4.** $\begin{pmatrix} 3x \\ 3y \end{pmatrix}$, Enlargement centre O, scale factor 3 (**b**) $\begin{pmatrix} y \\ x \end{pmatrix}$, Reflection in the line $y = x$

5.(a) 7 **(b)** 2 **6.(a)** −6 **(b)** −8 **7.(a)** 15
(b) 1:9 **8.(a)** True **(b)** True **(c)** False **(d)** True

9. Any matrix of form $\begin{pmatrix} a & a \\ b & b \end{pmatrix}$, (a, b) ≠ (0, 0)

10.(a) A30 = {1, 2, 3, 5, 6, 10, 15, 30}, A45 = {1, 3, 5, 9, 15, 45}
(b) (i) {1, 3, 5, 15} (ii) {1, 2, 3, 5, 6, 9, 10, 15, 30, 45}
(c) $r = 15$ **(d)** $s = 90$ **11.(a)** (3, 4) **(b)** $\begin{pmatrix} 6 \\ 4 \end{pmatrix}$
(c) DC = 2AB **12.(a)** 7 **(b)** 16 **(c)** 6
13. $b-a$, $a+\frac{1}{2}b$; $p=\frac{1}{3}$ $q=\frac{2}{3}$

14.(a) (i) $\begin{pmatrix} 5 & 2 \\ 4 & 4 \end{pmatrix}$ (ii) $\begin{pmatrix} 3 & 2 \\ 2 & 0 \end{pmatrix}$ (iii) $\begin{pmatrix} 22 & 12 \\ 18 & 10 \end{pmatrix}$
(iv) $\begin{pmatrix} 1 & 0 \\ 3 & 4 \end{pmatrix}$ **(b)** (i) $\begin{pmatrix} 1 & -1 \\ -1.5 & 2 \end{pmatrix}$
(ii) $\begin{pmatrix} 2.5 & -3 \\ -4.5 & 5.5 \end{pmatrix}$ **15.(a)** A = {3, −2},
B = {4, −2}, A ∩ B = {−2}, A ∪ B = {−2, 3, 4}
(b) $p = 1$ $q = -2$

Chapter 10

Page 138 Exercise 77

1.(a) right **(b)** acute **(c)** right **(d)** obtuse
(e) reflex **(f)** obtuse **2.(a)** (i) 50° (ii) 140°
(b) (i) 15° (ii) 105° **(c)** (i) 80° (ii) 170°
(d) (i) 1° (ii) 91° **(e)** (i) $44\frac{1}{2}$° (ii) $135\frac{1}{2}$°
(f) (i) $80\frac{1}{2}$° (ii) $170\frac{1}{2}$° **3.(a)** a,b; e, f; c, d; g, h
(b) a, e; b, f; c, g; d, h; a, f; e, b; c, h; g, d . . .
(c) Yes **(d)** Yes

Page 141 Exercise 78

1.(a) 720° **(b)** 1080° **(c)** 3240° **(d)** 1440°
2.(a) 150° **(b)** 156° **(c)** 140°
(d) 128° 34′ (128.6°) **3.(a)** 10 **(b)** 36 **(c)** 8
(d) 20 **4.(a)** 40° **(b)** 30° **(c)** 120° **(d)** 100°
(e) 50° **(f)** 80° **(g)** 120° **(h)** 150°

Page 143 Exercise 79

1. 1 **2.** No **3.** 3 **4.** order 3 **6.(a)** Yes
(b) Yes **7.(a)** 0 **(b)** No **8.** Point
10. 4 (if 'square') **11.** 0 **12.** 2 **13.** 120°
14.(a) 6 **(b)** 6 **15.** 108° **16.(a)** 5 **(b)** 5
17. 9 **18.** 4 max. **19.** 6 **20.** 3 max.

Page 145 Exercise 80

1. A **2.** 43° **3.** 82° **7.** 67° 30′
8. 50 **9.** $L\hat{M}N = 140°$; $L\hat{N}P = 120°$ **11.** 12
12.(a) 155° **(b)** 85° **13.(a)** 150° **(b)** 91°
14.(a) 18 **(b)** 132°

Chapter 11

Page 151 Exercise 81

The numbers given refer to the reasons **1.–4.** in the text preceding this exercise.
1. 1 **2.** 2 **3.** 3 **5.** 2 **6.** 2 **9.** 4
10. 1 and 3 **11.** 1 **14.** 3 **15.** 2 **16.** 1
17. $4\frac{1}{2}$, 6 **18.** $1\frac{1}{3}$, $6\frac{2}{3}$ **19.** $\frac{4}{5}$, $4\frac{4}{5}$

Page 154 Exercise 82

1. Parallelogram, Rhombus, Rectangle **2.** Square
3. Rhombus, Square, Kite (1 only)
4. Kite, Isosceles trapezium
5. Parallelogram, Rhombus, Rectangle, Square **6.** Kite
7. Parallelogram, Rhombus, Rectangle, Square, (Kite 1 only)
8. Rhombus, Square, Kite **9.** Rhombus, Square
10. Rhombus, Square, Kite.

Page 156 Exercise 83

1.(a) 5 **(b)** 13 **(c)** 8 **2.(a)** 8.1 **(b)** 6.3
(c) 15.5 **3.(a)** 15 **(b)** 9.5 **4.** 4.24 cm
5. 13 cm **6.** 9.66 cm **7.** 12 m **8.** 27.27 cm^2
9. 21 cm^2 **10.** 7 cm

Page 159 Exercise 84

1. 24 cm^2 **2.** 6 cm **3.** 27 cm^3 **4.** 5 cm
5. 29 700 cm^2 **6.** 5 cm^3 **7.** $4\frac{2}{3}$ cm **8.** $4\frac{1}{2}$ cm

Page 161 Exercise 85

3. 9 **4.(a)** ABD, ABC or ADC, BDC
(b) ABF DCF **5.** $\frac{3}{4}$ **6.** 2.4 cm **7.** 42.5°
8.(a) 5 cm **(b)** 20 cm **9.** 32 cm^2
11.(a) 10 000 cm^2 **(b)** 100 cm^3 **12.** 450 cm^3
13. 10 and 11 **14.(a)** 24 cm **(b)** 120 cm^2
15. B **16.** C **18.(a)** Parallelogram
(b) Kite or Isosceles Trapezium **(c)** Quadrilateral
19. Isosceles trapezium, any parallelogram which is not a square
20. 2.4 cm

Chapter 12

Page 167 Exercise 86

1. $a = b = 60°$ **2.** $c = 40°$ **3.** $d = 70°$, $e = 110°$
4. $f = 90°$, $g = 90°$ **5.** $h = j = 40°$ **6.** $k = 30°$
7. $l = 40°$ **8.** $m = 40°$, $n = 100°$
9. $p = 30°$, $q = 50°$, $r = 20°$ **10.(a)** $s = 35°$, $t = 70°$
(b) Congruent

Page 171 Exercise 87

1. 35° **2.** 4.58 cm **3.** $a = b = 55°$ **4.** $d = 70°$
5. 105° **6.** 1 cm **7.** 6 cm **8.** 4 cm **9.** 2 cm
10. 7 cm

Page 174 Exercise 88

1.(a) 0.3491 **(b)** 1.1432 **(c)** 63° 01′ **(d)** 40° 06′
2.(a) 7 cm **(b)** 35 cm^2 **3.(c)** 1.4 cm **(d)** 2.8 cm^2
4.(a) $\frac{3\pi}{2}$ cm **(b)** $\frac{9\pi}{2}$ cm^2 **5.** $\frac{\pi^c}{6}$

Page 176 Exercise 89

1. $x = 27$, $y = 63$ **2.(a)** $(180 - 7x)°$ **(b)** 15
3. 155° **4.** 40° **5.** 10°
6. x, x, $180 - 3x$, $360 - 6x$ **(a)** 10 cm
(b) 3.2 cm **(c)** $5\frac{5}{8} : 1$ **8.** $9\frac{4}{5}$ cm **9.(a)** $\frac{3}{2}$ **(b)** 12 cm^2

Chapter 13

Page 180 Exercise 90

1. $\frac{5}{12}$ **2.** $\frac{12}{5}$ **3.** $\frac{5}{13}$ **4.** $\frac{12}{13}$ **5.** $\frac{12}{13}$ **6.** $\frac{5}{13}$
7. $\frac{8}{6}$ **8.** $\frac{6}{10}$ **9.** $\frac{6}{10}$ **10.** $\frac{8}{10}$

Page 181 Exercise 91

1. 0.7071 **2.** 0.7108 **3.** 0.7112 **4.** 0.9751
5. 0.5774 **6.** 0.5938 **7.** 2.3183 **8.** 0.0192
9. 0.8192 **10.** 0.8161 **11.** 0.8158 **12.** 0.3420
13. 0.3330 **14.** 0.3324 **15.** 0 **16.** 0.711
17. 0.1998 **18.** 2.0278 **19.** 0.8638 **20.** 0.9810

Page 182 Exercise 92

1. 55° **2.** 85.4° **3.** 70.5° **4.** 40° 33′ **5.** 40°
6. 13.4° **7.** 74° 26′ **8.** 47° 35′ **9.** 5° 54′ **10.** 0° 10′

Page 182 Exercise 93

1. 38° 39′ **2.** 22° 01′ **3.** 51° 19′ **4.** 66° 25′

5. 74° 45′ **6.** 60° **7.** 19° 28′ **8.** 28° 47′
9. 65° 58′ **10.** 48° 32′

Page 184 Exercise 94

1. 3.36 **2.** 4.04 **3.** 8.19 **4.** 6.3 **5.** 10.04
6. 3.85 **7.** 5.39 **8.** 7.75 **9.** 7.52 **10.** 16.88

Page 187 Exercise 95

1. 0.9848 **2.** −0.1736 **3.** −5.6713 **4.** 0.2588
5. 0.0087 **6.** −0.2368 **7.** −0.9212 **8.** −57.29
9. −0.6346 **10.** −2.6051 **11.** $-\frac{3}{5}$ **12.** $\frac{8}{17}$
13. $\frac{3}{5}$ **14.** $-\frac{4}{3}$ **15.** $\frac{15}{8}$

Page 191 Exercise 96

1. 0.866 **2.** −0.6428 **3.** −0.866 **4.** −0.866
5. −0.866 **6.** 0.9848 **7.** −2.4751 **8.** 4.7046
9. −1.1918 **10.** −0.5

Page 192 Exercise 97

1. (2.6, 1.5) **2.** (1.5, 0.6) **3.** (0.8, 0.6)
4. (−1.5, 2.6) **5.** (−4.3, −2.5) **6.** (13, 67° 23′)
7. (2.8, 45°) **8.** (4.1, 14° 02′) **9.** (4.2, 135°)
10. (4.5, 333° 26′)

Page 193 Exercise 98

1. 18 cm **2.** 28 **3.** 13 **4.** 12 **5.** 16
6. 29° 45′ **7.** 32° 28′ **8.** 37° 40′
9. 49° 10′ **10.** 57° 43′

Page 195 Exercise 99

1. 15.09 miles **2.** 95.63 km **3.** 46.13 km
4. 36.86 km **5.** 70.71 km **6.** 3.61 km bearing 236° 18′
7. 4.24 km bearing 135° **8.** 5 km bearing 36° 52′
9. 5 km bearing 233° 08′ **10.** 4.3 km bearing 99° 28′

Page 198 Exercise 100

1. 0.917 **2.** −0.64 **3.** $-\frac{15}{17}$ **4.** 130° **5.(a)** $\frac{12}{13}$
(b) $-\frac{5}{13}$ **6.** +0.64 **7.** 13.99 m **8.** 65°, 295°
9.(a) 153 **(b)** 207 **10.(a)** (3, 90°) **(b)** $r = 5$, $\theta = 36° 52'$
11. 45° 14′ **12.** 113.4 m **13.** 92.3 m **14.** 4 m
15. 9.96 km **16.** 104° 02′ **17.(a)** 3750 m **(b)** 135°
18. 24.22 m **19.** CD = 16 km,
AD = 21 km **20.** 1 hour, 5 mins, 45 seconds from start.

Chapter 14

Page 202 Exercise 101

(a) 18.03 m **(b)** 10.44 m **(c)** 9° 27′ **(d)** 11° 19′
(e) 16° 42′ **(f)** 78° 41′

Page 204 Exercise 102

(a) 7.1 cm **(b)** 13.2 cm **(c)** 61° 52′ **(d)** 69° 18′

Page 209 Exercise 103

1. AC = 8.4 cm, BC = 10.03 cm **2.** AB = 3.877 cm, BC = 6.089 cm **3.** AB = 6.337 cm, AC = 11.62 cm
4. BC = 4.582 cm **5.** AC = 1.889 cm **6.** AC = 7.708 cm, $B\hat{A}C$ = 30° 05′ **7.** 107° 07′ **8.** 5.33 cm
9. 2.93 cm **10.** 4.14 cm

Page 211 Exercise 104

1.(a) 64°, 116° **(b)** 90° **2.(a)** −53°, 53°
(b) 0° **3.(a)** 45° **(b)** 0°, 45° **4.** 6°, 122°; (±2°) **5.** 7°, 65°; (±2°).

Page 215 Exercise 105

1. 3000 n.m., 10 hours **2.** 828 n.m., 2 hrs 22 mins.
3. 4447 km **4.** 6670 km **5.** 4380 km **6.** 2221 km
7. 20 012 km **8.** 40 024 km **9.** 30 660 km
10. 1546 km.

Page 218 Exercise 106

1.(a) 6.403 m **(b)** 27° 56′ **(c)** 36° 52′
2. **(a)** (i) $V\hat{A}C$ (ii) $V\hat{P}R$ **(b)** $\frac{1}{4}$ **3.(a)** 30
(b) 38° 56′ **(c)** 26° 34′ **4.(a)** 12 m
(b) 67° 23′ **(c)** 7.071 m **(d)** 31° 34′
5. 113° 35′ **6.** 33° 20′
7. 59.95 km, 20° 42′, 270° 42′, 77.47 km
8. 30.5, 149.5 **9.(a)** 0, 60 **(b)** 0.37, 0.13
10.(a) (i) 1539 n.m. **(b)** 7670 km **11.(a)** Yes
(b) No **(c)** No **(d)** Yes **(e)** Yes
12.(a) 3244 m **(b)** 4247 m **(c)** 3950 m **(d)** 59° 50′

Chapter 15

Page 229 Exercise 109

2. AD = 6.0 − 6.4 cm, AP = 1.1 − 1.5 cm or 11.1 − 11.4 cm
4. 4 cm, 6.4 cm **5.** 2.1 cm **6.** 5 cm **7.** 8.04 cm
9. 11.10 cm, 7.60 cm **10.(c)** 2 **11.** AX = 4.2 cm, BX = 8.2 cm **12.** 3.3 cm **13.** 9.9 cm
14.(a) 56.3 cm^2 **(b)** 48 m **15.(a)** 6.53 cm **(b)** 6.6 cm

Page 237 Exercise 110

7. 9, 9, 9. **10.(a)** 7 marks, 11 marks,
(b) (i) 12 marks, (ii) 4 marks (iii) 2 marks.

Page 240 Exercise 111

1. $\frac{4}{5}$ **2.** $\frac{1}{52}$ **3.** $\frac{1}{4}$ $\frac{1}{169}$ **5.** $\frac{1}{221}$ **6.** $\frac{1}{36}$
7. $\frac{1}{12}$ **8.** $\frac{5}{18}$ **9.** $\frac{1}{4}$ **10.** $\frac{1}{32}$ **11.(a)** $\frac{1}{2}$ **(b)** $\frac{3}{14}$
(c) $\frac{3}{14}$ **(d)** $\frac{3}{14}$ **12.** $\frac{5}{12}$ **13.** $\frac{11}{850}$ **14.** $\frac{1}{6}$
15.(a) $\frac{8}{27}$ **(b)** $\frac{1}{27}$

Page 242 Exercise 112

1.(a) -4 **(b)** 12.5% **2.(a)** 40 – 49 **(b)** 47.5
(c) 31 **3.(a)** 30 **(b)** 20
4.(b) (i) 31 (ii) 30 (iii) 30 **5.(a)** 175.6 cm
(b) 175.9 cm **6.(c)** 3.84 kg **(d)** 91
7. 100.1, greater; 0.8 and 0.5 approximately.
8.(a) median = 52, mode = 54 **(b)** 49.5 **(c)** 49.6
9.(a) 1.992 **(b)** (i) 0.054 (ii) 0.26 **(c)** (i) 0.0625 (ii) 0.25
10. Mode = 3, median = 4, mean = 3.7, probability = $\frac{2}{33}$
11.(a) (i) $\frac{1}{6}$ (ii) $\frac{1}{4}$ **(b)** (i) $\frac{1}{506}$ (ii) $\frac{3}{253}$
(iii) $\frac{5}{506}$ (iv) $\frac{10}{253}$ **12.** $\frac{1}{12}$ **13.(a)** $\frac{3}{5}$ **(b)** $\frac{1}{5}$
14.(a) 10 **(b)** $\frac{2}{5}$ **(c)** $\frac{1}{10}$ **(d)** $\frac{3}{10}$ **15.** $\frac{4}{9}$
16.(a) $\frac{12}{20}$ **(b)** $\frac{7}{20}$ **17.(a)** $\frac{12}{100}$ **(b)** $\frac{88}{100}$
18.(a) (i) $\frac{1}{9}$ (ii) $\frac{8}{9}$; 180; **(b)** (i) $\frac{1}{52}$ (ii) $\frac{12}{51}$
19.(b) Even number **(d)** $\frac{7}{192}$
20. $\frac{3}{200}$, $\frac{1}{120}$, $\frac{419}{600}$, $\frac{59}{60}$